Springer Proceedings in Energy

More information about this series at http://www.springer.com/series/13370

Tanay Sidki Uyar
Editor

Towards 100% Renewable Energy

Techniques, Costs and Regional Case-Studies

Springer

Editor
Tanay Sidki Uyar
Department of Mechanical Engineering
Marmara University
Istanbul, Turkey

ISSN 2352-2534 ISSN 2352-2542 (electronic)
Springer Proceedings in Energy
ISBN 978-3-319-45658-4 ISBN 978-3-319-45659-1 (eBook)
DOI 10.1007/978-3-319-45659-1

Library of Congress Control Number: 2016960199

Printed on acid-free paper

This Springer imprint is published by Springer Nature
The registered company is Springer International Publishing AG
The registered company address is: Gewerbestrasse 11, 6330 Cham, Switzerland

Preface

We, as inhabitants of this planet, are tired of nuclear weapons and atmospheric pollution from petroleum, natural gas, and coal combustion. We desire independence from fossil fuel and nuclear wars, and we want greater employment. A solution is urgently needed to make continued life on this planet viable, and we believe this means living in harmony with nature. Ultimately, the Sun is the only energy source available for human beings, plants and animals. The Sun and greenhouse gases work in concert to bring the average temperature of the atmosphere to 16 °C.

Humanity is trying to recover from the damage initiated by the Industrial Revolution, which supplied energy needs by polluting the environment, causing death of millions of human beings and destroying natural resources and agricultural products. The journey from coal to fuel oil and to nuclear waste heat and finally to energy efficiency, directed by the external costs of energy production, ended up with the internalisation of external costs (in the 1980s for the USA and in 1993 for Europe by the work of environmental economists from Germany and the UK). Tax Credits for Renewable Energy Investments in California and the results of the ExternE Project of the European Union are the main reasons for the initial steps for the transition to renewable energy. Support from governments for clean coal technologies and nuclear energy through harmful subsidies have delayed the transition to renewable energy and end use efficiency. By learning from the mistakes of energy policies supporting polluting and inefficient technologies, and after becoming aware of the social costs of energy production and consumption, countries which industrialized early have started to support development and implementation of renewable energy technologies and end use efficiency.

Wind and solar electricity, electric vehicles and LED lighting technologies have become commercially available and competitive. Beginning in the 2010s, electricity production from renewables, even without internalisation of externalities of conventional energy production (health problems, agricultural product losses, natural capital destruction and climate change effects), has become the cheapest option.

Due to the high entropy nature of renewable fuels (kinetic energy of the moving air, potential energy of the water molecules, chemical energy stored in biomass and the heat stored in geothermal resources) renewable energy technologies can supply process heat and electricity locally. Efficient management and supply of the locally produced energy to other consumers requires construction of microgrids and smart grids with storage options. The intermittent nature of renewable energies requires storage of renewable energies for backup when there is no wind or sun available.

Chemical energy stored in biomass and renewable energy stored as process heat or electricity and their conversion to liquid and gases fuels make the transition to 100% renewable energy possible without relying on conventional fossil fuels. This makes renewables independent from fossil fuels, or, the energy solution independent from the energy problem.

Extensive availability of renewable energies locally necessitates the involvement of communities, municipalities and cooperatives as part of the decision making process, and as owners of their renewable energy investments for sustainable energy solutions.

The implementation of renewable energy as community power would mean equity, freedom, peace and local employment. Increasing the number of such applications will result in renewable energy regions and, finally, with 100% Renewable Energy Cities.

We hope that this book provides an academic platform for the interaction of experts and scientists working in different fields of renewable energy research, development and implementation.

Handling global problems threatening human life and our common living space requires that human activities be carried out in harmony with nature. We must understand and digest the reality that living in harmony with nature is not a favour for nature, but a prerequisite for human beings and nature to coexist in peace on this planet. Since nature cannot negotiate, the only solution available for human beings is to take into account natural constraints on human activities.

Relations and interaction of human beings with nature also teach us to respect the information, expectations and demands of other individuals in our society, in order to find the correct definition of our problems and to find solutions that can be implemented. Ecologically-sound societies can only evolve from democratic societies where everyone struggles together in solidarity.

Istanbul, Turkey Tanay Sidki Uyar

Contents

Part I
Basics, Fundamentals, Concepts

Efficient Use of Energy in the Industry

Hasan A. Heperkan

1 Introduction

Although Turkey has several energy sources, she imports more than half of her energy production. In 2006 the total energy production was 31.7 million TOE and the consumption 101.7 million TOE [1]. The energy deficiency was 70 million TOE. Predictions indicate that this figure will exceed 200 million TOE by 2020. Import of oil, gas, and coal will continue to cover the energy demand in the future.

Energy sources are decreasing rapidly with the steady increase in the demand for energy, elevating the price. This on the other hand has increased the price of goods, causing problems in the developing markets. As a result energy has to be utilized rationally.

Table 1 presents that the energy demand of Turkey is provided by 30.9 % coal, 13.1 % natural gas, 40.9 % oil, 4.9 % hydraulic, 0.1 % solar, and 9.5 % wood, animal, and plant waste. The same picture for the different sectors is 25.3 % buildings, 29 % industry, 14.5 % transportation, 3.6 % agriculture, and 3.1 % other [2]. Forty-six percent of Turkey's energy consumption depends on oil (Table 1).

Energy supply of Turkey, according to 2012 figures, depends on 30.7 % coal (14.8 % lignite); 25.3 % oil; 4 % hydraulic; 30.9 % natural gas; 6.3 % solar, geothermal, wind, and other renewables; and 2.8 % noncommercial sources like wood, animal, and plant waste [3]. Seventy-two percent of Turkey's energy consumption depends on imports.

H.A. Heperkan (✉)
Mechanical Engineering Faculty, Yıldız Technical University, Beşiktaş, İstanbul
e-mail: heperkan@yildiz.edu.tr

T.S. Uyar (ed.), *Towards 100% Renewable Energy*, Springer Proceedings
in Energy, DOI 10.1007/978-3-319-45659-1_1

Table 1 Share of the primary energy sources in the total energy consumption (%)

Years	1970	1975	1980	1985	1990	1995	2000	2001	2002	2003	2004	2005	2006
Anthracite	15.4	11.0	8.9	9.8	12.2	10.6	12.3	9.3	11.3	13.4	14.0	13.7	14.8
Lignite	9.2	9.8	12.4	20.3	18.3	16.8	15.5	15.2	13.3	11.3	10.8	10.2	11.2
Asphaltite	0.1	0.7	0.8	0.6	0.2	0.0	0.0	0.0	0.0	0.1	0.4	0.4	0.3
Total coal	24.7	21.5	22.1	30.7	30.7	27.4	27.9	24.5	24.6	24.7	25.2	24.3	26.2
Natural gas	–	–	0.1	0.2	5.8	10.0	17.1	19.7	20.6	23.2	23.3	27.2	28.9
Oil	42.2	51.8	50.4	46.3	44.8	46.4	40.1	41.0	39.5	37.9	37.5	35.4	32.6
Hydraulic	1.4	1.9	3.1	2.6	3.7	4.8	3.3	2.7	3.7	3.6	4.5	3.8	3.9
Geothermal													
Electricity	–	–	–	–	0.1	0.1	0.1	0.1	0.1	0.1	0.1	0.1	0.1
Heat	–	–	–	–	–	0.1	0.8	0.9	0.9	0.9	0.9	1.0	1.1
Solar	–	–	–	–	–	0.1	1.0	1.2	1.3	1.5	1.5	1.6	1.5
Total commercial	68.3	75.2	75.9	80.2	85.1	88.9	90.2	90.2	90.7	92.0	93.0	93.3	94.3
Wood	20.4	16.0	14.8	13.3	10.1	8.7	6.3	6.5	6.0	5.4	4.9	4.6	4.0
Animal waste	11.3	8.8	9.3	6.5	4.8	2.5	1.7	1.8	1.7	1.5	1.4	1.3	1.2
Total noncommercial	31.7	24.8	24.1	19.8	14.9	11.2	8.0	8.2	7.6	6.9	6.3	5.8	5.2
Electricity ımport	–	–	–	–	–	–0.1	0.3	0.3	0.4	0.8	1.1	1.3	1.4
Total	100	100	100	100	100	100	100	100	100	100	100	100	100

Table 2 Share of the primary energy sources in the total energy production (%)

Years	1970	1975	1980	1985	1990	1995	2000	2001	2002	2003	2004	2005	2006
Anthracite	19.2	17.9	12.7	10.1	8.1	5.0	4.1	4.7	4.3	4.8	4.4	4.8	5.0
Lignite	12.0	16.7	21.6	37.8	36.9	40.9	43.8	45.3	42.5	40.0	37.6	39.3	43.1
Asphaltite	0.1	1.2	1.4	1.0	0.5	0.1	0.0	0.1	0.0	0.6	1.3	1.6	0.7
Total coal	31.3	35.8	35.7	48.9	45.5	46.0	47.9	50.0	46.8	45.3	43.3	45.7	48.9
Natural gas	–	–	0.1	0.3	0.7	0.6	2.2	1.2	1.4	2.2	2.7	3.3	3.1
Oil	25.7	19.8	14.1	10.2	15.1	14.1	11.1	10.9	10.6	10.5	9.8	9.8	8.5
Hydraulic	1.8	3.1	5.6	4.8	7.7	11.6	10.2	8.4	11.9	12.8	16.3	13.8	14.2
Geothermal													
Electricity	–	–	–	–	0.3	0.3	0.3	0.3	0.4	0.3	0.3	0.4	0.3
Heat	–	–	–	–	0.1	0.2	2.5	2.8	3.0	3.3	3.3	3.8	4.0
Solar	–	–	–	–	0.1	0.2	1.0	1.2	1.3	1.5	1.5	1.6	1.5
Total commercial	52.8	58.7	55.6	64.3	69.3	73.1	75.2	74.7	75.5	75.8	77.2	78.3	80.7
Wood	26.5	26.0	27.3	24.0	20.8	21.0	19.5	19.9	19.3	18.9	17.8	16.9	15.0
Animal waste	14.7	26.6	17.1	11.7	9.9	5.9	5.3	5.4	5.3	5.3	5.0	4.8	4.3
Total noncommercial	41.2	14.7	44.4	35.7	30.7	26.9	24.8	25.3	24.6	24.2	22.7	21.7	19.3
Total	100	100	100	100	100	100	100	100	100	100	100	100	100

2 Importance of Energy Efficiency for the Industry

Table 3 shows that the share of the energy demand for the industrial sector has increased from 21 % in 1970 to 29 % in 1998. Oil is the most important primary energy source with a share of 35.5 % [4].

If we follow the energy prices in the world and Turkey, we see that the price of a barrel of oil was $3 in 1970. After the raise by the OPEC countries in 1973, it became $12, the Iran revolution $28, and the Iran–Iraq war $34. Dropping to $16 later showed a sharp peak again when Iraq invaded Kuwait, $40. Today it is around $30–40. Gasoline sells for around $1.5/L.

The above picture indicates that more than a half of the energy used by the industrial sector is imported and its price fluctuates drastically according to sociopolitical reasons. To reduce the foreign dependence of our energy character, alternatives such as

- Utilizing new energy sources (solar, wind, biogas, wastes, etc.)
- Applying new technologies (energy storage, hybrid systems, fluidized bed power plants, heat pumps, etc.)

could be considered. But we should not forget that the most effective means is to reduce the amount of energy used and energy savings. In fact, when we remember the environmental impact (CO_2 and greenhouse gases) of the problem, we will deduce that the best solution is the rational use of energy.

3 Planning Energy Savings in the Industry

The first step towards the energy analysis of a facility is to establish the energy balance. Points of energy consumption, their energy demand, and the load variations have to be assessed carefully. The distribution of the energy bought from the market (like electricity), the efficiencies at the consumption points, and the losses are determined. Not only the momentary values, but also the daily, weekly, and annual variations have to be considered. From the data, information like the average energy consumption, deviation from these values, and the energy per product can be calculated.

In the second stage, energy rejected from each point is recorded. The waste is than compared with the energy demand of other points to find possible utilization. The synchronization of the available energy and the need is also very important; energy storage should also be considered if possible.

During the third stage, the technology needed to accomplish the energy transfer is investigated; relevant methods and equipment are selected. Feasibility studies are carried out and the return periods are calculated.

Table 3 Energy consumption of different sectors (%)

	1990	%	1995	%	2000	%	2005	%	2006	%	2009	%
Industry	14,542	35	17,372	35	24,501	40	28,282	39	30,974	40	25,966	32
Buildings	15,358	37	17,596	35	20,058	33	23,013	32	23,726	31	29,466	37
Transportation	8723	21	11,066	22	12,008	20	13,849	19	14,884	19	15,916	20
Agriculture	1956	5	2556	5	3073	5	3359	5	3610	5	5073	6
Other	1031	2	1386	3	1915	3	3296	5	4163	5	4153	5
Total	41,610	100	49,976	100	61,555	100	71,799	100	77,357	100	80,574	100

The last period of the process is to follow up the performance of the improvements. This stage is also very important because it is the dissemination of the gathered information and the experience to parties of interest.

There are several important elements related to the rational use of energy in industrial facilities:

- Electrical energy savings
- Thermal energy savings
- Mechanical energy savings
- Process energy savings
- Reclamation of materials
- Freshwater supply

3.1 Electrical Energy

Electrical energy is used in the industry for heating, as in ovens, to drive electric motors to move machinery and in lighting of production areas, administration buildings, warehouses, and other special areas.

The lighting need of office quarters and production areas should be determined correctly during the design process. The most important point is to avoid unnecessary lighting and give up the strategy for general lighting; instead local lighting in accordance with the specific requirements should be adopted.

The type of the bulbs used in the illuminated areas is also effective on the overall energy consumption. Wolfram filament, fluorescent, mercury, and sodium vapor lamps are most commonly used in Turkey. Lamps with high luminescence efficiency should be selected. This value is 15 lm/W for wolfram filament, 35–40 lm/W for fluorescent, and 140 lm/W for high-pressure vapor lamps. Screens, covers, and shutters reducing light should not be used and care should be given to cleaning. Special lamps with high luminary flux available in the market also pose an interesting alternative.

The illumination plan of the facility is as important as the type of lamps used. The lamps should become operational stage by stage as it gets dark and unoccupied areas should not be illuminated except for security reasons. Seasonal variations of sunlight should be considered. Passive measures like the location of windows and the selection of the work areas can also be effective.

The highest fraction of the electric energy used in the industry is consumed by electric motors. Therefore it is of prime importance to keep electric motors under continuous surveillance and regular maintenance to guarantee their operation around their optimum design point.

If motors are operated with low voltage, the current through the coils increases and overheats the system reducing its life cycle and promoting losses. The balanced current distribution among the phases should also be followed carefully to avoid heating; this is an indication of probable failure in the future.

Transformers, generators, and electric motors operate according to the induction principle and draw magnetization currents to form the magnetic field. This is a reactive current. This current is not transformed into useful power but is necessary for magnetization. Current transformed into light in lamps, into motion in motors, and into heat in wires is an active current. It is recorded by the meters and is the basis of the invoice. The grid current falls behind the active current by a phase angle of ϕ due to this reactive current. Cos ϕ is known as the power factor. When reactive current is used in a facility, ϕ increases and cos ϕ reduces and hence the active current also reduces. Therefore power companies limit the value of the power factor. The limit is set as 0.9 in Turkey. This allows the utilization of half the active current as reactive current. Phase shifters and capacitors can be used to exploit reactive power.

Another effective measure is the staged tariffs used in Turkey recently, which has been effective in Western countries since a long period. The timing of the electric power usage can be scheduled to avoid going into a higher tariff zone.

3.2 *Thermal Energy*

Thermal energy requirement at the consumption points can show different characters in energy-intensive industrial plants. Some processes require high-pressure superheated steam, whereas some processes can take place with saturated low-pressure steam. Thermal energy for space heating and social needs can be supplied by hot water and not steam. Hence the correct distribution of thermal energy within the plant could play an important role in the overall energy consumption of the facility.

Thermal energy optimization and saving studies can be collected under three main topics:

1. Thermal energy production (boiler)
2. Thermal energy distribution (insulation)
3. Thermal energy utilization (condensate)

The industry uses low-, medium-, or high-pressure saturated or superheated steam, pressurized hot water, hot water, and hot oil boilers depending on capacity and type. It is important to select the right boiler to meet the capacity of the process. Production is not continuous in all processes around the year. Moreover the production of certain goods varies according to the demand from the market. Process changes can also take place and the production of thermal energy does not show a steady regime. Boilers are often operated under part-load conditions. The losses of a classical boiler change with the load; the losses increase [5]. Boiler efficiencies are measured under full-load conditions and these values do not represent the real performance and fuel consumption at part-load operation. If the boiler is operable under part load during most of the time, the annual operational thermal efficiency should be considered [6]. The annual operational thermal efficiency is the ratio of the useful energy obtained to the total fuel consumed annually. To improve the efficiency, an effective control system to balance the energy supply and demand is required as well as special technology and construction (e.g., modulating burner) [7].

Table 4 Excess air, boiler efficiency, and specific consumption at different operating conditions (flue gas temperature 230 °C, steam pressure 9 kg/cm^2, boiler feed temperature 63 °C)

	O_2	Excess air	Load	Specific steam prod.	Specific fuel consum.
Information	%	%	%	kg steam/kg F.O.	kg F.O./ton steam
Normal operating condition	15.5	300	72	11.66	85.76
Optimum operating condition	13.0	174	83	13.44	74.40
Maximum efficiency condition	3.9	25	90	14.57	68.63

Another topic important for energy saving during energy production in a boiler is the control of combustion and excess air. Extra air compared to stoichiometric air is supplied to ensure complete combustion. However if the excess air increases too much, flue gas losses increase to reduce the overall efficiency of the system. It is important to provide the minimum air to the system that would not cause condensation in the chimney. Condensing boilers available in the market in the recent years condense the water vapor in the flue gases to exploit the latent heat. The best way to assess the optimum combustion conditions for a boiler is to monitor and analyze the flue gases. O_2 and CO must always be measured. O_2 gives the excess air and CO indicates whether the combustion is complete or not. Table 4 shows excess air, efficiency, and steam–fuel ratio values of a boiler under different operating conditions [8]. A good control system assesses the energy demand of the system and modulates the generator without turning it off.

It is useful to care for the insulation of valves and flanges used in the distribution network of thermal energy. Although these components are generally overlooked, the heat losses when calculated can reach considerable values, especially for high-pressure steam lines. Locating distribution lines in closed enclosures helps to prevent the adverse influence of varying outside temperatures on heat losses. Accessibility is also important for maintenance and repair activities.

Other components that have significant effect on energy saving are fuel tanks and stack connections; they should be properly insulated. Facilities using heavy oil must heat the fuel to reduce the viscosity for pumping. At least the bottom sections of the fuel storage tank have to be kept at 50–60 °C. The losses are elevated during winter operation. The heat lost through the stack connection could be recovered utilizing an economizer if insulated effectively.

Steam produced in the boiler condenses at the consumption points and the condensate is collected and fed back into the boiler. In the distribution lines however, the steam that condenses due to heat losses, together with the dissolved air and CO_2, causes corrosion. Steam traps are used to evacuate the mixture, which is usually discarded. Moreover the condensate from the process lines is also discarded thinking that they are contaminated. Flow rates that are small individually can add up to considerable amounts in larger facilities. When collected in a tank in the plant, material and heat recovery is possible.

Table 5 Thermal energy losses

Losses		Heat loss kcal × 10^7	Fuel-oil Equivalent kg	Fuel loss %	Heat loss cost MTL/month (1986)
Flue gas	Optimum conditions	86.00	88,704	13.20	11.0
	Maximum conditions	130.00	134,400	20.00	16.8
Insulation	Fuel-oil tank	7.83	10,087	1.50	1.3
	Stack connection duct	8.17	10,540	1.57	1.0
Condensate	Escort line	16.10	20,714	3.08	2.6
	Other	52.80	68,069	10.13	8.5
Total[a]		171.00	198,114	29.48	24.4

[a] Optimum flue gas conditions were assumed during total heat loss calculations

Table 5 reflects the energy optimization studies carried out for a chemical plant. Recovery potentials are also given. It should be kept in mind that the total energy savings after applying all the measures will be less than the total from the table. If the boiler is operated at optimum conditions, the flow rate of the flue gases through the stack will reduce and the economizer load will decrease. The recovered energy from the insulation will also reduce. Proper insulation on the other hand will reduce steam consumption, decreasing the condensate flow rate.

3.3 *Mechanical Energy*

The potential fields in energy savings in the industry are high-pressure air lines, pumps, and fans.

It is important to determine the correct need of the system during the design stage while selecting the operating pressure and the compressor; a large compressor for safety purposes is not recommended. Such a selection would increase the investment cost, and increase the leakages and losses in the distribution lines and hence the operational cost as well. The pipes should be smooth, components creating pressure losses such as reductions and bends should be kept at a minimum number, and the pressure loss coefficients of valves should be examined carefully.

Air is heated during compression. Heat is a by-product of the process and reduces the efficiency of the compressor. Using this energy will increase the overall performance of the system. Controlling the inlet air temperature is also helpful. If we reduce the inlet air temperature, we can increase the mass flow rate by increasing the density. This will also increase the power consumption but the rate is not as high as the density [9].

Another potential component is the pump. Process technique requires that the flow lines should be designed with minimum friction, without armatures and bends, keeping the velocities low to reduce losses, hence improving the performance of the pump. During operation, the valves should be fully open and flow through bypass lines should be avoided. Pumps should not be kept under idle conditions and care should be given to operate at the most efficient point of the pump curve. The pump characteristic and the system load curve should intersect at the maximum of the efficiency curve.

As in almost every field of engineering, recent advances in electronics have brought innovative improvements to pump technology. It is possible to control the speed by changing the frequency of the voltage. Pumps have a range of operation freedom for pressure drop, flow rate, and speed to achieve optimum conditions. The performance declines at other points. Systems usually operate under off-design conditions. Part-load operation causes the specific consumption of pumps to rise. With an appropriate control system, a frequency converter enables the pump to operate at the optimum point even under part-load conditions by adjusting the speed. When a thermostatic valve throttles the flow by 15 %, the heating efficiency reduces only 3 %; on the other hand adjusting the speed with a converter decreases the power consumption by 50 % [10].

4 Energy Analysis of an Industrial Facility

The measurement parameters and measurement points have been established. Nine energy zones have been chosen; the zones are summarized in Table 6. Steam flow rates and temperatures have been used to assess the energy consumption of the processes. Temperatures were recorded continuously and the steam flow rates in groups to minimize the number of flow meters. Temperatures and production programs were used to determine the operation periods.

Steam flow rates were measured using 22 orifice plates and five flow meters. Two electronic flow meters were mounted on the two main steam lines and recorded continuously. The meters were calibrated at the beginning and end of the measurement periods. Temperatures were measured with J- and T-type thermocouples and recorded by 12 data loggers. The measurements were taken every 20–30 min. Data was processed with Excel-Macros. Flue gas analysis and fuel meter readings were also recorded regularly. Winter measurements were taken in February and March, and summer readings in August and September. Production was 8 h a day.

Tables 7 and 8 summarize the results of the measurements both for winter and summer. The total steam production calculated from the fuel meter readings has been distributed among the two major lines according to the measurements recorded by the flow meters. This consumption was then distributed among the various processes and equipment. The actual reading from the flow meters, the real values (*values of fuel meters*), and the total found as a result of the redistribution were compared for checking. It can be observed that the difference between the values is well within the error limits typical for such on-site studies.

Table 6 Energy zones for steam consumption

Number	1	2	3	4	5	6	7	8	9
Energy zones	Boiler house	Pen. process	Pen. space heating	• Office heating	Non-pen. process	Non-pen. space heating	Ceph. space heating	Ceph. process	Diesel room + distillation
				• Old office building	Liquids manufacturing				
				• Solids manufacturing	Solids manufacturing				

Table 7 Distribution of steam consumption between processes and heating in the winter (production, night, weekend)

	Production			Night			Weekend		
		Heating and ventilation (%)			Heating and ventilation (%)			Heating and ventilation (%)	
	Process (%)	Process area	Office	Process (%)	Process area	Office	Process (%)	Process area	Office
Line-1	19.1	11.0	12.3	10.4	14.5	20.0	13.4	9.7	18.8
Line-2	14.1	42.2	1.4	8.5	45.4	1.6	13.7	42.6	1.6
Total	33.2	53.2	13.7	18.9	59.9	21.6	27.1	52.3	20.4

Table 8 Distribution of steam consumption between processes and heating in the summer (production, night, weekend)

	Production			Night			Weekend		
		Heating and ventilation (%)			Heating and ventilation (%)			Heating and ventilation (%)	
	Process (%)	Process area	Office	Process (%)	Process area	Office	Process (%)	Process area	r
Line-1	25.3	5.5	0.2	40.4	1.7	0.0	38.8	0.9	0.4
Line-2	30.2	38.8	0.0	22.1	35.7	0.0	26.6	34.4	0.0
Total	55.5	44.3	0.2	62.5	37.4	0.0	65.4	35.3	0.4

Sankey diagrams were prepared to illustrate the share of each process or unit group in the total energy consumption. As an example, Fig. 1 shows this distribution for the total steam consumption of the plant for the weekend operation period for winter. Figure 2 presents the same information for summer.

5 Conclusion

The final distribution of the total steam consumption between the various processes and heating in winter and summer, respectively, is summarized in Tables 6 and 7 for production, night, and weekend operations. The share of heating has been given for the production areas and the offices separately. It can be observed that roughly 30 % of the total consumption is consumed for the processes, 50 % for heating and conditioning of the production areas, and 20 % for office space heating in the winter. In the summer roughly 55 % of the total consumption is consumed for the processes, and 45 % for heating and conditioning of the production areas. Office space heating isn't applicable during the summer. Only domestic hot water is consumed but the relative share is negligible.

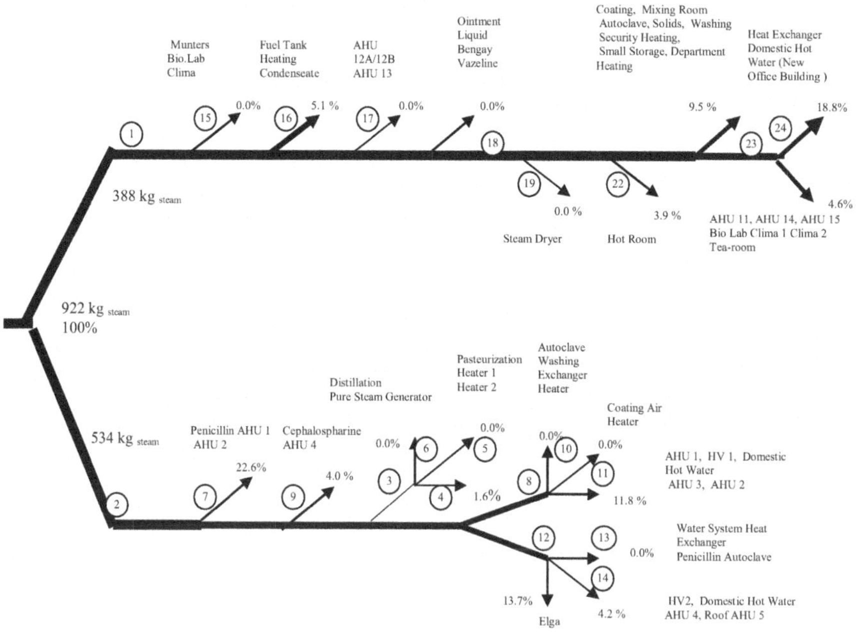

Fig. 1 Total weekend (winter)

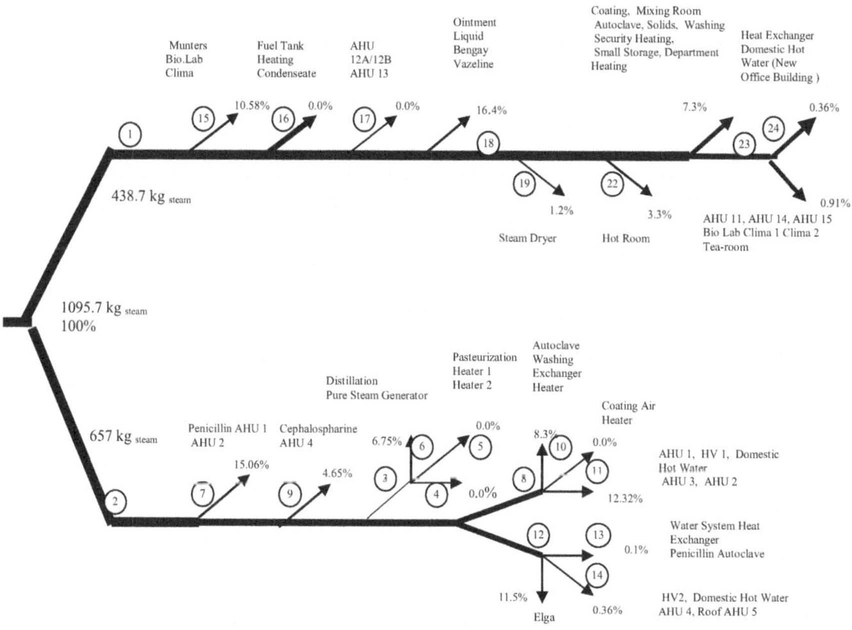

Fig. 2 Total weekend (summer)

References

1. 11. National Energy Congress final report, 21–23 October 2009, İzmir
2. 11. National Energy Congress Turkey's Energy Sector Development and Supply Demand Projections, 21–23 October 2009, İzmir
3. Ministry of Energy and Natural Resources (2013), www.enerji.gov.tr/en-US
4. 11. National Energy Congress Energy Statistics, 21–23 October 2009, İzmir
5. Heperkan H, Baloğlu F, Karahan A (1994) Modern heating systems (1). Install Eng 2(16):53–55
6. DIN 4701-2:1983-03, Regeln für die Berechnung des Waermebedarfs von Gebaeuden-Tabellen, Bilder, Algorithmen (has been replaced by DIN EN 12831:2003-08, Heizsysteme in Gebaeuden-Verfahren zur Berechnung der Norm-Heizlast)
7. Heperkan H, Baloğlu F, Karahan A (1994) Modern heating systems (2). Install Eng 2(17):36–39
8. Heperkan H, Atakan S, Altıntaş N (June 1987) Klor-Alkali Tesislerde energy optimization for chlorine-alkali facilities. TÜBİTAK—MBEAE report
9. Dağsöz AK (1991) Energy savings in the industry. In: ALP technical books, www.nadirkitap.com/article #2849919
10. Pumpenleistungsregelung, Grundlagen und Anwendungen (August 1992), Wilo Werk GmbH W887/7.000/9208/D/StD.

Sustainable Energy Transition: Local Governments as Key Actors

Maryke van Staden

1 Introduction

Alarming trends such as increasing energy consumption, declining fossil fuel resources, the visible negative impact caused by burning fossil fuels (on air quality, health, and the environment), rising energy prices, and grave future development scenarios impacting on energy-economy-society-security require a global spotlight on the interplay between energy, energy security, and climate change.

Accelerated climate change and declining natural resources are two important issues facing the world today. These do not only pose environmental challenges. They also have significant socio-economic, security, and political impacts and present a range of challenges that will impact everyone. Yet, it will have a significant larger impact on the poorer section of the global population, who have fewer resources to prepare and respond to the unfolding effects thereof.

The local impact *on* climate change is clear. Greenhouse gas emissions are being emitted by burning fossil fuels, mostly for energy use in urban areas. This in turn accelerates (the natural phenomenon of) climate change—and causes extensive problems for the global ecosystem. Climate scientists and experts in many disciplines are exploring these issues, building a valuable knowledge base which can help decision-makers guide us towards a sustainable future. Further, many local and national leaders are also already tackling climate change mitigation, approaching it also from the perspective of achieving a myriad of co-benefits that can result from well-planned and -implemented action.

The local impact *of* climate change is visible around the world: from changing rainfall patterns and rainfall intensity, with more storms, heavy downpours, rising sea levels, as well as droughts, shifts in growing seasons, and disease distributions, to

M. van Staden (✉)
World Secretariat, ICLEI—Local Governments for Sustainability,
Kaiser-Friedrich-Str. 7, 53113 Bonn, Germany
e-mail: maryke.van.staden@iclei.org; http://www.iclei.org/gcc

T.S. Uyar (ed.), *Towards 100% Renewable Energy*, Springer Proceedings
in Energy, DOI 10.1007/978-3-319-45659-1_2

mention but a few impacts. All these aspects directly influence humans and their environment, quite often, but not exclusively in urban areas. The Intergovernmental Panel on Climate Change (IPCC) Summary for Policy Makers of 2013 [1] offers a useful outline of a consensus-based report on research and the relevance of extreme weather events for society and sustainable development. There are also many other reports that show alarming trends, and draw attention to the need for accelerated action.

2 Context for Local Action

There can be no solution to climate change without local climate and energy action. A summary for decision-makers of the Fifth Assessment Report of the Intergovernmental Panel on Climate Change (IPCC) was provided to specifically address cities [2], which among others highlights that cities account for 37–49 % of global greenhouse gas emissions and urban infrastructure accounts for over 70 % of global energy use. It also outlines the main climate impacts on urban spaces, where already more than 50 % of the global population now live.

Effective local action requires a supportive and enabling national framework, which will allow, guide, and actively aid local and subnational governments in comprehensively tackling climate change mitigation and adaptation. Further, citizens, business, and industry look towards (all levels of) their government to deal with the situation, and to guide and protect them. These elements are highly relevant to the local community level, where the impact *on* and *of* climate change can, and should, be addressed. Communities may find it easier to engage, as local decision-making processes and implementation can be faster and action can have immediate impact, compared to action at other levels of government. This does not however mean that other levels do not need to engage.

Increasingly local governments are engaging, where they have recognised the need and benefits of action. Energy conservation and using improved energy efficiency technologies are key action areas to reduce energy demand and save costs. Immediate or fast financial benefits can be gained; for example where more efficient technologies are used the payback time on investment (i.e. amortisation) will lead to mid- to long-term savings. This can easily be done for example by installing energy-efficient lighting, but also with more expensive action areas such as complete building renovation to achieve modern energy standards (low energy, passive house, or zero energy).

However, the importance of switching to renewable energy sources should not be underestimated in this transition discussion. We need energy alternatives to fossil and nuclear energy, where fuels are safe and not so easily depleted. The wide range of renewable energy sources available implies that nearly every location on earth has some form of renewable energy that could be used—ideally in a mix: from solar—and wind energy to waste-to-energy and tidal power solutions, and many others. The renewable energy mix and using effective energy storage solutions are key to planning and rolling out a well-managed energy transition that is low-carbon and resilient.

2.1 Motivation for Action

People, money, energy, and environment—these issues are drawing the attention of modern society. They are also interconnected to climate change and our response to this global phenomenon. It is well known that global resources are limited. This requires us to reduce waste, optimise resource efficiency, and generally be cleverer in applications that require energy and resources. The human impact on climate change is acknowledged, and the solutions are in many cases clear and readily available. Our response requires a combination of applying technology and effective policy, as well as changing behaviour, with the latter being the hardest angle to address in many cases.

Increasingly local communities—cities and towns of all sizes—are responding to these challenges by engaging in local climate and sustainable energy action. Why do local governments engage? A starting point is often an exploration on how government operations can be optimised to reduce costs and yet remain efficient. Rising energy costs make it challenging to budget effectively. Yet, recognition of responsibility is typically also a driver for action. Local leaders and municipal staff explore how local climate and sustainable energy action can lead to local (co)benefits, how this can be financed, and how to retain or further develop "good quality of life" living and working space for citizens, businesses, and industries.

This local process is typically led by local governments, but also requires the involvement of many stakeholders: citizens, business and industry, and non-profit organisations. The active support and engagement of many different groups is a proven key to success. In practice this means especially engaging with people in these stakeholder groups: municipal staff, councillors, chief executive officers (CEOs), etc. The value of individual "champions" in each of these stakeholder groups is a necessity to help drive the agenda, and ensure that it remains "on the agenda".

Municipalities understand the local context and local needs. They are responsible for good local governance. Engaging in local climate and energy action seems to be a logical choice—but this choice clearly implies change. It requires change in the way energy is generated and distributed (moving away from centralised systems to decentralised, smaller interconnected, and safer systems). It also requires change in the way energy used. Energy has become a valuable commodity! Yet, this is part of the transition challenge, as local governments often do not own energy utilities and typically do not have a mandate to address energy—but they understand the need for acting in this area. The leaders have started, and use diverse reasons for this.

Municipalities have different (often multiple) motivations to engage and reap multiple local benefits, which include the following:

- Exploring approaches for improving air quality and associated improved health
- Urban infrastructure improvement: anticipating the impact of climate change and adapting to this. "Local to global" responsibility: protecting people and the environment in a changing climate where resources are becoming scarcer, addressing the global common good

- Keeping money in the local economy: local energy production, local energy use, generating local profits, and local taxes
- Green economy development: improvement by saving energy and reducing energy bills, making money for local sustainable energy generation, local job creation, stimulating the small- and medium-sized enterprise sector
- Social upliftment of poorer residents: by reducing their need for energy, e.g. through energy-efficient housing renovation, and offering effective public transport options
- Energy security: securing stable energy supply, ensuring that energy remains affordable for citizens

Some local governments have been engaging in climate action since the 1990s, when ICLEI started its Cities for Climate Protection (CCP) Campaign. Many more have started since then, with the CCP Campaign participants numbering over 1000 cities and towns in the early 2000s. Since 2013 ICLEI has transitioned to the GreenClimateCities programme, offering an updated process with practical guidance to local governments of cities of all shapes and sizes and at varying levels of development—both in the Global South and North.

2.2 *Role of Local Governments*

The role of local government (also referred to as local authority, municipality, council, administration, etc.)—as the level of government closest to citizens—is critical in the context of climate protection, climate change adaptation and resilience, and transition to sustainable energy.

Local governments are usually responsible for defining *strategy*, implementing local *policy and regulations*, developing and maintaining *structures* that handle administration and provide services, and providing a range of *services* to local inhabitants and businesses. This role differs from country to country but can include policing; health services; education; social services; transport, water, and sanitation services; and sometimes also energy services (e.g. local sales from national grid). Further to this they often *own or manage infrastructure* such as buildings, roads, and even electricity grids in their geographical territory.

In all of these cases local governments can thus shape and guide change among inhabitants, businesses, and industry as well as their own municipal operations. They can motivate and lead a change of direction in the whole community, to benefit the community. These are areas where local climate action is possible, with vast potential for reducing emissions, and improving overall efficiency and quality of life.

Those cities and towns that are achieving success usually:

- Have one or more local champions who can motivate people, draw attention to these issues, and constantly make sure that climate/energy is on the political agenda. These include political representatives and senior municipal staff—highly recommended as a valuable driving force for local action.

- Have a comprehensive and regularly updated (climate or energy) action plan which is being implemented and monitored.
- Understand where challenges come from—i.e. where energy is being used and emissions released, and conduct regular greenhouse gas (GHG) inventories.
- Review local renewable energy resources to assess where energy imports can be reduced, and how local resources can be optimised.
- Conduct a community SWOT analysis to identify strengths, weaknesses, opportunities, and threats in the relevant sectors and areas.

These actions and processes help to coherently address climate change mitigation and adaptation at the community level.

3 Current Developments

3.1 Global Developments

ICLEI—Local Governments for Sustainability (ICLEI)—addresses these and many other issues impacting on sustainable urban development, working with municipalities and their partners around the globe. The organisation's approach is that many global challenges require local solutions but global governance, for example when addressing climate, air quality, biodiversity, and freshwater resources. This needs international commitment and processes, as well as cooperation between all levels of government. The local government level is essentially an implementing level, and thus highly impacted on by global, regional, and national planning and processes.

Yet, they typically have no capacity to monitor or influence these levels individually, looking to city networks such as ICLEI to engage in advocacy on their behalf. When national governments decided to craft a new global climate regime, ICLEI started a parallel global process called the Local Government Climate Roadmap. This was done as a collaborative effort with all leading global city networks to ensure engagement, recognition, and empowerment of local governments in this new climate regime.

Further, the carbon*n* Center[1]—the Bonn Centre for Local Climate Action and Reporting—was established as an international centre of excellence, as a joint United Nations Environmental Programme (UNEP) and ICLEI initiative. It manages the global reporting platform for local and subnational governments on local climate action: mitigation and adaptation [3].

Table 1 provides an overview of this and other key developments relevant to climate protection that also impact on local governments.

[1] carbon*n* (http://carbonn.org) is a centre for excellence that supports local governments around the globe with reporting through the carbon*n* Cities Climate Registry—http://carbonn.org/carbonn-cities-climate-registry.

Below three key developments that have impacted on the advocacy process for local governments are briefly presented, with the timeline outlined up to 2014.

The *Local Government Climate Roadmap* was an advocacy journey that began at the Local Government Climate Sessions, held in parallel to the COP13 in Bali in 2007. It provided a voice to local governments worldwide, mirroring the launch of the United Nations Climate Change Conference Roadmap, designed for nations, in

Table 1 Chronology of relevant efforts in the global response to climate change (non-exhaustive)

Year	Chronology of climate relevant efforts
1990	• First Assessment Report of IPCC (FAR)
1992	• Adoption of UNFCCC (*no specific definition of greenhouse gases*)
1993	• Launch of ICLEI's Cities for Climate Protection (CCP) Campaign
1995	• Second Assessment Report of IPCC (SAR)
	• IPCC 1995 Guidelines for GHG Inventories
	• Local authorities as observers at global climate negotiations
1996	• IPCC Guidelines and Good Practice Guidance for GHG Inventories
1997	• Adoption of Kyoto Protocol (Annex-A lists specific GHGs and sectors)
	• First Global IEA Report on CO_2 Emissions from Fuel Combustion
1998	• Launch of the Greenhouse Gas Protocol
2001	• Marrakech Accords
	• GHG Protocol Corporate Accounting and Reporting Standard (First Edition) Third Assessment Report of IPCC (TAR)
2004	• GHG Protocol Corporate Accounting and Reporting Standard (Revised)
2005	• Formation of the C40 Cities Climate Leadership Group by London Mayor Ken Livingstone (C20 then)
	• Launch of European Emissions Trading Scheme ETS
	• Launch of the US Conference of Mayors Climate Protection Agreement
2006	• Release of ISO14064 Standard
	• IPCC Revised Guidelines
	• First Global Report of the Cities for Climate Protection Campaign
	• Partnership between C40 and Clinton Climate Initiative (CCI) announced
2007	• Fourth Assessment Report of IPCC
	• Launch of Global City Indicators Facility and The Climate Registry in the USA
	• COP 13 in Bali: Launch of the Local Government Climate Roadmap[a]
2008	• Launch of European Covenant of Mayors[b]
	• Release of ICLEI's US Local Government Operations Protocol (LGOP)
	• Launch of ICLEI-US/CDP Cities Pilot Project
	• COP14 in Poznan: Local Government Climate Sessions
2009	• Launch of the first global protocol for local governments on GHGs—the International Local Government GHG Emissions Analysis Protocol (IEAP)
	• Launch of Greenhouse Gas Regional Inventory Process (GRIP)
	• Launch of Bonn Center for Local Climate Action and Reporting (carbon*n* Center)
	• COP15 in Copenhagen: Local Government Climate Lounge

(continued)

Table 1 (continued)

Year	Chronology of climate relevant efforts
2010	• Kick-off for ISO/TR14069
	• Launch of WB/UNEP/UNHABITAT Draft Standard
	• Launch of the Global Cities Covenant on Climate—the Mexico City Pact
	• Launch of carbon*n* Cities Climate Registry (cCCR) as a global reporting platform[c] (renamed carbonn Climate Registry in 2014)
	• Kick-off for drafting of US Community GHG Protocol
	• Start of Resilient Cities Congress series in Bonn
	• COP 16 in Cancun: Recognition of local governments as "governmental stakeholders"
2011	• Release of C40/CDP Cities Report
	• ICLEI-C40 MoU to design and develop the GPC
	• Release of GHG Protocol Corporate Value Chain (Scope 3) Accounting and Reporting Standard
	• Release of 2011 Annual Report of carbon*n* Cities Climate Registry
	• COP 17 in Durban: Launch of the Durban Adaptation Charter
2012	• Launch of the draft Global Protocol for Community-scale Greenhouse Gas Emissions Inventories (GPC)—developed by WRI, C40, and ICLEI
2013	• Local Government Climate Roadmap Phase II started
	• COP19 in Warsaw: First "Cities Day" organized at a COP
2014	• UN Special Envoy for Cities and Climate Change appointed, namely Michael R. Bloomberg
	• Launch of the Compact of Mayors
	• COP20 in Lima: Launch of GPC

[a]http://www.iclei.org/climate-roadmap
[b]http://eumayors.eu
[c]http://carbonn.org

determining a global action plan towards a post-Kyoto framework on climate change for the period after 2012. Partners have tirelessly campaigned to mobilise local governments, also to obtain referencing to subnational and local governments by national negotiators in Cancun, Mexico, in 2010. This in turn was used to step up activities to also ensure success in "empowerment" of local and subnational governments (empowerment implying enabling frameworks that support and assist effective local climate action).

The *Compact of Mayors*[2] was launched at the UN Climate Summit in September 2014 by C40, ICLEI, and UCLG, in partnership with UN Secretary-General Ban Ki-moon, and his Special Envoy for Cities and Climate Change, Michael R. Bloomberg, and UN-Habitat. As a global initiative, it aims at recognising new and existing local climate commitments, and make sure that these are recognised globally. Two recognised reporting platforms—carbon*n* Climate Registry and

[2] http://compactofmayors.org.

Table 2 Reporting elements in the carbonn Climate Registry [2]

	Performance (GHG emissions)	Local climate actions
• CO_2	• Local government emissions inventory	• Mitigation actions
• CO_2e	• Community emissions inventory	• Adaptation actions
• Carbon intensity		• Action plans
• Renewable energy		
• Energy efficiency		
Government and/or community level	Guided by the International Local Government GHG Emissions Analysis Protocol (IEAP)	For completed, planned, or ongoing actions
Absolute or business-as-usual		

CDP—allow cities and towns of all sizes to share their commitments publicly and to report consistently on targets, activities, and impact, on the platform of their choice. This has helped to create a global political movement, with political commitments to local climate action combined with global accountability.

Commitments and reporting are monitored by the *carbonn Climate Registry (cCR)*. The cCR is operated by the carbon*n* Center based in the City of Bonn and hosted by ICLEI. It is a mechanism created for cities and local governments to ensure transparency and accountability of local climate action through a commitment of regular reporting. By reporting on the cCR, cities and towns demonstrate leadership, contributing to transparency and accountability of local climate action. They will thus be better prepared for verification of their commitments, performance, and actions, which in turn should facilitate or ease access to global climate funds (Table 2).

4 Conclusion

Developments around the globe show that it is possible to mobilise many local governments to engage in local climate and energy action—especially when focusing on the clear benefit for their respective communities and for the world, as a local-global win-win approach.

All levels of government need to engage when dealing with climate and energy issues. These are not purely national issues, despite the strategic nature of energy. Synergy and cooperation between different levels of government are key to optimise processes and approaches. This is increasingly taking place in different countries, and is showing positive results—for example in the Nordic countries, where advanced climate action discussions are taking place, even exploring climate neutrality in some cases. However, political leadership is only one element. This should also be linked to engagement of other key stakeholders, to effectively implement comprehensive programmes that deal with these major challenges in modern life.

Processes are in place that mobilise, support, offer guidance, and address advocacy with and for local governments around the globe. Global developments were informed and guided by the Local Government Climate Roadmap, the Compact of Mayors which helped to collect political commitments and establish a harmonised reporting system, with the carbon*n* Climate Registry established as a voluntary reporting and monitoring platform for local climate action. Combined, these three elements have helped achieve progress also at the global level to draw attention to the need for, and interest in, accelerating local action. ICLEIs GreenClimateCities programme is available to local governments around the globe, developed to guide and support them in scaling up local climate action—both mitigation and adaptation.

Every local government can and should combine energy conservation, energy efficiency, and renewable energy—moving towards 100 % renewable energy communities that optimise policy, technology, and people power, to ensure that zero-emission, resilient communities are created around the globe.

References

1. IPCC (2013) Summary for policy makers. In: Stocker TF, Qin D, Plattner G-K, Tignor M, Allen SK, Boschung J, Nauels A, Xia Y, Bex V, Midgley PM (eds) Climate change 2013: the physical science basis. Contribution of working group I to the fifth assessment report of the Intergovernmental Panel on Climate Change. Cambridge University Press, Cambridge, United Kingdom and New York, NY, USA. https://www.ipcc.ch/report/ar5/wg1/
2. van Staden R (2014) Climate change: implications for cities—key findings from the Intergovernmental Panel on Climate Change fifth assessment report. University of Cambridge and ICLEI. www.iclei.org/fileadmin/PUBLICATIONS/Brochures/IPCC_AR5_Cities_Summary_FINAL_Web.pdf
3. ICLEI (2015) carbon*n* Climate Registry—five year overview report (2010–2015). http://e-lib.iclei.org/wp-content/uploads/2015/12/cCR2015_5Year_Report.pdf

Institutionalization of Solar Cities Development

Jong-dall Kim

1 Key Sectors of Solar Cities

Solar cities program is designed to systematically introduce solar and other renewable energy sources, technologies, and industries in cities, within the context of other measures such as environmental, economic, and spatial planning programs. The purpose of solar cities is to find a pathway towards economically and ecologically appropriate energy system while also taking into account limited financial and human resources as well as incomplete insight into the future development of economic and technical conditions.

In the increasing demand to solve environmental problems and introduce renewable energies in urban dimension, International Solar Cities Initiative (ISCI) has started as IEA/ISES task since 1999. Since the first ISCI Congress hosted in Daegu, 2004, ISCI Congress has been recognized as a major international event in the field of green urban energy policy, consisting of mayors meeting, academic conference, business/citizen's forum, and exhibition.

This chapter introduces how to institutionalize solar cities programs from renewable energy and urban policies to practices that reduce city greenhouse gas emissions. The case study of Daegu city is also introduced.

How can we achieve solar cities development? First of all, six main sectors can be summarized in diverse discussions for green development as follows.

J.-d. Kim (✉)
International Solar Cities Initiative (ISCI), Kyungpook National University,
Daegu, South Korea

School of Economics and Trade, Kyungpook National University, Daegu, South Korea

Research Institute for Energy, Environment and Economy (RIEEE),
Kyungpook National University, Daegu, South Korea
e-mail: jdkim@knu.ac.kr

T.S. Uyar (ed.), *Towards 100% Renewable Energy*, Springer Proceedings
in Energy, DOI 10.1007/978-3-319-45659-1_3

- Renewable energy (solar, wind, biomass, fuel cell), green buildings (retrofit, green products and materials, LEED construction)
- Clean transportation (alternative fuels, public transit, hybrid and electric vehicles), water management (water reclamation, grey water and rainwater systems, low-water landscaping, water purification, and management)
- Waste management (recycling, municipal solid waste salvage, brownfield and remediation)
- Land management (organic agriculture, habitat conservation and restoration, urban forestry and parts, reforestation and afforestation)

Among these sectors, sustainable energy is central to addressing the current concerns about climate change and economic development. Environmental risks resulting from obsolescent technology and outdated models of economic growth constitute one of the major challenges facing the energy sector. The era of large-scale power plants, high-voltage transmission systems, massive oil cracking and refining complexes, and huge coal mining is being forced to end to curb the effects of fossil fuel combustion.

Since the early 1970s, Asian countries have registered the most rapid economic growth of any region in the world, but we have achieved this with the lowest ecological carrying capacity. Due to rising energy demand, greenhouse gas (GHG) emissions (around 30 % share of GHG) and other environmental pressures are testing the limits of this capacity. In this respect, the key question facing Asian countries is whether this increase will occur in a sustainable manner, or whether it will reproduce the patterns that industrialized countries witnessed in the past. There is increasing evidence that low carbon development hinges upon this choice [3] (Fig. 1).

Throughout the sectors, development and deployment of technologies is the engine of solar cities development. Technologies such as renewable and low-carbon energy (supply and storage), advanced water treatment, LED application, and high-tech green city are representatives in current green economy. Second, high future commercial rates and associated high initial costs for research, development, and marketing of renewable technologies and products prevent firms from being voluntarily interested in green future investment. Thus, the green economy may need government subsidies as market incentives to motivate firms to invest and produce green products and services. And last, solar cities development requires to promote public awareness and participation as important because public/business participation in production and consumption of green products is essential. Solar cities development programs need a broad and deep support from both the highest policy-making level and at the grassroots level, for which institutional capacities need to be formed. Partnership model including local government, business, experts, and social groups is an important actor in green development. Further, public interest groups are also powerful sources of support for this path. The government can provide financial assistance for the development of public interest groups and professional societies and associations. It will be green development regime that can help to overcome the most powerful institutional barriers and to implement new path.

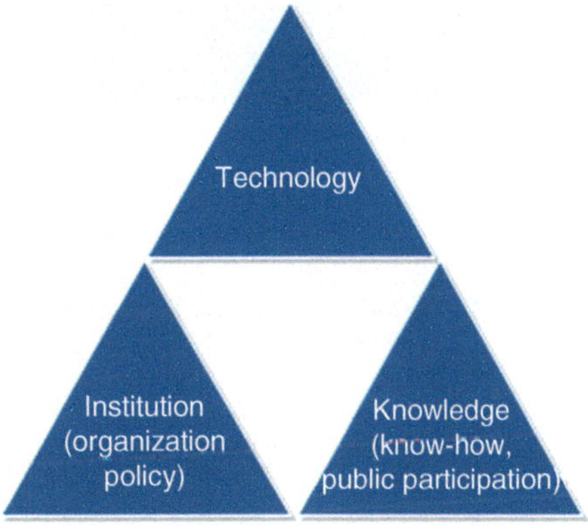

Fig. 1 Policy areas for solar cities development

In order to realize policy measures at the top of the city agenda or national agenda, the strong green leadership is required with new institutions such as Committee on Green Development composed of key decision makers from different sectors and excellent advisory groups.

2 Institutionalization of Solar Cities Development

In order to sustain solar cities development policy, three actions may be needed. First action is to lay the legislative foundation like the Basic Act on Green Development. Second one is to have a long-term plan and consecutive implementation plans for green growth. Back casting method will be very effective for reviewing long-term plan and revise its course efficiently and effectively. Third, the creation and expansion of the green budget is finally an essential part for the green development which may require reorienting public and private finance toward green policy, R&D, and SOC. Through these actions, renewable policy and technology development will be sustained (Fig. 2).

I want to introduce my own experience in pursuing green development in my city for 12 years. The Solar City Daegu Project started in 2000 and was designed to systematically introduce solar and other renewable energy sources, technologies, and industries on a large scale, within the context of other measures such as environmental, economic, and spatial planning programs. The project aims at sustainable emissions of greenhouse gases and focuses on urban business and industry in solar and other renewable energy technology systems while also establishing a new urban development of employment base for the future. Daegu city gives priority to fulfilling its policies for the development of an environment-friendly city, one of which is the use of alternative fuels to reduce emissions of exhaust gases by automobiles.

The vision of Solar City Daegu 2050 is to "Create a Green and a New Economy of Solar City Daegu." Through systematic implementations of solar and other renewable energy projects, Solar City Daegu 2050 foresees sustainable development

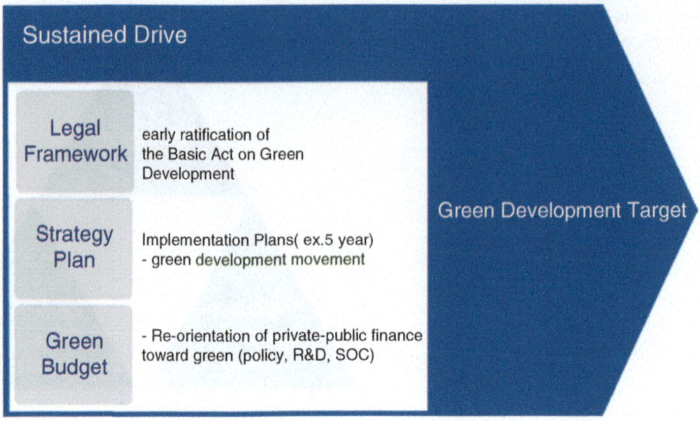

Fig. 2 Actions

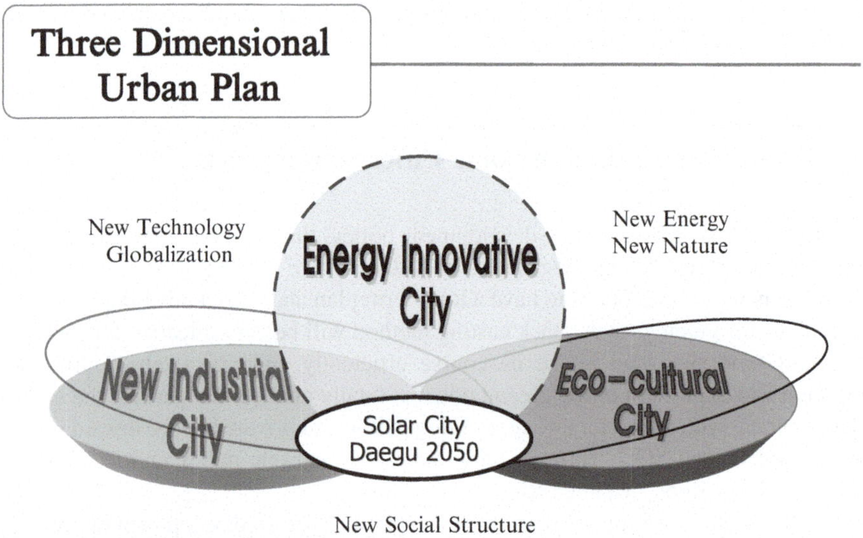

Fig. 3 The three pillars of green development

at the municipal level, which can balance between environmental preservation and economic development. In order to achieve this vision, three dimensions of key urban plans are developed. Those are energy-innovative, new industrial, and eco-cultural cities (Fig. 3).

The objective of energy-innovative city is the development of an innovative implementation system through which energy demand-side management and renewable energy will be intensively and systematically introduced in the city. The objective of new industrial city is the development of new industries and new employment opportunities through intensive promotion of the solar and hydrogen economy. The objective of eco-cultural city is the provision of a clean nature and advanced healthy culture.

Using a back casting methodology, the visions and key target projects are being used for strategic action plans. The 50-year plan is being evaluated and modified every 5 years. Implementation plan will be developed each 10 years through which the solar city project will be active in a number of planning processes for sustainable urban design in the city of Daegu, including energy and environmental planning, economic and technology planning, planning for sustainable transportation futures, and urban planning.

3 Conclusions

Based on theoretical analysis and experiences, I would suggest some ideas for institutionalization of solar cities development.

First, energy and environmental policy needs to be integrated with economic and technology policy. Green development is related with many other socioeconomic and spatial sectors in nature. And these programs involve the interaction or dominance of interrelated forces and also include various means to ensure that all regulatory, market, and legislative forces work together. This means that an efficient implementation of a number of programs requires a comprehensive strategy with a coordinated manner. All sectors' efforts need to be concentrated for green development (industry, transportation, residential, and commercial and public sectors). Region and citizens should be producers and cooperators beyond simple consumers (high-efficiency equipment, green home and building, green transportation, and green energy are closely linked with technology, policy, and citizen's life).

Second, the government needs to design coordination channels for solar cities development. Coordination area is planning language, planning scope, DB, indicators, and integrated evaluation model.

Third, guidelines for institutionalization of solar cities development are also needed to be prepared. There are many conflicting relations, understanding about green development among generations, among regions and there are sustainable/multiuse of ecological resources.

References

1. Climate Change Ambassador, ROK, New paradigm of low carbon development, 2009
2. Droege P (2006) The renewable city: a comprehensive guide to an urban revolution. Wiley, Chichester
3. Kim J (2010) Energy transition issues in South Korea's sustainable development. Harvard Asia Quarterly, Spring

State-of-the-Art and New Technologies of Direct Drive Wind Turbines

Klinger Friedrich and Müller Lukas

1 Advantages of Direct Drive Wind Turbines

Gearbox problems are responsible for many turbine failures. According to data collected by Germanischer Lloyd (GL), 26 % of turbine downtime is due to the gearbox, another 13 % to the shaft and couplings, and 17 % to the generator. Lifetime of the gearboxes sometimes does not reach 5 years and its replacement creates cost of up to 200.000,00 € for a 1.5 MW turbine in Germany which is more than 10 % of the turbine investment cost. Figure 1 shows a damaged gearbox.

On the right side of Fig. 2 a conventional turbine concept is shown. The torque produced by the rotor is reduced by a gearbox at a very high gear ratio of 1:100. At the same time, the speed at the output shaft is increased by the same ratio adjusted to the high-speed generator that feeds electric power in the grid. On the left side, a direct drive system is shown using a high-torque generator that is directly coupled to the hub with the rotor blades. In both cases, converters enable variable speed of the rotor in order to increase the aerodynamic efficiency.

Gearless turbines have important advantages compared to turbines with the conventional drive train:

- Maintenance requirements are reduced close to zero.
- Downtime due to gearbox failure does not exist.
- Lifetime for bearings is increased and lubrication requirements are reduced due to low-speed operation.
- Therefore the technology availability of direct drive is very high.

K. Friedrich (✉) • M. Lukas
INNOWIND Forschungsgesellschaft mbH in cooperation with Fraunhofer IWES, Altenkesseler Str. 17/D2, Saarbrücken 66115, Germany
e-mail: f.klinger@wind-energy-research.de

T.S. Uyar (ed.), *Towards 100% Renewable Energy*, Springer Proceedings in Energy, DOI 10.1007/978-3-319-45659-1_4

Fig. 1 Gearbox failure

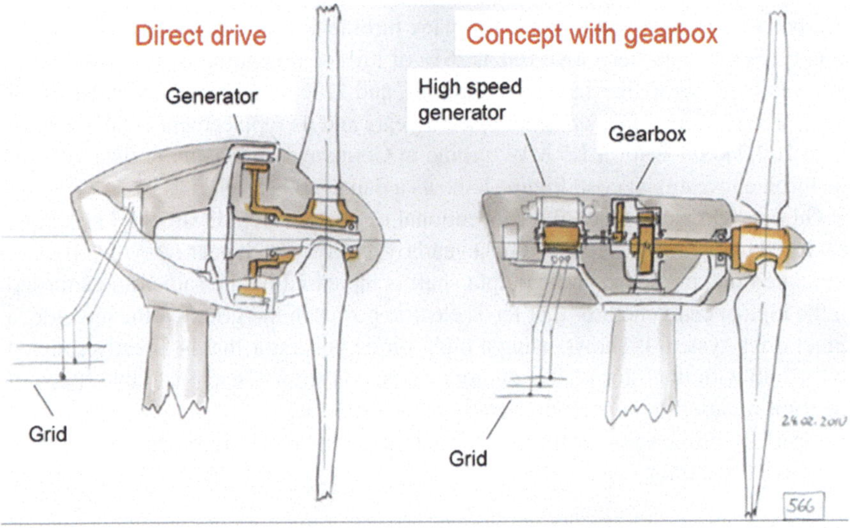

Fig. 2 Comparison of wind turbine with and without gearbox (by author)

- Energy yield increases about 10 %.
- The overall efficiency is about 10 % higher as there are no losses in the gearbox.

Figure 3 presents the power train efficiency for different drive train concepts over a range of wind speed from 20 to 100 % of rated wind speed.

The direct drive concept with permanent magnet excitation (PMDD) shows important advantages in efficiency at lower wind speeds, upper line in orange color.

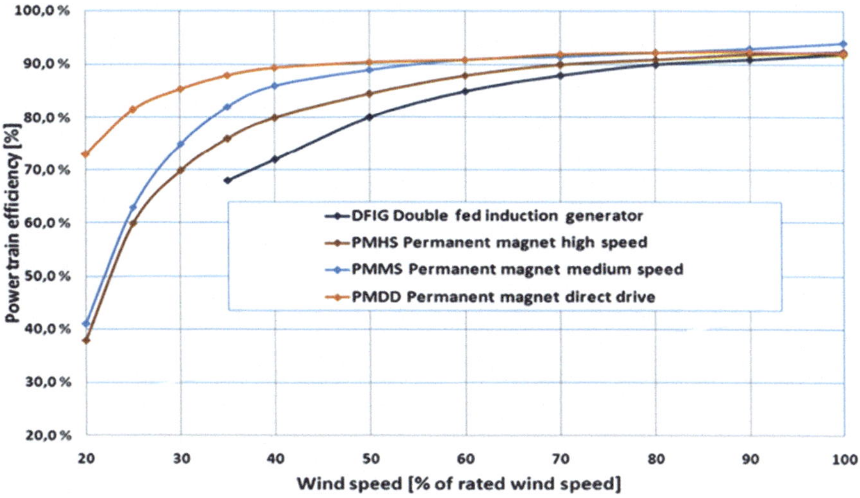

Fig. 3 Efficiency comparison for different drive train concepts (by author)

The lower line in black color represents the efficiency of the drive train of a conventional Danish turbine concept with a gearbox and a high-speed double-fed induction generator followed by a high-speed synchronous generator with permanent magnet excitation and gearbox. The blue line represents the concept with low-ratio gearbox 1:10 and a medium-speed synchronous generator with permanent magnet excitation.

2 Bearing Concepts

The rotor of electric machines is usually located inside the stator as demonstrated on the left side of Fig. 4. The innovative generator concept by VENSYS Energy AG is using outer rotor and permanent magnet excitation on the inner side of the rotor as shown on the right side of Fig. 4. In case the rotor uses permanent magnets, the outside diameter DA2 of such machines is smaller compared to DA1 of inner rotor concepts for the same air gap diameter D.

An innovative and robust bearing concept with only a large-diameter slewing bearing is used for Vensys 100, Leitwind, and XEMC turbines as shown in Fig. 5 for inner rotor and Fig. 6 for outer rotor turbines. A further innovation is shown in Fig. 6 on the right side, where the generator is integrated into the rotor hub.

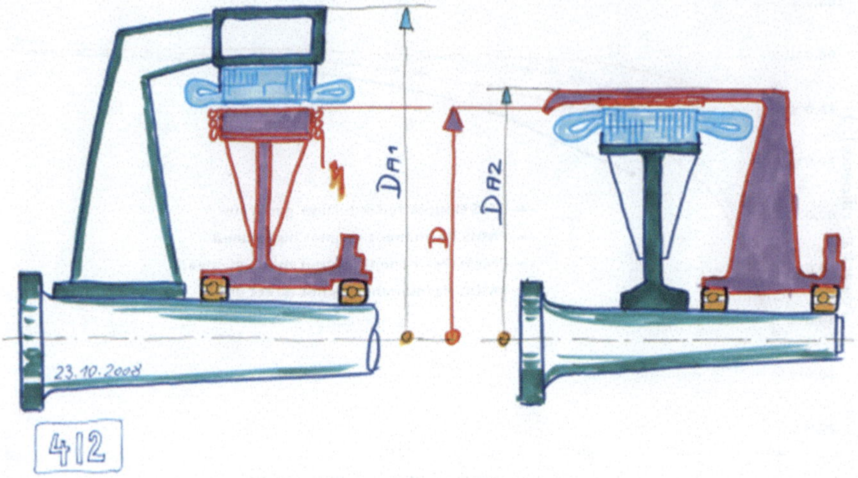

Fig. 4 Comparison of outer and inner rotor (by author)

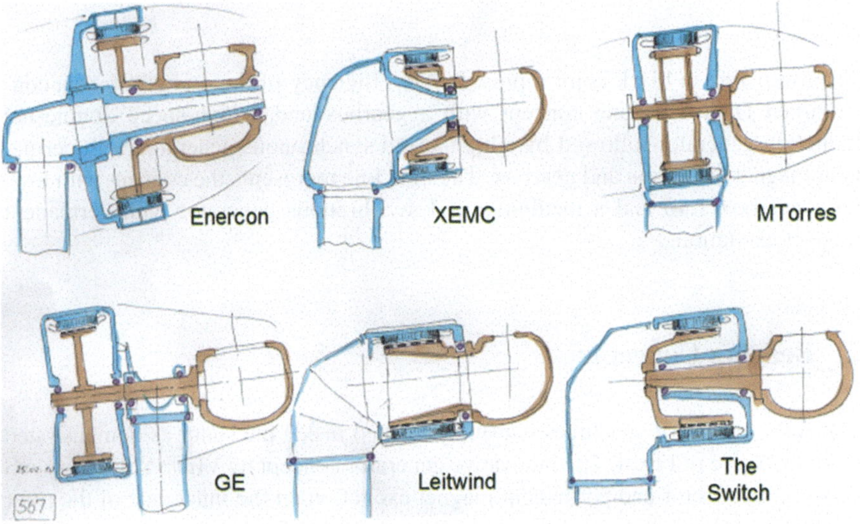

Fig. 5 Bearing concept with inner rotor (by author)

3 Generator Design Concept: State of the Art

Worldwide there are several manufacturers for direct drive wind turbines, namely Enercon, Vensys as original designer and his license holder Goldwind, and Zephyros as original designer and his license holder XEMC (Table 1).

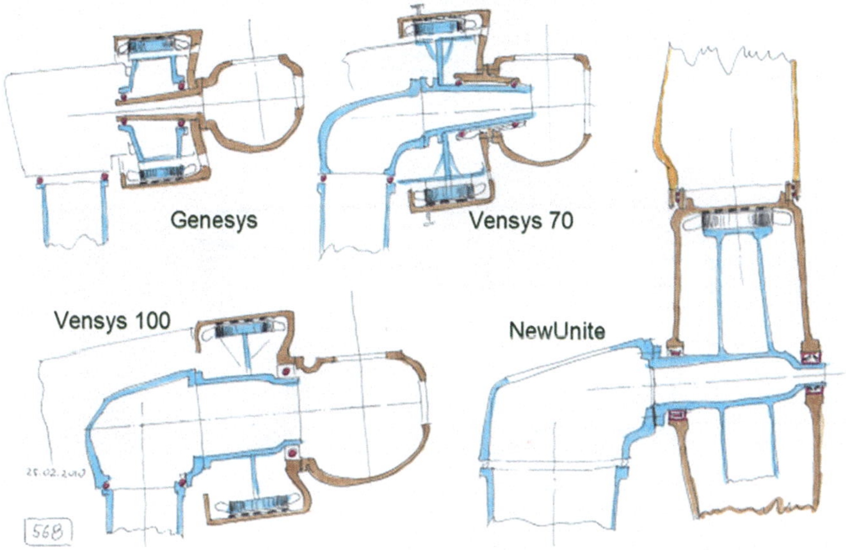

Fig. 6 Bearing concept with outer rotor (by author)

Table 1 State-of-the-art wind turbines

Manufacturer	Power (MW)	Rotor arrangement	Number of blades
Enercon	3	Internal	3
Genesys	0.6	External	3
Vensys	1.5	External	3
Vensys	2.5	External	3
Harakosan	2	Internal	3

3.1 Enercon

Enercon's E 82 (Figs. 7 and 8) is the most important direct drive product worldwide with 82 m rotor diameter and 2 MW rated power following a simple concept to have only a few slowly rotating parts. The generator has an inner wound rotor with external excitation and an outer stator of about 5 m diameter. Two bearings on a rigid shaft carry the hub and the generator rotor. The cooling system uses fresh air from outside. Enercon has sold different turbines using this concept with more than 24,000 MW worldwide. Their market share in Germany is 60 %. In 2011 Enercon launched a 3 MW Turbine E-101 that is using combined cooling by air and liquid.

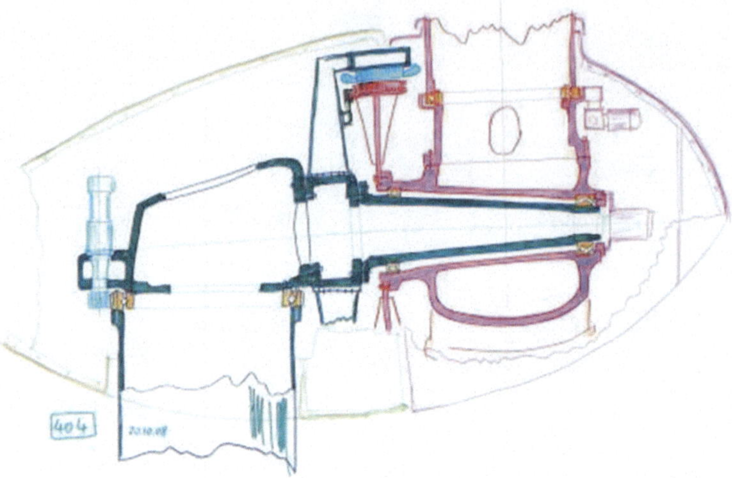

Fig. 7 Design concept of Enercon E82 (by author)

Fig. 8 Enercon (*origin*: www.enercon.de)

3.2 *Vensys and Goldwind*

Vensys has a similar design (Figs. 9 and 10), but compared to Enercon the generator has a different layout with a permanent magnet excitation system on the outer rotor. This leads to a smaller outer diameter and higher efficiency. The iron core is air cooled by the outside airflow through a channel on the stator. Goldwind in China, Eozen in Spain, CKD in the Czech Republic, and Impsa in Brazil are using this concept by license. The rotor is 70, 77, and 82 m, and the rated power is 1.5 MW.

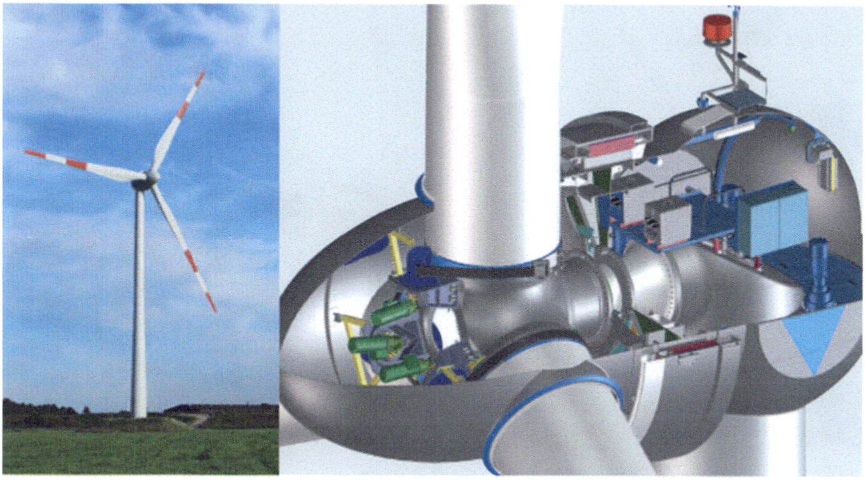

Fig. 9 Vensys V70 (*origin*: Vensys brochure)

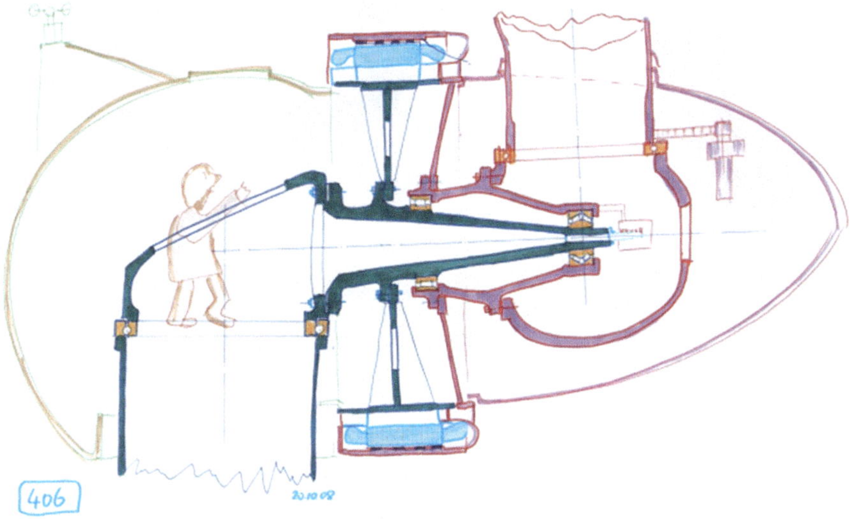

Fig. 10 Design concept 1.5 MW of Vensys/Goldwind V70 (by author)

The Vensys/Goldwind wind turbine V90 (Figs. 11 and 12) has 2.5 MW rated power that is 66 % more than his forerunner. And thanks to the active cooling system this turbine is only 10 % heavier than its forerunner.

Fig. 11 Vensys V90 (*origin*: Vensys brochure)

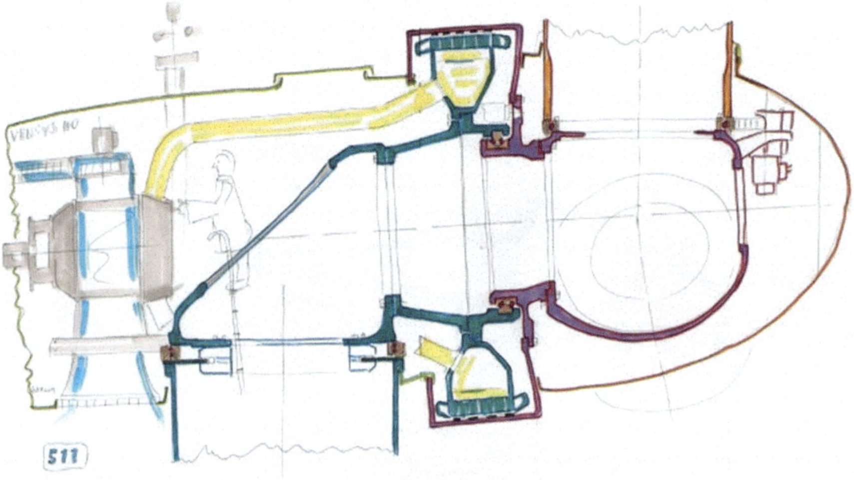

Fig. 12 Design concept 2.5 MW of Vensys/Goldwind V90 (by author)

3.3 *Zephyros and XEMC*

The Harakosan Turbine Z72 (see Figs. 13 and 14) is based on the design of the Zephyros project from the Netherlands. The generator is a multipole permanent magnet generator that is directly mounted to the hub. The magnets take care of the rotor excitation, so there is no external field excitation needed. This implies that the generator losses are reduced by 25 % compared to a wound rotor. The stator, equipped with cooling fins, is located on the outside and is cooled by the outside air.

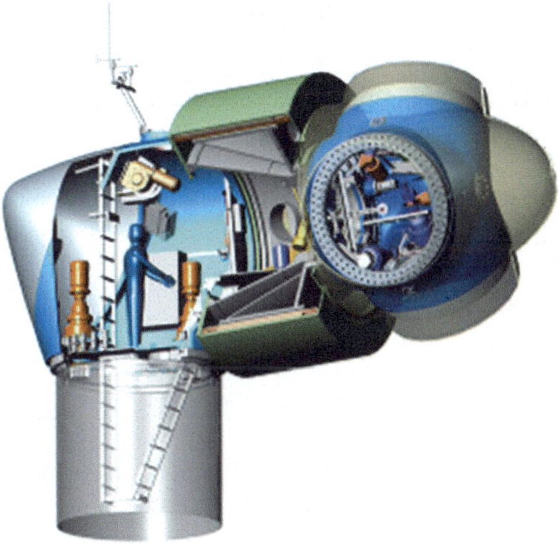

Fig. 13 Harakosan Z72 (*origin*: Harakosan brochure)

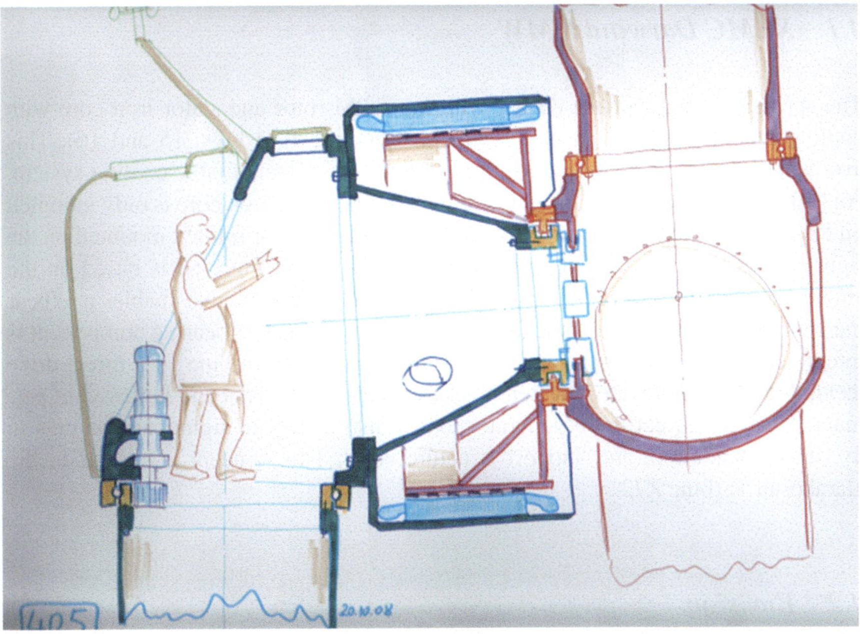

Fig. 14 Design concept of Harakosan Z72 (by author)

The stator windings have a 7 kV insulation and are connected to an ABB ACS 1000 medium-voltage converter. The generator, designed and delivered first by ABB Finland and XEMC China, has a weight of 49 t and the dimensions are 3.9 m x 2.0 m. A specially designed cylindrical roller bearing of large diameter carries rotor and hub.

4 Generator Design Concept: Future Concepts

In the following pages, several future concepts, which are in development or prototype phase, are shown and described.

Manufacturer	Power (MW)	Rotor arrangement	Number of blades
Siemens	3	External	3
Alstom	6	Internal	3
XEMC Darwind	5	Internal	3
Envision	3.6	Internal	2
Sway	10	External	3
InnoWind	3.2	External	3
(The Switch)	3	External	–

4.1 *XEMC Darwind 5 MW*

The specialty is a generator design with inner PM rotor and stator iron core with cooling fins at the outer surface for passive cooling (see Figs. 15 and 16). This avoids the need of an energy-consuming and potentially unreliable cooling system. An additional inner cooling circuit is provided. The stator iron core is only mounted on the left side and the bearing of the hub and the generator rotor is mounted on the right side of the coned inner shaft. The XEMC-Darwind turbine is based on the single main bearing concept. The bearing has a large diameter that enables it to bear the enormous force of the rotor: there is no central shaft. The bearing component is integrated into the generator and is possible because the turbine uses direct drive technology. This innovative design not only creates a compact and lightweight permanent magnet direct drive generator, but it also results in significant savings in logistics, foundation, and support structure costs. This concept is similar to the Harakosan Turbine Z72.

4.2 *Envision*

Chinese-owned Envision Energy has launched the E-128 3.6 MW (see Figs. 17 and 18) as a new concept. The prototype is announced to be erected in summer 2012 in Denmark where the R&D Team (Danish Global Innovation Centre) is based. Two bearings and a stationary main pin on the right side carry the rotor which is a well-known bearing concept similar to Enercon. The wind turbine uses a flexible torque

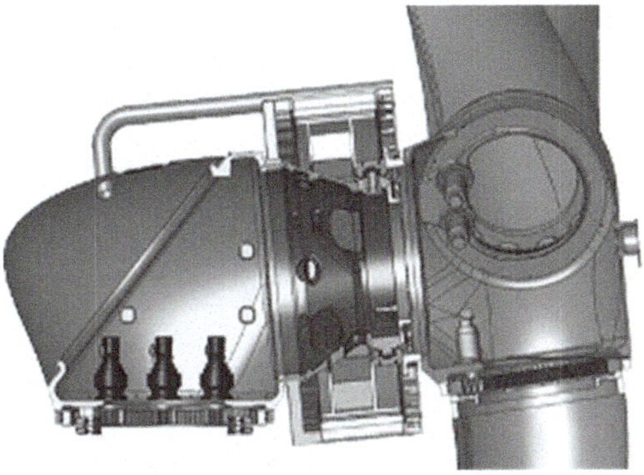

Fig. 15 XEMC Darwind (*origin*: www.xemc-darwind.com)

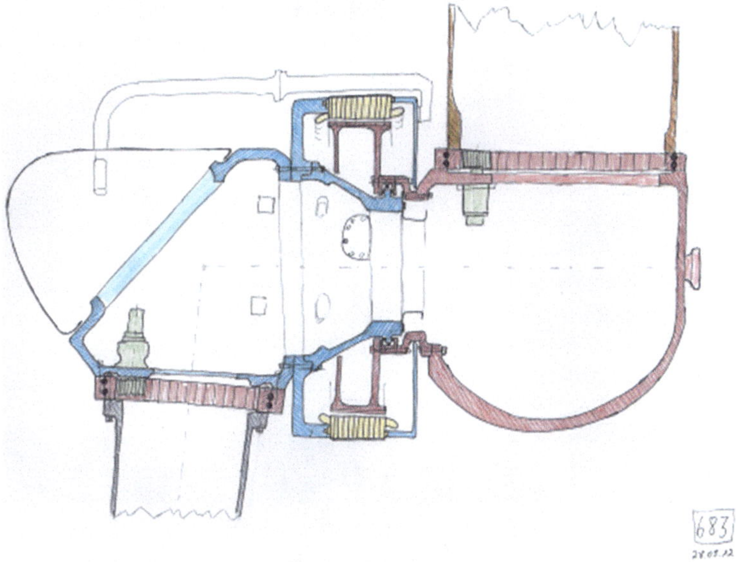

Fig. 16 Design concept of Zephyros/XEMC (by author)

shaft to connect the hub on the right side of the tower with the generator on the left side. The torque shaft is out of carbon fiber and is inspired by Spanish pure torque concepts. The idea is to separate the deformation of the rotor bearing system by wind loads from the bearing system of the generator. Each 62 m blade consists of a 20 m inner section with a fixed blade angle and a pitchable 42 m outer blade, manufactured by LM. The pitch bearings and electric pitch mechanism are located in between inner and outer blade part.

Fig. 17 Envision (*origin*: Envision brochure)

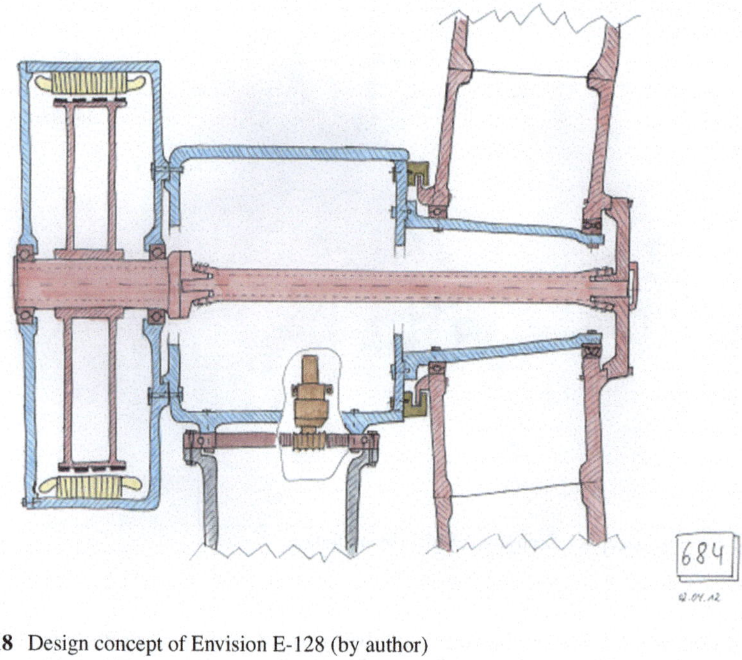

Fig. 18 Design concept of Envision E-128 (by author)

4.3 *Sway AS*

The 10 MW turbine is designed for both fixed seabed and floating installations and is scheduled to be commercially available before 2015 (see Figs. 19 and 20). The generator rotor resembles a bicycle rim with tension members connected to the

Fig. 19 Sway AS (*origin*: www.turbosquid.com/3d-models/3d-model-sway-wind-turbine/563463)

Fig. 20 Sway perspective and cross-sectional view (by author)

bearings on a rigid shaft (blue colored). The generator stator is rigidly connected to the shaft by tension members, as well. The rotor blades form forks which are rotational mounted on a fixed shaft with additional tension members to transfer the torque to the generator rotor. Compared to the in-hub wind turbine by InnoWind (see Fig. 27) SWAY has an open hub with large diameters suited for excellent passive cooling.

Fig. 21 The Switch PMG3200-12 (*origin*: www.theswitch.com)

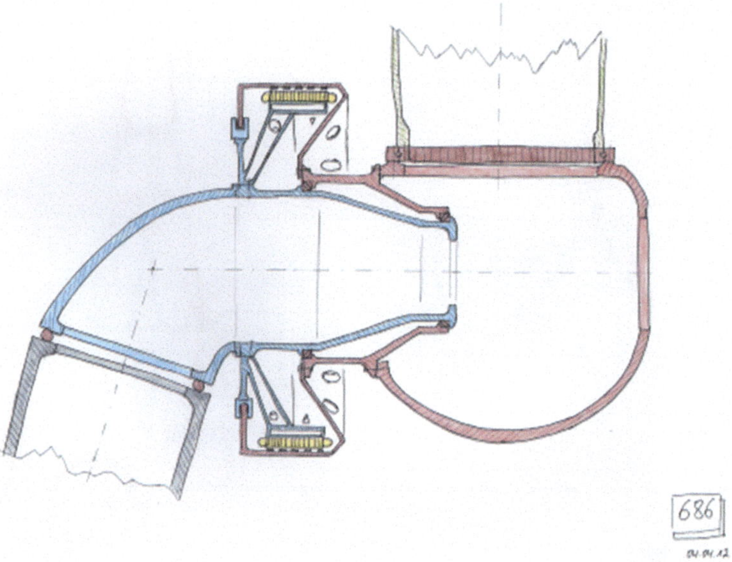

Fig. 22 The Switch PMG3200-12 generator integrated (by author)

4.4 *The Switch*

The Switch is fully committed to wind power generation as manufacturer of generators and convertors (see Figs. 21 and 22). The purpose-built permanent magnet generators (PMG) cover all wind power applications. Each PMG is designed with special magnet shapes and arrangements to match specific wind conditions for smooth operation and maximum efficiency. The Switch PMGs provide excellent availability and productivity. By eliminating cogging, the overall mechanical stress is reduced. This improves reliability and extends the turbine's lifetime. The Switch low-speed, direct-drive PMGs operate without any gearbox to create superior all-round drive train efficiency. Torque variations are effectively filtered to eliminate undesirable vibration. Direct drive PMG versions are available with either inner or outer rotors from 1.6 to 6 MW rated power.

4.5 Siemens

The SWT-3.0-101 rotor is a three-bladed cantilevered construction, mounted upwind of the tower. The power output is controlled by pitch regulation. The rotor speed is variable and designed to maximize the aerodynamic efficiency (see Figs. 23 and 24).

The B49 blades are made of fiberglass-reinforced epoxy in Siemens' proprietary IntegralBlade® manufacturing process. In this process the blades are cast in one piece to eliminate weaker areas at glue joints. The blades are mounted on pitch bearings and can be feathered 80° for shutdown purposes. Each blade has its own independent pitching mechanism capable of feathering the blade under any operating condition. The blade pitch arrangement allows for optimization of the power output throughout the operating range, and the blades are feathered during standstill to minimize wind loads. The rotor hub is cast in nodular cast iron and is fitted to the generator rotor with a flange connection. The hub is sufficiently large to provide a comfortable working environment for service technicians during maintenance of blade roots and pitch bearings from inside the structure. A cast, hollow, and fixed main shaft ensures a comfortable internal access from the canopy to the hub. The rotating parts of the wind turbine are supported by a single, double-tapered roller bearing. The bearing is grease lubricated. The generator is a fully enclosed synchronous generator with permanent magnet excitation. The generator rotor construction and stator winding are designed for high efficiency at partial loads. The generator is positioned between the tower and the hub producing a comfortably lean arrangement of the internals in the nacelle. The mechanical brake is fitted to the non-drive end of the generator rotor and has three hydraulic calipers. A cast bed frame connects the shaft to the tower. The yaw bearing is an externally geared ring with a friction bearing. A series of electric planetary gear motors drive the yawing. The weather screen and housing around the machinery in the nacelle are made of fiberglass-reinforced laminated panels with multiple fire-protecting properties. The design implies fully integrated lightning and EMC protection. The SWT-3.0-101 wind turbine is mounted on a tapered tubular steel tower. The tower has internal ascent and direct access to the yaw system and nacelle. It is equipped with platforms and internal electric lighting [origin: www.energy.siemens.com/hq/en/power-generation/renewables/wind-power/].

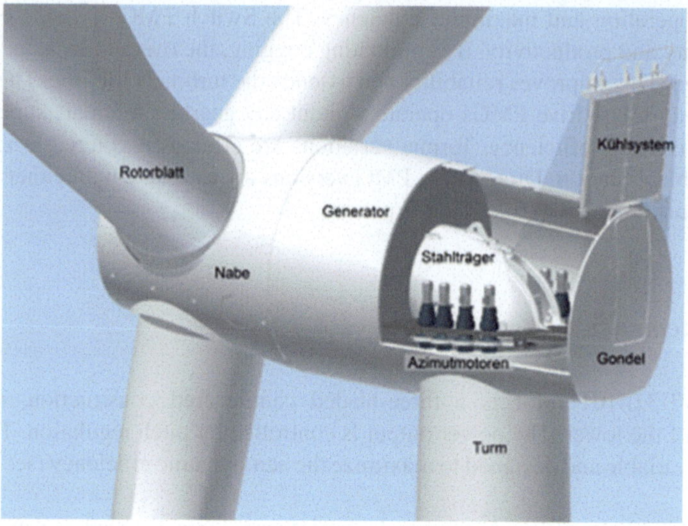

Fig. 23 Siemens SWT-3.0-101 (*origin*: www.energy.siemens.com/hq/en/power-generation/renewables/wind-power/)

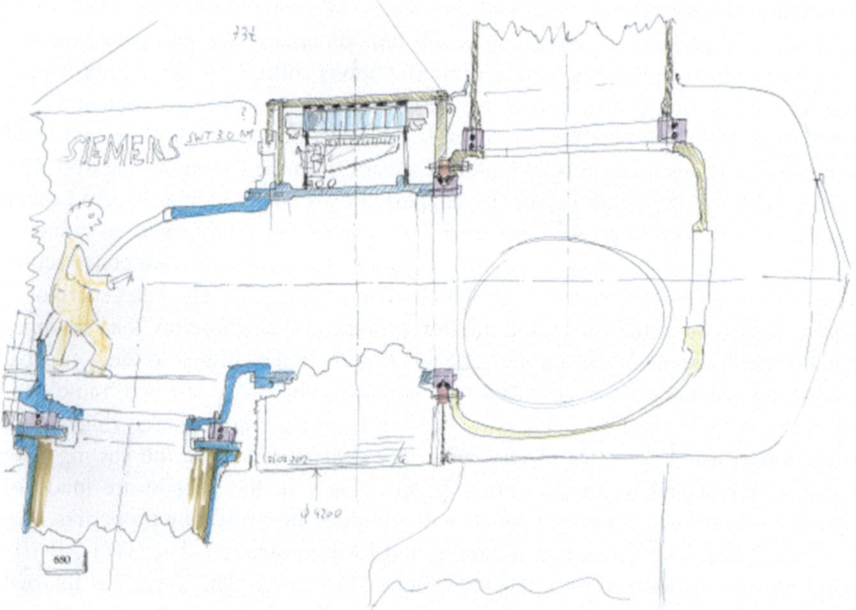

Fig. 24 Design concept of Siemens SWT-3.0-101 (by author)

Fig. 25 Alstom (*origin*: www.alstom.com)

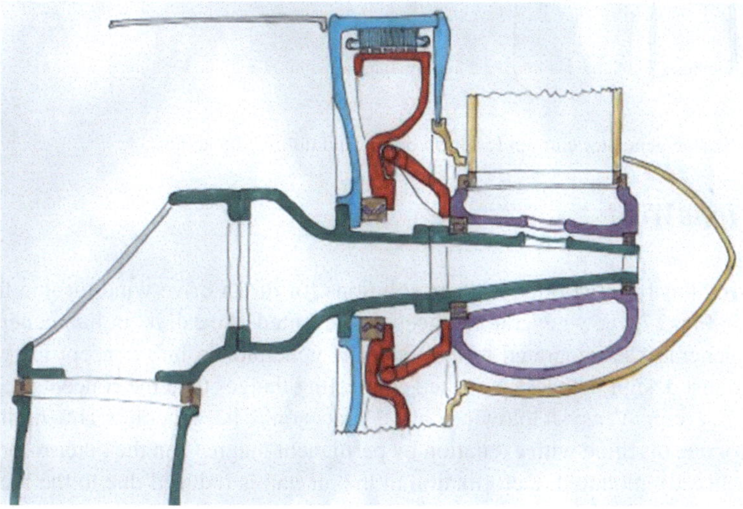

Fig. 26 Design concept of 6 MW Haliade 150 (by author)

4.6 *Alstom Haliade*

The 6 MW Haliade 150 wind turbine has been developed in response to a call for tenders launched by the French Government. It is the largest turbine prototype so far with a 150 m rotor diameter (see Figs. 25 and 26). The ALSTOM PURE TORQUE™ design protects the generator and improves performance by diverting unwanted stresses from the wind safely to the turbine's tower through the main frame.

The design separates the turbine rotor and generator to ensure that only torque is transferred to the generator. This allows the minimum sufficient air gap to be maintained between the generator rotor and stator. The "Advanced High Density" direct drive PMG, supplied by power conversion specialist Converteam, is a more compact and lightweight design compared to earlier generation direct drive systems [origin: www.alstom.de].

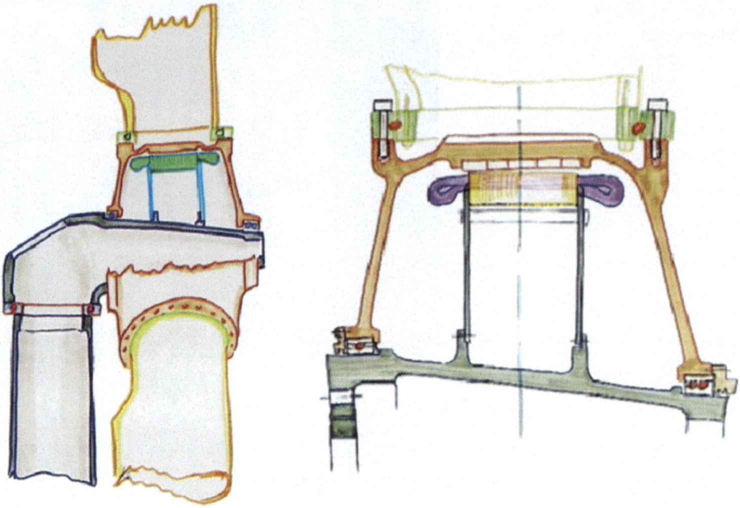

Fig. 27 In-hub generator concept for direct drive wind turbine (by author)

4.7 InnoWind

InnoWind has been working on new solutions for direct drive wind turbine for 20 years. In Fig. 27 a new generator concept is presented. We call it "in-hub generator" as the generator is integrated to the hub. The generator in hub concept leads to a more robust design with less parts and connecting flanges than the concepts according to Enercon, Vensys/Goldwind, and Haracoson. The generator is a multipole synchronous machine with excitation by permanent magnets on the outer rotor with blades directly mounted. Deformation in the air gap is reduced due to the location of the main rotor bearings on both sides of the generator.

5 Conclusion

Direct drive wind turbine concepts with gearless drive trains have been developed and improved in the last 20 years. Compared to turbines with gearboxes, reliability and efficiency are increased and cost for maintenance and downtime is reduced. Their design is more robust with half of the parts and components. The newest turbine designs in the range of 2–3 MW are proving that the tower head mass is at the same level as of gearbox turbines. Offshore turbines are under development for 5–10 MW and the trend for onshore turbines is also in the direction of direct drive.

State of the Art in Polycrystalline Compound Thin-Film Photovoltaics

Bülent M. Başol

1 Introduction

Global annual PV installations have grown from about 0.25 to over 20 GW/year between 2000 and 2011. This rapid growth in the market was accompanied by impressive reductions in manufacturing costs that brought down the world average of the installed PV system price from over $7/W in 2007 to close to $3.50/W in 2011 [1]. The corresponding levelized cost of energy (LCOE) in 2011 was in the range of 14–23 cents/kWh, which is close to the cost of retail electricity in certain parts of the world. Despite these developments the goal of grid parity for solar electricity still remains to materialize. Accomplishment of grid parity requires PV-generated electricity costs to reduce down to the 5–6 cents/kWh range, which in turn calls for an installed system price of about $1/W in the USA.

Thin-film PV has attracted much attention and considerable R&D effort since early 1980s because of its low cost potential. Some of the attractive features of thin-film devices can be summarized as follows:

1. Thin-film solar cells employ device structures with only 1–10 μm thick active layers.
2. Thin-film cells can be processed employing cost-effective film deposition approaches.
3. Devices can be formed over low-cost substrates which may be rigid or flexible.
4. Thin-film cells may be monolithically integrated and interconnected using cost-effective laser scribing approaches to form modules.
5. Because of their high bandgap, thin-film PV modules provide high energy yield (kWh of energy generated per peak−kW of modules installed), especially in locations with hot climate.

B.M. Başol (✉)
EncoreSolar, 6541 Via Del Oro, Unit: B, San Jose, CA 95119, USA
e-mail: bbasol@encore-solar.com

T.S. Uyar (ed.), *Towards 100% Renewable Energy*, Springer Proceedings in Energy, DOI 10.1007/978-3-319-45659-1_5

6. Thin-film modules, in general, have shorter energy payback periods compared to the crystalline Si products, the lowest payback times being in the order of a few months.

The share of thin-film products in the global PV market had reached about 17 % in 2009. CdTe modules dominated this market share with about 13 % and the balance was distributed between CIGS and amorphous Si (a-Si) products. Although the overall shipments increased, the market share of thin-film PV declined in 2010 and 2011 due to the precipitous drop in the wafer Si module prices and the rapid growth of large-scale PV installations. Presently, thin-film PV's share in the global market is in the range of 13 %, and these technologies are getting steep competition from low-cost wafer Si modules manufactured in China. In this chapter we present a brief review and the current status of two important polycrystalline thin-film PV technologies, one based on CdTe and the other on CIGS-type absorbers. Both of these technologies are in large-scale (>1 GW/year) manufacturing at this time.

2 Device Structures

The two configurations employed in the fabrication of CIGS- and CdTe-based solar cells are shown in Fig. 1. In the "substrate" structure used for CIGS solar cells, device fabrication involves the steps of (1) back contact deposition onto a substrate, which may be a sheet of glass or a metallic or polymeric foil; (2) formation of the absorber layer (p-type CIGS) over the back contact layer; (3) deposition of a junction partner layer (also called a buffer layer); and (4) deposition of a transparent top contact through which light enters the device. There may also be a collection grid deposited on the top surface of the device to reduce the series resistance.

In CIGS cells the back contact typically comprises Mo since this material is relatively stable in the high-temperature CIGS growth environments containing Se and/or S species. Two of the successfully used buffer layers in CIGS solar cells are sub-100 nm thick films of CdS and Zn(O,S,OH). The transparent conductor is a transparent conductive oxide (TCO) stack comprising i-ZnO/doped ZnO, i-ZnO being a high-resistance undoped layer and the doped ZnO film is doped with Al (if sputtering is used to deposit the film) or B (if MOCVD method is employed). Sputter-deposited indium tin oxide (ITO) may also be used in place of the doped ZnO layer in the stack.

In the "superstrate" configuration depicted in Fig. 1 the order of deposition for the various layers in the solar cell structure is reversed. In this case the superstrate needs to be transparent. Glass is the most commonly used material for this purpose. Solar cell fabrication involves the steps of (1) deposition of a transparent top contact over the superstrate, (2) deposition of a junction partner layer (also called a buffer layer), (3) formation of the absorber layer (p-type CdTe) over the junction partner layer, and (4) deposition of a back contact.

In CdTe cells the TCO layer is typically a fluorine-doped tin oxide (SnO_2:F) film. The junction partner layer is CdS and the back contact may comprise an inert material such as C and/or a refractory metal as well as Al as the current carrying layer.

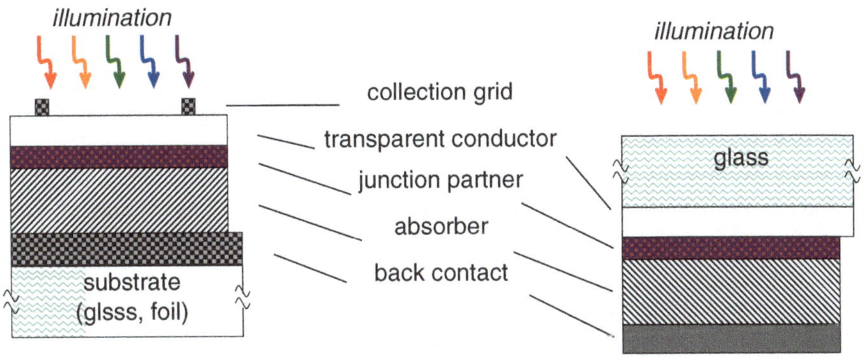

Fig. 1 "Substrate" (*left*) and "superstrate" (*right*) device configurations employed for the fabrication of CIGS and CdTe solar cells, respectively

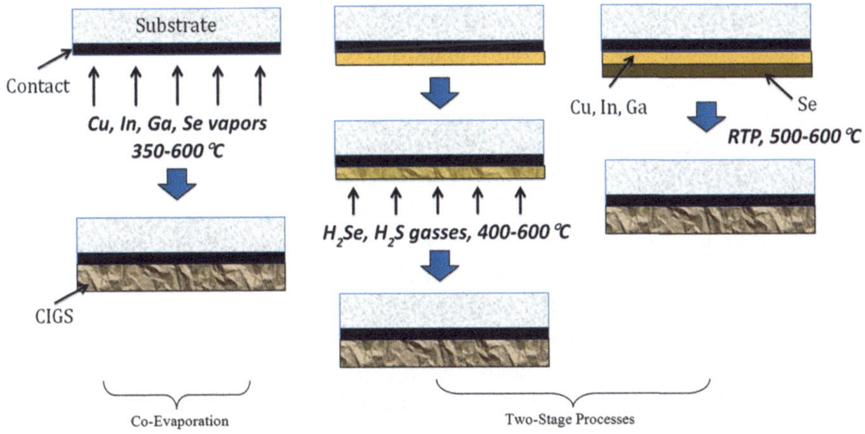

Fig. 2 CIGS solar cell manufacturing approaches

3 CIGS Technologies

Copper–indium–gallium–selenide/sulfide (CIGS) group of materials are chalcopyrite semiconductors that have been developed as thin-film solar cell absorber layers since early 1980s. CIGS layers are commonly grown by co-evaporation or two-stage techniques. As can be seen from the sketch of Fig. 2, co-evaporation method involves delivery of the elemental components of the absorber layer, i.e., Cu, In, Ga, and Se, onto a heated substrate from separate evaporation sources in a vacuum system. The vapors react and form a film of the compound on the substrate surface. The substrate temperature is usually varied in the range of 350–600 °C during the process and the composition of the incoming vapor is closely controlled to vary the composition of the deposited species. The most successful co-evaporation approach to date utilized a three-step process which involves deposition of In, Ga, and Se during the first step of the process, forming a (In,Ga)–Se layer. This is then followed by the deposition of Cu and Se and conversion of the underlying (In,Ga)–Se layer into a Cu-rich large-grain CIGS film. In the third and final step of the process In, Ga, and Se are co-evaporated over the Cu-rich CIGS layer to adjust its overall composition so that a Cu/(In + Ga) molar ratio of less than one is obtained. Films obtained by this method yielded the highest efficiency devices to date with conversion efficiencies near 20 % [2, 3].

In the two-stage techniques schematically shown in Fig. 2, a precursor layer is first deposited, typically at room temperature, on the contact layer previously coated on a substrate. Then the precursor layer is treated at an elevated temperature to convert it into a solar cell grade p-type CIGS layer. The techniques that have been employed for precursor deposition include sputtering, electrodeposition, and ink printing. Sputter-deposited precursors have by far been the most popular in manufacturing and they are metallic stacks comprising Cu–Ga alloys and In. The thicknesses of the alloy layer(s) and the In layer are adjusted to yield the desired Cu/(In + Ga) and Ga/(In + Ga) molar ratios in the active regions of the final films, which are typically <1, and in the range of 0.2–0.3, respectively.

Heating step of the two-stage processes involves either a slow furnace process or rapid thermal processing. The slow furnace process is a batch approach where many substrates with metallic precursor layers are loaded in a furnace and a temperature profile in the 400–600 °C range is applied, for durations measured in hours, in the presence of hydride gases, H_2Se and H_2S. Hydride gases convert the precursor layers into device-quality CIGS films that contain Se as well as S [4]. This technique is commercialized by Solar Frontier and is considered to be the front-runner in terms of manufacturing capacity. RTP process is typically applied to precursor layers containing Se. For example, the precursor layer treated by RTP may be a metal/Se stack where the metal layer may contain Cu, In, and Ga. The temperature of the stack is raised at a rate of 5–10 °C and the precursor is converted into a device-quality CIGS film in a matter of minutes. This approach is preferred by companies employing in-line processing approaches, such as roll-to-roll processing of flexible devices.

Besides the sputtering technique, precursor layer preparation step may employ non-vacuum approaches such as electrodeposition [5, 6] and ink deposition [7]. Electrodeposition has been adapted by SoloPower to manufacture flexible CIGS solar cells using flexible stainless steel foil as the substrate [8]. Recent results from this technology include flexible cells and modules with efficiencies of 14.7 % [9] and 13.4 % [10], respectively. Ink deposition technologies for CIGS are being developed by International Solar Electric Technology (ISET) [11] and Nanosolar [12]. Nanosolar recently reported 17.1 % confirmed solar cell efficiency using this approach [13].

Another Silicon Valley company, Miasole, has been developing a sputtering-based roll-to-roll process for manufacturing flexible CIGS solar cells. These devices can then be packaged in rigid or flexible materials to fabricate rigid or flexible modules. A recent announcement from the company states that a large-area flexible module with 15.5 % efficiency has been demonstrated [14].

4 CdTe Technologies

The most commonly used methods of forming CdTe layers for solar cell applications involve vacuum processes such as vapor transport (VT) and close-spaced sublimation (CSS), although non-vacuum approaches such as electrodeposition [6] and ink printing [15] are also being investigated. In the VT technique, which is employed by First Solar, CdTe powder source is heated saturating an inert carrier gas with Cd and Te vapor species, which are then directed onto a moving substrate through a nozzle that is designed to assure uniform injection onto the substrate surface. Process is carried out at a typical pressure of 10–100 Torr, although higher pressures may also be utilized. The substrate temperature is in the 550–650 °C range, higher temperatures requiring the use of specially designed glass sheets. In the CSS method, the CdTe source material is placed in close proximity of the substrate surface and heated up causing sublimation of the Cd and Te species at a pressure of 1–50 Torr. Sublimed species then react and condense on the substrate surface in the form of a thin CdTe layer. These vacuum techniques are being employed by most of the CdTe PV companies including Abound Solar, GE Energy, Calyxo, and Antec Solar.

Success of the above-described vapor deposition techniques for CdTe film growth is partially due to the fact that the Cd and Te species sublime congruently from the CdTe sources, i.e., the composition of the vapor is basically the same as the source. Furthermore, use of high substrate temperatures assures that no secondary phases of Cd or Te may be included in the deposited layer since both of these substances are rather volatile in their elemental form. Therefore, formation of stoichiometric CdTe film on the substrate is rather straightforward. Non-vacuum approaches deposit the CdTe films or precursors at low temperatures and cannot utilize the mechanisms described above. In the electrodeposition process an under-potential deposition mechanism assures the stoichiometric nature of the films grown at slow rates [6, 16]. This technique was used for large-area module fabrication by BP Solar and a 10.6 % efficiency was demonstrated over a decade ago [17]. In ink deposition methods, stoichiometry is fixed at the source which is typically a nano-CdTe ink. For both of the abovementioned non-vacuum approaches, deposited films have to be treated at temperatures above 400 °C to improve their structural and electronic properties. All CdTe layers, whether they are grown by vacuum or non-vacuum methods, are also subjected to a treatment in Cl- and oxygen-containing atmosphere at temperatures above 350 °C. Importance and details of these post-deposition treatments for CdTe have been widely published in the literature [18, 19].

5 Status

Table 1 below shows the status of the CdTe and CIGS technologies. CdTe, due to its ease of processing and low cost, has the larger market presence, which is near 2 GW by First Solar. CIGS manufacturing capacity is nearing 1 GW, most of it by Solar

Table 1 Champion cell and module efficiencies demonstrated for CdTe and CIGS technologies

Technology	Champion cell efficiency	Champion module efficiency
CdTe	16.7 % NREL (2001)[a]	12.8 % GE Energy (2011)[a]
	17.4 % First Solar (2011)	**14.4 %** First Solar (2012)
		12.4 % Abound Solar (2012)
CIGS	19.6 % NREL (2009)[a]	**15.7 %** Miasole (2010)[a]
	20.3 % ZSW (2010)[a]	14.5 % ShowaShell (2012)
	17.3 % Miasole (2012)	13.4 % SoloPower (2012)

[a]denotes results reported in [20]

Frontier. Champion cell efficiencies in Table 1 represent the efficiencies obtained in the laboratory for small-area (near 1 cm^2) devices.

One has to take into account several factors while evaluating the data presented in Table 1. For example, the champion cell efficiencies reported for CIGS are from devices with finger patterns and antireflective (AR) coatings. In monolithically integrated CIGS module structures there is no AR coating and finger pattern, but there is a cover glass. Therefore, while the lack of AR coating and the presence of cover glass may reduce the short-circuit current density of a module with respect to a solar cell, lack of a finger pattern may work in the opposite direction. In CdTe solar cell structures since light enters through the glass superstrate and since finger patterns are not employed in the champion cells, it is easier to translate the cell efficiencies into module efficiencies, except that some specialized processes (high temperature, special expensive glass, etc.) may have been utilized in the fabrication of a champion cell. Although such specialized processes can also be applied to the fabrication of modules, they may increase the cost and may not be practical for large-scale manufacturing. One other point to keep in mind is the fact that although all reported CdTe modules are monolithically integrated, CIGS modules utilizing flexible cells fabricated on metallic foils (e.g., devices produced by Miasole, SoloPower, Nanosolar, and Global Solar Energy) are manufactured by interconnecting individual cells, which may be premeasured, classified according to efficiency, and then interconnected to yield the highest efficiency champion module.

6 Addendum—June 2016

Since the presentation of these results in 2012 there has been much development in this field. Solar Frontier announced a 22.3 % CIGS solar cell with 0.5 cm×0.5 cm area. First Solar reported a small-area 22.1 % efficient CdTe device. The champion module efficiencies for the CdTe and CIGS technologies are now at 18.6 % and 17.9 %, respectively.

References

1. Pernick R, Wilder C, Winnie T (2012) Clean energy trends 2012. Clean edge report
2. Repins I, Contreras M, Egaas B, DeHart C, Scharf J, Perkins C, To B, Noufi R (2008) 19.9% efficient ZnO/CdS/CIGS solar cell with 81.2% fill factor. Prog Photovolt Res Appl 16(3):235–239
3. Jackson P, Hariskos D, Lotter E, Paetel S, Wuertz R, Menner R, Wischmann W, Powalla M (2011) New world record efficiency for CIGS thin film solar cells beyond 20 %. Prog Photovolt Res Appl 19:894–897
4. Goushi Y, Hakuma H, Tabuchi K, Kijima S, Kushiya K (2009) Fabrication of pentanary $Cu(InGa)(SeS)_2$ absorbers by selenization and sulfurization. Sol Energy Mater Sol Cells 93:1318–1320
5. Kapur VK, Basol BM, Tseng ES (1987) Low cost methods for the production of semiconductor films for CIS/CdS solar cells. Sol Cells 21:65–72
6. Basol BM (2011) Application of electrochemical deposition techniques to thin film solar cell processing. In: Proceedings of the SPIE optics and photonics conference, San Diego, California, USA, 21–25 August, p 8110: 81100Q1–81100Q9
7. Basol BM (2000) Low cost techniques for the preparation of Cu(In, Ga)(Se, S)2 absorber layers. Thin Solid Films 361–362:514–519
8. Basol BM, Pinarbasi M, Aksu S, Freitag J, Gonzalez P, Johnson T, Matus Y, Metin B, Narasimhan M, Nayak D, Norsworthy G, Soltz D, Wang J, Wang T, Zolla H (2009) Status of electroplating based CIGS technology development. In: Proceedings of the 34th IEEE PV specialists conference, Philadelphia, PA, USA, 7–12 June, p 231
9. Aksu S, Pinarbasi M (2012) An overview of the roll-to-roll electrodeposition-based CIGS technology. In: Proceedings of China PV technology international conference (CPTIC), Shanghai, China, 21–23 March
10. SoloPower (2012) SoloPower sets record for flexible CIGS solar panel efficiency. http://www.solopower.com/2012/03/solopower-sets-record-for-flexible-cigs-solar-panel-efficiency/. Accessed 16 June 2012 from the World Wide Web
11. Kapur VK, Bansal A, Le P, Asensio O (2003) Non-vacuum processing of CIGS solar cells on rigid and flexible substrates using nanoparticle precursor inks. Thin Solid Films 431–432:53–57
12. Nanosolar (2012) Production process. http://www.nanosolar.com/technology/production-process. Accessed 16 June 2012 from the World Wide Web
13. Nanosolar (2012) Blog. http://www.nanosolar.com/company/blog/nrel-certifies-164-nanosolar-foil-efficiency. Accessed 16 June 2012 from the World Wide Web
14. Miasole (2012) Miasole sets flexible PV efficiency world record at 15.5 percent. http://www.miasole.com/press-releases. Accessed 16 June 2012 from the World Wide Web.
15. http://www.solexant.com
16. Basol BM (1988) Electrodeposited CdTe and HgCdTe solar cells. Sol Cells 23:69–88
17. Cunningham D, Rubcich M, Skinner D (2002) Cadmium telluride PV module manufacturing at BP solar. Prog Photovolt Res Appl 10:159–168
18. Basol BM (1992) Processing high efficiency CdTe solar cells. Int J Sol Energy 12:25–35
19. McCandless B, Sites J (2003) Cadmium telluride solar cells. In: Luque A, Hegedus S (eds) Handbook of photovoltaic science and engineering. Wiley, West Sussex
20. Green M, Nopiah ZM, Yusof AM, Sopian K (2012) Solar cell efficiency tables (version 39). Prog Photovolt Res Appl 20:12–20

Bülent M. Başol Ph.D., Electrical Engineering (Solid State Electronics), University of California, Los Angeles (UCLA), 1980; M.S., Electrical Engineering (Electronic Circuits), UCLA, 1974; B.S., Electrical Engineering, Boğaziçi University, Istanbul, Turkey, 1973.

He has been active in the fields of thin-film photovoltaics and semiconductor wafer processing since 1978. He has been a Founder/Cofounder/Chief Technology Officer (CTO) of several start-up companies in California, including International Solar Electric Technology (ISET), NuTool, SoloPower, and EncoreSolar. At NuTool he led the development of copper electrodeposition and polishing technologies for integrated circuit interconnect fabrication. At the other companies he developed several novel technologies with application in thin-film CdTe and CIGS solar module manufacturing. These technologies include electrodeposited CdTe, ink-deposited CIGS devices, electrodeposited CIGS technologies, and flexible CIGS solar cell structures. He authored over 100 publications and holds 167 US patents in the fields of semiconductor manufacturing methods and equipment, and PV materials and device processing.

How Can We Include External Costs in the Price of Energy?

Ian Dickie and Lawrie Harper-Simmonds

1 Introduction

Chapters 1 and 2 discussed the concept of external costs, and how they compare between renewable and non-renewable methods of energy generation, respectively. However, it is critical that the evidence highlighted in these chapters is incorporated into policymaking. The aim of this chapter is to discuss the methods by which external costs can be included in the price of energy, as well as related issues such as who should bear those price increases, and whether such policies are (politically) feasible. Five tools are discussed in this chapter, and they can be broadly divided into incentive-based instruments (taxes, permit trading, and subsidies), and non-incentive-based instruments (command and control and investment appraisal).

2 Concepts

There are two general concepts that form part of the basis by which these tools are assessed. These are the polluter pays principle (PPP) and the least cost principle (LCP).

2.1 Polluter Pays Principle

The PPP states that the party responsible for producing pollution is responsible for paying for the damage that this pollution causes. In effect, it states that the

I. Dickie (✉)
London, UK
e-mail: ian@eftec.co.uk

L. Harper-Simmonds

T.S. Uyar (ed.), *Towards 100% Renewable Energy*, Springer Proceedings in Energy, DOI 10.1007/978-3-319-45659-1_6

externalities caused by a firm in generating energy should be internalised to them. It is mentioned under Principle 16 of the Rio Declaration on Environment and Development.[1]

2.2 *Least Cost Principle*

The LCP provides guidance on attaining a specific reduction in emissions at least overall cost—in economic terms, it is efficient. Generally, this refers to the case when there is more than one source to a form of pollution (i.e. more than one firm) and these firms have different abatement costs. The principle says that the firm with the lowest abatement costs should undertake the majority of the abatement, as it is more economically efficient to do so.

The attainment of these principles is seen as being a key aim of dealing with externalities. These principles are viewed within the overarching objective of correcting the externality and achieving the socially optimal level of production/consumption of the good in question.

3 Taxation

The use of taxation corrects an externality by internalising it. This process is illustrated in Fig. 1:

The effect of a unit tax is to raise the marginal private cost of producing each unit of a good, so the PMC curve shifts to PMC1. A unit tax of size t consequently raises the costs to producers and reduces emissions to the socially optimal level (q^*, e^*); this is known as a Pigouvian tax. Such a tax adheres to the PPP, because the tax targets the cost to the producer. The tax t is equal to the social marginal damage (SMD), i.e. the external cost. In the example above, both q and e fall, as it is implicitly assumed that there is a linear relationship between the two; this, however, will not always be the case as the tax makes polluting more expensive, and abatement relatively cheaper. This can encourage new technologies, which allow the same level of q, but with lower emissions (e).

A Pigouvian tax is also economically efficient; that is, it adheres to the LCP. This is because, by setting a uniform price signal to all producers, those who have the lowest marginal abatement costs are incentivised to reduce their emissions the most, as it is cheaper for them to do so than pay the tax to keep emitting.

In reality, however, the knowledge requirements necessary for imposing a Pigouvian tax may be too strict. The policymaker needs to know the exact size of the externality to get the tax right; this would most likely involve an iterative process of tax-setting to reach the socially optimal solution. A more reasonable approach may be to set a tax that is expected to achieve a given reduction in emissions and stick to it.

[1] http://www.un-documents.net/rio-dec.htm.

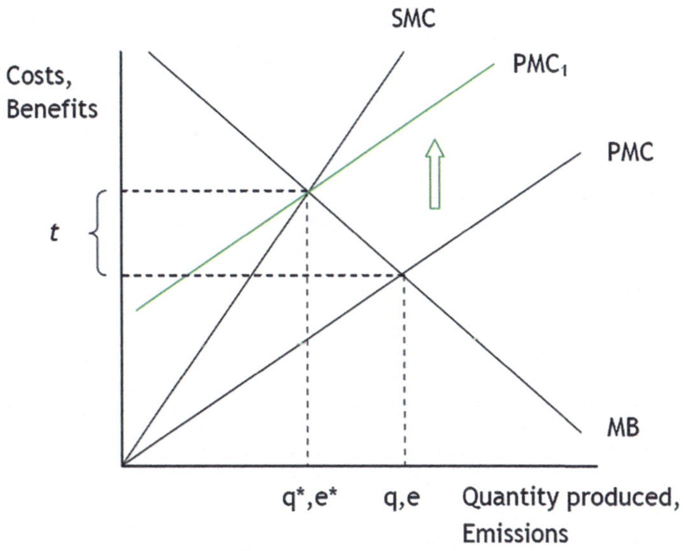

Fig. 1 Use of a Pigouvian tax to correct an externality

Where it is not possible to estimate the damage and/or measure the pollutant, it is not possible to implement a Pigouvian tax in its theoretical form. In these cases, a more blunt (or indirect) tax can be used, where a sufficiently related good (to the pollutant) is taxed. The success of the tax is dependent on the extent of the relationship between the pollutant and the chosen good; the stronger the relation is, the more accurate and efficient the tax is. Using the example of a tax on carbon emissions from electricity generation, these can be graded as follows, in terms of the most preferable to the least:

1. Pigouvian tax on actual carbon emissions resulting from electricity generation.
2. Indirect tax on the carbon content of the fuel used: This would likely be enacted by monitoring the amount and type of fuel used by producers, and calculating an average amount of fuel used per unit of electricity (kWh) produced.
3. A per unit tax on each kWh of electricity generated when the type of fuel used (and thus carbon content) cannot be measured or when monitoring is too costly.

Taxes may also be used as a revenue-raising instrument, to the extent that they are set lower than the cost of abatement. This does not discourage production, and hence emissions, but raises revenue that could be ring-fenced for mitigation of the external costs incurred.

3.1 Permit Trading

A permit trading system essentially creates a market where one previously lacked. It sets a cap on the total amount of emissions (which corresponds to the socially optimal level), and then distributes permits to polluters, who are free to trade these

Box 3.1: Climate Change Levy

The climate change levy (CCL) was introduced in the UK in April 2001. It is effectively a tax on the use of energy in industry, commerce, and the public sector, which had the original aims of promoting energy efficiency and reducing carbon emissions in these sectors. Initially, the levy was intended to be revenue neutral, so that any revenues raised were recycled back to those paying the tax; this was to be achieved through a 0.3 % point cut in employers' national insurance contributions (NIC).

However, the tax was in fact revenue-reducing for the government. The elasticity to the tax of those energy consumers targeted by it was greater than expected; that is, they reduced their energy consumption by more than expected. Consequently, the tax raised less revenue than the NIC cut cost to implement. In 2002, the NIC rate was increased 1 % point, and so the scheme no longer had the explicit aim of being revenue neutral.

The tax also failed to directly target carbon emissions. Although renewable energy sources were exempt, nuclear was not, despite being carbon free during the actual process of electricity generation. The size of the tax was also fixed for all electricity, so there was no incentive to move from a more polluting source to a (non-renewable) less polluting source. That the levy resulted in a reduction in emissions does represent a success, however, and research by Cambridge Econometrics and the Policy Studies Institute [1] illustrated that simply the announcement of the levy resulted in a permanent reduction in energy use.

permits. This results in a similar outcome to that in Fig. 1; the scarcity of the permits, capped to the socially optimal level, creates a market price, which settles at t.

Broadly speaking, permits may be distributed in two ways: auctioning and free distribution, an example of which is grandfathering. When auctioning permits, the supply of permits is announced and each individual permit is sold to the highest bidder, with the revenue accruing to the government. Under a free distribution, these permits are handed out to producers with no charge; the grandfathering method is an example under which producers are given permits that are proportional to their historical emissions. Under the assumption of competitive markets, the distribution of permits does not affect the efficiency of the permit trading system; it still adheres to the LCP. This is because a firm may choose to sell a permit if the market price for the permit is greater than the marginal abatement cost it faces to reduce its emissions by that unit. A permit trading system also adheres to the PPP, which is also not affected by the method of permit distribution, because the external costs are then reflected in the producer's own private cost function.

The method of distribution does have repercussions, however; if permits are auctioned, then the permit trading system acts as a revenue-raising instrument for the policymaker (government) just like the tax. Under some system of free distribution (for example, grandfathering, under which permits are allocated based on past emission levels) the permit trading system is not revenue raising, as payments will simply be between firms. Consequently, the method of distribution can have equity implications: when the permits are auctioned, revenue is raised by the government and can benefit the wider population, while when permits are feely distributed, they can increase profits for large producers.

Box 3.2: EU Emissions Trading Scheme

The EU ETS requires that large emitters of CO_2 within the EU monitor and report their annual emissions. They must subsequently return a number of emission allowances, or permits, that correspond to their emissions of CO_2 within that year. Allowances are allocated to those firms in the energy and industrial sectors that arc above the relevant limit and distributed on the basis of past emissions; this is known as grandfathering. The allowances are allocated in blocks that cover several years at a time; these are known as trading periods. The EU Emissions Trading Scheme (ETS) covered almost half the EU's CO_2 emissions and 40 % of its GHG emissions in 2008.

The first trading period, or Phase 1, ran from 2005 to 2007. It actually oversaw an increase in member states' verified emissions by 1.9 % over the 3 years, and fell victim to criticism as the allowance prices crashed to zero towards the end of the period. This crash was due to an oversupply of permits.

Phase 2 runs from 2008 to 2012, and has attempted to ensure a tighter national cap for emissions. Allocations have been set to represent an average reduction of almost 7 % on 2005 levels; however, there is concern that this phase will also represent an oversupply in permits. Grubb et al. [2] notes that carbon prices are being maintained by the potential to bank allowances for use in the third phase, which is expected to be significantly tougher than the previous phases. It should be noted that this 7 % does not represent the benefit of the trading scheme; emissions may have fallen anyway in the period without the cap. Jones et al. [3] estimate that the cap contributed to emission reduction 2.4 % compared to expected emissions without the cap, although this research was conducted before the global downturn.

Phase 3 will run from 2013 to 2020. Along with aviation emissions being included from 2012, the EU is aiming for a tighter cap on total emissions through Phase 3, with the ultimate aim of including all greenhouse gases and all sectors, and eventually moving to an auctioning system. However, the ability to bank allowances from Phase 2, while sustaining the current price, means that realised emission reductions in Phase 3 may not be as severe as the cap should dictate.

The prospective use of offsets is also of concern to some groups, especially those which originatc outside of the EU. While it has been proposed as a measure to help those producers recover from the economic downturn, the burden of reduction is effectively shifted away from the EU and into other, often less developed economies. Furthermore, their full effect can often be very difficult to quantify, along with their additionality, which is a requirement of such offsets.

The EU ETS has demonstrated how a permit trading scheme can be implemented, and the potential problems associated with it. While its introduction has been accompanied by reductions in emissions, it is difficult to attribute the

(continued)

Box 3.2 (continued)

whole of these reductions to the lower cap imposed by the ETS itself. Furthermore, the decision to distribute the permits as allowances, while in all likelihood making the system politically acceptable, has resulted in windfall profits for producers, who have benefitted from their oversupply, and represents a missed opportunity for government revenue. Newbery [4] has criticised this, and also noted that the costs have been passed on to consumers. Newbery has also promoted the institution of a price ceiling and price floor, in order to deliver a stable price signal to producers regarding the promotion of long-term investment in low-carbon technologies.

3.2 *Command and Control*

Command and control instruments are those which set particular standards which technologies must meet; a hypothetical example would be that, in producing a kWh of electricity, a generation technology cannot produce more than a particular amount of CO_2. Such an instrument can guarantee that the socially optimal level of production is achieved in the short term. It also adheres to the PPP, because the firms face the costs of reducing their emissions.

However, command and control instruments are not economically efficient. That is because all producers have to meet the same particular standard regardless of the cost associated with it. Once that standard is met, they have no further incentive to reduce their emissions. As such, producers with relatively large marginal abatement costs have to reduce their emissions by a disproportionately large amount relative to taxes and trading. Command and control instruments are generally thought of as fair, as they both target polluters and are politically acceptable, and ensure that a minimum standard is met, but are neither economically efficient nor dynamic; they will need to be revised as they are based on currently available technology.

Box 3.3: Renewable Obligation Certificates

Renewable Obligation Certificates (ROCs) are issued for renewable electricity generated within the UK. They are related to the Renewables Obligation Order, introduced in 2000, which permits the government to require electricity suppliers to supply a certain proportion of their total electricity sales in the UK from particular renewable sources. These orders are issued annually. The default is for a renewable generation technology to receive on ROC per MWh of output; however, this varies by technology. For example, offshore wind installations receive 2 ROCs per MWh, while onshore wind installations receive 1 ROC per MWh. In order to meet their obligations, producers can also purchase these ROCs from other producers; in essence the scheme is a mixture of command and control and permit-based trading measures.

3.3 Investment Appraisal

Investment appraisal is a somewhat different approach to the previous four outlined. In essence, it requires that decision-makers conduct a social cost-benefit analysis illustrating that their project is socially beneficial (i.e. that the benefits accruing to society from the project exceed the costs). Within the context of energy generation, this could relate to the construction of a new power plant; when benefits cannot be shown to exceed costs, then either the project is not allowed to go ahead or amendments are required, so as to increase the benefits or reduce the costs, accordingly. There is also the benefit of being able to take into account site specifics when judging a policy; as previously discussed, some burdens will have widely varying impacts depending on their location.

It should be noted that investment appraisal cannot be applied in an ex post context to make a proposal account for its external costs, even if these are still ongoing. Investment appraisal is often required outright of large projects, and the necessary impact assessments undertaken by offshore wind companies looking to construct turbines are evidence of attempts to incorporate such external costs. However, it is a relatively expensive instrument; the costs of engaging in investment appraisal fall on producers, and it may rule out relatively small-scale developments. It is also true that existing uncertainties may mean that some impacts cannot be properly accounted for.

3.4 Instrument Summary

The choice of instrument depends on the burden that is to be addressed. The quantifiable emissions, as valued in the ExternE work, are best addressed using the tax and tradable permit systems. They ensure that the PPP and the LCP are adhered to, because they specifically target the burden, and provide incentives for it to be tackled. In comparison, command and control approaches impose inefficiencies on the economy, and theoretically worse for firms than taxes and trading, while renewable technology subsidies do not directly target the externalities of concern.

However, they cannot be realistically used for other burdens, such as visual amenity and health risks, as these are not classic goods that can be easily measured and monitored. To incorporate these external costs, which are relatively more important for renewable than fossil fuel technologies, investment appraisal is required. Such a technique will ensure that as full an account of external costs is given as possible, but it is expensive in specific to each project.

It should be noted that depending on the uncertainties surrounding the externality, taxes or permit trading may be preferred. When there is uncertainty regarding the social marginal damage arising from a pollutant, or potential for strong threshold effects, it may be best to use a trading system which can definitely cap emissions at a particular level (although the costs to producers of adjusting to this may be

Table 1 Summary of instruments

Instrument	PPP	Least cost	Equity	Dynamic
Taxation	✓	✓	Polluters pay; depends if revenue is recycled to benefit society (other lower tax rates/ring-fenced for environmental projects), or if is given back to producers	Ongoing incentive, but may require updating if doesn't meet required emission reductions or new information comes to light
Permit trading	✓	✓	Depends if revenue-raising or not. If permits are auctioned, then further benefits to society	Requires a series of trading periods and rules regarding transferability of permits. If these are stable then ongoing incentive to innovate
Command and control	✓	X	Burden lies with producers, but not revenue-raising	Less incentive to innovate because it is based on current technologies
Investment appraisal	N/A	N/A	Ensures that external costs are considered as far as possible, but does not dictate where burden lies	Incentive to innovate is based on approach after appraisal conducted. Can generate information which helps future innovation

high). In contrast, when there is considerable uncertainty relating to the costs to producers of abatement, relative to the social damages, a tax system which allows producers more flexibility may be preferred (Table 1).

4 Conclusion

This chapter has demonstrated that choosing a policy instrument to correct for negative externalities requires the consideration of many issues. We defined two principles by which instruments can be assessed; the PPP and the LCP. The only instruments that conform to both these principles are Pigouvian taxes and permit trading, and their appropriateness is defined by the relative uncertainties surrounding damage and abatement costs. However, they are not appropriate for all externalities in their current form.

Even for those externalities for which taxes and permit trading are applicable, they may be eschewed in favour of other instruments whose implementation is more feasible. This is because they may require less of an upfront cost for producers, and so do not affect existing producers to the same degree. Investment appraisal has become frequently adopted to demonstrate the viability of a proposal regarding energy generation, and can theoretically include all relevant externalities, although data is lacking in some instances. Subsidies are not discussed in this chapter because they are not readily used to target or incorporate external costs into decision-making, but have been used as a transitional policy to stimulate the development of new sectors. Under the current environment of spending cuts, subsidies are not particularly popular.

References

1. Cambridge Econometrics and the Policy Studies Institute (2005) Modelling the initial effects of the climate change levy. Report to HMCE. http://www.enagri.info/gold/knowledge_base/reports_local/070115/Modelling+the+initial+effects+of+Climate+Change+Levy.pdf
2. Grubb M, Brewer TL, Sato M, Heilmayr R, Fazekas D (2009) Climate policy and industrial competitiveness: ten insights from Europe on the EU emissions trading system. http://www.climatestrategies.org/component/reports/category/61/204.html
3. Jones B, Keen M, Norregaard J, Strand J (2007) Appendix 1.2 climate change: economic impact and policy responses. In: World economic outlook, October 2007: globalization and inequality. International Monetary Fund, Washington, DC
4. Newbery D (2009) Memorandum submitted by David Newbery, Research Director, Electric Policy Research Group, University of Cambridge. The role of carbon markets in preventing dangerous climate change. The fourth report of the 2009–2010 session. UK Parliament House of Commons Environmental Audit Select Committee. http://www.publications.parliament.uk/pa/cm200910/cmselect/cmenvaud/290/290we33.htm

References

Evaluation of Renewable and Conventional Ammonia as a Potential Solution

Yusuf Bicer and Ibrahim Dincer

Nomenclature

ADF	Abiotic depletion factor
ASU	Air separation unit
CFBG	Circulating fluidized bed gasifier
CCS	Carbon capture storage
CML	Center of Environmental Science of Leiden University
CFC	Chlorofluorocarbons
DB	Dichlorobenzene
DG	Downdraft gasifier
FFE	Feed plus fuel energy
GHG	Greenhouse gas
GWP	Global warming potential
HB	Haber-Bosch
HFC	Hydrofluorocarbons
HTP	Human toxicity potentials
ICE	Internal combustion engine
IPCC	Intergovernmental panel on climate change
LCA	Life cycle assessment
LCI	Life cycle inventory
LPG	Liquefied petroleum gas
SB EQ.	Antimony equivalents
SCR	Selective catalytic reduction
SMR	Steam methane reforming

Y. Bicer (✉) • I. Dincer
Clean Energy Research Laboratory (CERL), Faculty of Engineering and Applied Science, University of Ontario Institute of Technology, 2000 Simcoe Street North, Oshawa, ON, Canada L1H 7K4
e-mail: yusuf.bicer@uoit.ca; ibrahim.dincer@uoit.ca

T.S. Uyar (ed.), *Towards 100% Renewable Energy*, Springer Proceedings in Energy, DOI 10.1007/978-3-319-45659-1_7

STP Standard temperature and pressure
UCG Underground coal gasification
WMO World Meteorological Organization

1 Introduction

Ammonia is projected to be a potential hydrogen carrier with high hydrogen content in the near future. In recent years, expectations are rising for hydrogen and hydrogen carriers as a medium for storage and transportation of energy in the mass introduction and use of renewable energy. Both storage and transport of hydrogen are considered an important issue since hydrogen is a gas under normal temperature and pressure. Hydrogen carriers are mediums that convert hydrogen into chemical substances containing large amounts of hydrogen, to simplify storage and transport processes. Hydrogen carriers include ammonia synthesized from nitrogen and hydrogen that can be used for direct combustion. Ammonia becomes an important hydrogen carrier that does not contain any carbon atoms and has a high hydrogen ratio. Therefore, it is evaluated as a power-generating fuel. Since ammonia produces mainly water and nitrogen on combustion, replacing a part of conventional fuel with ammonia will have a large effect in reducing carbon dioxide emissions. The following advantages can be listed for ammonia usage.

In this regard, ammonia (NH_3):

- Consists of one nitrogen atom from air separation and three hydrogen atoms from any conventional or renewable resources
- Is the second largest synthesized industrial chemical in the world
- Is a significant hydrogen carrier and transportation fuel that does not contain any carbon atoms and has a high hydrogen ratio
- Does not emit direct greenhouse gas emission during utilization
- Can be used as solid and/or liquid for many purposes
- Can be stored and transported under relatively lower pressures
- Can be produced from various type of resources ranging from oil sands to renewables
- Is a suitable fuel to be transferred using steel pipelines with minor modifications which are currently used for natural gas and oil
- Can be used in all types of combustion engines, gas turbines, and burners as a sustainable fuel with only small modifications and directly in fuel cells which is a very important advantage compared to other type of fuels
- Brings a non-centralized power generation via fuel cells, stationary generators, and furnaces/boilers and enables smart grid applications
- Can be used as a refrigerant for cooling in the car

2 Ammonia Production and Transportation

Ammonia is one of the largest synthesized industrial chemicals on Earth. However, production of ammonia consumes almost 1.2 % of total primary energy and contributes about 1 % of GHG emissions in the world. Approximately 1.5 t of CO_2 is released to the environment during the production of 1 t of ammonia because 72 % of current ammonia production is carried out by natural gas in the world [1, 2]. The annual increase in ammonia production covering the period from 2002 to 2015 is shown in Fig. 1.

A most common ammonia synthesis technique is recognized as Haber-Bosch process in the world. In this process, nitrogen is supplied through air separation process mainly as cryogenic. Cryogenic air separation is presently one of the most efficient and cost-effective technologies for generating large amount of oxygen, nitrogen, and argon [3]. Using cryogenic technology, nitrogen can also be produced in high purity which can further be utilized as a useful by-product. Among other air separation processes, cryogenic air separation has the most mature and developed technology. The required electricity could be supplied either from conventional or alternative sources. The Haber-Bosch is an exothermic process that combines hydrogen and nitrogen in 3:1 ratio to produce ammonia. The reaction is facilitated by catalyst and the optimal temperature range is 500–600 °C [4].

In terms of conventional resources, naphtha, heavy fuel oil, coal, natural gas coke oven gas, and refinery gas can be used as feedstock in ammonia production. Natural gas is the primary feedstock used for producing ammonia in worldwide as shown in Fig. 2. There are 11 ammonia plants operating in Canada, producing an average of 4–5 million metric tonnes per plant annually [5]. Ammonia production plants in Canada are ranked internationally as having the highest feed plus fuel energy (FFE) plant efficiency which consume a typical of 33.8 GJ natural gas per tonne of produced ammonia [5]. In comparison, the world FFE average is 38.6 GJ/t NH_3. FFE is related

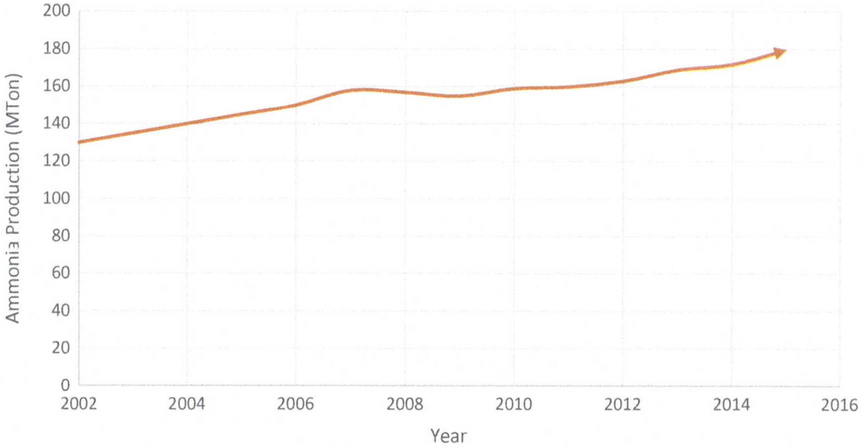

Fig. 1 World ammonia production growth (data from [16])

to the CO_2 generated within the ammonia plant, whereas net energy efficiencies contain electrical consumption and modifications for other energy debits and credits, which may have connected offsite CO_2 emissions not directly from the ammonia plant [5]. In China, coal is intensively used and is generally characterized by high energy requirements [1]. In 2010, ammonia plants in China had a projected average energy intensity of 49.1 GJ/t NH_3. About 75 % of the ammonia in China was produced with the coal-based method. The average energy intensity of the Chinese coal-based ammonia-producing plants is 54 GJ/t NH_3 [1]. Natural gas costs constitute 70–90 % of the production cost of ammonia. Since ammonia production is based on natural gas in SMR method, if natural gas prices rise, production costs for ammonia increase in parallel. For the Haber-Bosch process, production of ammonia is based on various hydrogen production techniques as shown in Fig. 3.

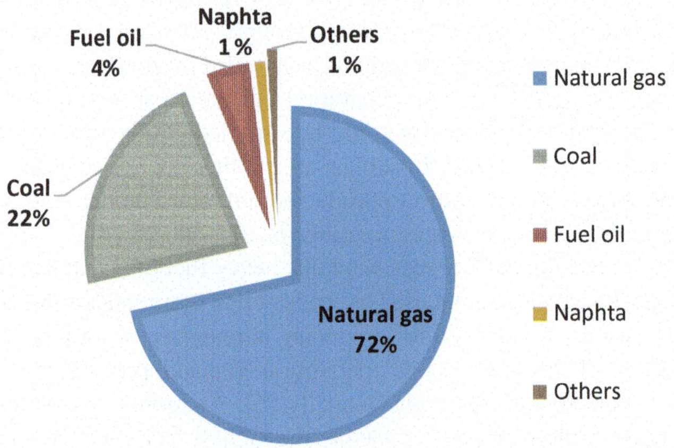

Fig. 2 Sources of global ammonia production based on feedstock use (data from [2])

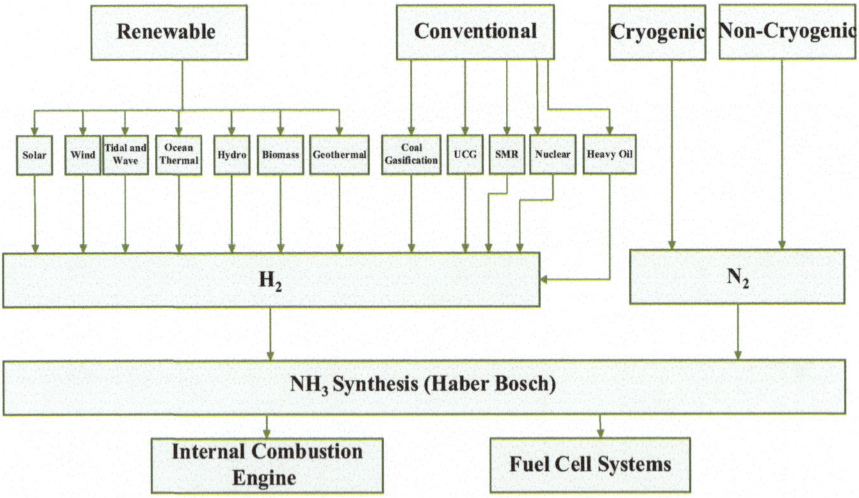

Fig. 3 Ammonia production and usage routes

The storage and delivery infrastructure of ammonia are similar to liquefied petroleum gas (LPG) process. Under medium pressures (5–15 bars), both of the substances are in liquid form which brings the significant advantage because of storage benefits. Today, vehicles running with propane are mostly accepted and used by the public since their onboard storage is possible and it is a good example for ammonia-fueled vehicle opportunities since the storage and risk characteristics of both substances are similar to each other. An ammonia pipeline from the Gulf of Mexico to Minnesota and with divisions to Ohio and Texas has served the ammonia industry for many years. It indicates that there is a working ammonia pipeline transportation which can be spread overall the world. The potential of ammonia usage in many applications will be dependent on the availability of ammonia in the cities. Ammonia is a suitable substance to be transferred using steel pipelines with minor modifications which are currently used for natural gas and oil. In this way, the problem of availability of ammonia will be eliminated. A pipeline may deliver almost 50 % more energy when transporting liquid ammonia than carrying compressed natural gas because of the volumetric energy densities [6].

3 Ammonia Utilization

Ammonia as a sustainable fuel can be used in all types of combustion engines, gas turbines, and burners with only small modifications and directly in fuel cells which is a very important advantage compared to other type of fuels [7]. In an ammonia economy, the availability of a pipeline to the residential area could supply ammonia to fuel cells, stationary generators, furnaces/boilers, and even vehicles which will bring a non-centralized power generation and enable smart grid applications [6]. Ammonia can be reformed to hydrogen for any application because of very low energy requirement of reforming (46.22 kJ/mol) although the temperature required for efficient cracking depends on the catalyst [8, 9]. Ammonia is at the same time a very appropriate fuel for solid oxide fuel cells and direct ammonia fuel cells. These medium-temperature fuel cells promise to be of low cost, highly efficient, and very robust [6].

It is emphasized that the physical characteristics of ammonia are close to propane. The capability to convert a liquid at adequate pressure permits ammonia to store more hydrogen per unit volume than compressed hydrogen/cryogenic liquid hydrogen. Besides having a significant advantage in storing and transporting hydrogen, ammonia may also be burned directly in internal combustion engine (ICE). Compared to gasoline vehicles, ammonia-fueled vehicles do not produce direct CO_2 emission during operation. However, it is important to determine not only direct emissions associated with vehicle operation, but also complete energy-cycle emissions related with fueling the vehicles. Furthermore, ammonia can be produced at locations where oil and natural gas extraction wells are located. In this way, generated CO_2 can be reinjected into the ground for sequestration. Ammonia can then be

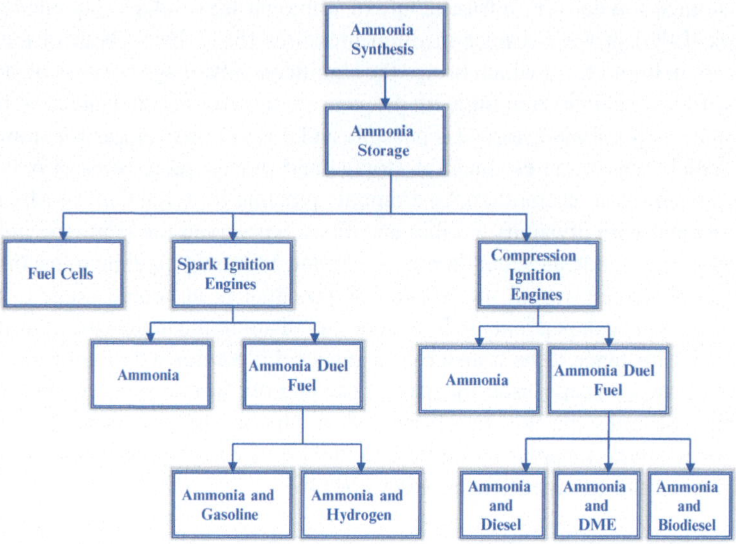

Fig. 4 Direct ammonia utilization pathways for the transportation sector

easily transferred through pipelines, railway cars, and ships by delivering to consumption area where it may be utilized as a source of hydrogen, chemical substance, and fertilizer for agriculture, fuel for the transportation and power generation sectors, working fluid, or refrigerant. Ammonia can be utilized in many transportation applications as shown in Fig. 4.

A few of the following alternatives are listed for direct ammonia usage in various applications:

- Spark-ignited ICE
- Diesel ICE with H_2 or diesel "spike"
- Combustion turbines
- Gasoline or ethanol mixture ICEs
- Transformed biogas generators
- Direct ammonia fuel cells

For power generation systems, for which the storage space is readily available, the energy density is not the determining factor for the fuel selection, as the cost per MJ and emission levels are typically the important parameters. With the new energy-efficient methods of producing ammonia on the cost per MJ basis, ammonia produced using renewable energy sources would be competitive with the fossil-based fuels. The toxicity issue is also not as critical for power generation systems since the fuel will be handled by professionals following well-established handling procedures.

Note that ammonia has been recognized and employed as a leading refrigerant in the industrial sector due to its outstanding thermal properties, zero ozone depletion, and global warming potential (GWP). Ammonia has the highest refrigerating effect per unit mass compared to all the refrigerants being used including the halocarbons. The remarkable advantages of ammonia over R-134a could be lower overall operating costs of ammonia systems, flexibility in meeting complex and several refrigeration needs, and lower initial costs for numerous applications [10]. Ammonia is available almost everywhere and is the cheapest of all the commonly used refrigerants. The remarkable differences favoring the choice of ammonia over R-134a are going to be lower overall operating costs of ammonia systems, flexibility in meeting complex and multiple refrigeration needs, and, for many applications, lower initial costs [10]. Ammonia has better heat transfer properties than most of the chemical refrigerants and consequently allow for the use of equipment with a smaller heat transfer area. Thereby plant construction cost will be lower. But as these properties also benefit the thermodynamic efficiency in the system, it also reduces the operating costs of the system. In many countries the cost of ammonia per mass is considerably lower than the cost of HFCs. This kind of advantage is even multiplied by the fact that ammonia has a lower density in liquid phase. Modern ammonia systems are fully contained closed-loop systems with fully integrated controls, which regulate pressures throughout the system. Additionally, every refrigeration system is required by codes, which are effective, mature, and constantly updated and revised, to have safety relief valves to protect the system and its pressure vessels from overpressurization and possible failure.

4 Environmental Impact Assessment

Ammonia has significant potential as an alternative fuel to further the sustainable development of the transportation sector. Currently, the majority of the locomotive fleet is made up of diesel–electric locomotives, operating with either two-stroke or four-stroke prime mover diesel engines that are coupled to an electric generator. Application of ammonia fuel for ICE with the alternative locomotive configuration direct feed, or a combination of direct feed and decomposition subcategory options, will bring more sustainable solutions. Additionally, fuel cell-driven locomotives may contribute to solve the associated matters of urban air superiority and national energy security influencing the rail and transportation sector. The matters are connected by the point that about 97 % of the energy for the transportation segment is dependent on oil, and beyond 60 % in the USA is imported. Therefore, the investigation of environmental effect of various ammonia production options needs to be carried out. Table 1 illustrates the average GHG emissions in different regions of the world for ammonia production processes. The European average value is 1.9 kg CO_2/kg of ammonia. On the other hand, ammonia plants located in the USA have lower emission values corresponding to about 1.3 kg CO_2/kg of ammonia. The estimates in the table were normally based on the amount of energy disbursed and the suitable emission aspects for the particular fossil fuel contributions. Natural gas is the main fossil fuel for entire estimations given in Table 1.

Table 1 Greenhouse gas emission factors for ammonia production in various countries (data from [11, 17])

Country	g CO_2/kg ammonia produced
Norway	1500
The Netherlands	2163
Europe	1711.3
Europe average	1910
Europe modern	1660
West Europe	1550–1300
Canada	1600
USA	1260
Australia (integrated ammonia/urea plant)	1250–1800

Table 2 The average energy use of ammonia production in different global regions (data from [11])

Region	MJ/t NH_3
Western Europe	41.6
North America	45.5
Russia and Central Europe	58.9
China and India	64.3
Rest of the world	43.7
World average	52.8

Furthermore, the average energy use of various ammonia plants in the world is illustrated in Table 2. It is noted that counting the extra emissions for making and transport of fossil fuels leads to an important growth of the influence of ammonia produced. Besides, a difference in global districts consuming diverse fossil fuel mixtures and different ammonia production efficiencies yields a more explicit vision in influences per ton ammonia produced [11].

4.1 *Life Cycle Assessment*

Life cycle assessment (LCA) presents a methodical set of processes for assembling and investigating the inputs and outputs of materials and energy, and the related environmental impacts, directly assignable to a product, system, or service during the course of its life cycle. LCA is principally a cradle to grave analysis method to examine environmental impacts of a system or process or product based on the specific steps illustrated in Fig. 5. A life cycle is the set of phases of a product or service system, from the extraction of natural resources to last removal. Overall environmental impact of any process is not complete if only operation is considered; all the life steps from resource extraction to disposal during the lifetime of a product or process should be considered. Mass and energy flows and environmental impacts related to plant construction, utilization, and dismantling stages are taken into account in LCA analysis (ISO 14044).

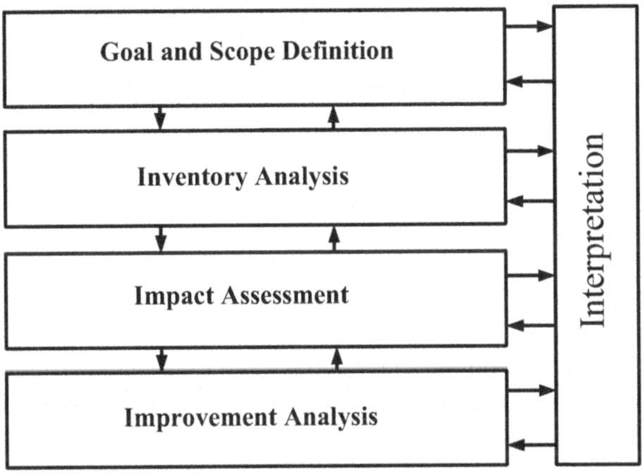

Fig. 5 Framework of LCA analysis

The goal definition and scoping describe the product, process, or activity. It identifies the boundaries and environmental effects to be studied for the assessment. Inventory section classifies and quantifies energy, water, and material usage and environmental discharges. Impact assessment step evaluates the human and ecological effects of energy, water, and material usage and the environmental releases identified in the inventory analysis. Interpretation step evaluates the results of the inventory analysis and impact assessment to select the chosen product, process, or service. The procedures for performing of an LCA have been defined by such International Organization for Standardization (ISO) based on ISO 14040—environmental management—life cycle assessment—principles and framework and ISO 14044: [12]—environmental management—life cycle assessment—requirements and guidelines. The four steps are explained as follows:

4.1.1 Goal and Scope Definition

This is the first step of an LCA study. This step defines the objectives of study and also the range of activities under investigation. The utmost care and detail are required to define the goals and scope of study. The LCA is an iterative process; therefore the feedback consideration should be kept in definition of systems.

4.1.2 Inventory Analysis

In this step, raw material and energy, emissions, and waste data are collected. This data is used to calculate the total emissions from the system. The mass and energy balances are used at each step to calculate the life cycle inventory of the system.

The life cycle inventory needs to include every possible energy and material input and all possible emissions to establish credible results. Data quality is an important aspect of LCA. During inventory analysis the standards are followed for maintaining the data quality.

4.1.3 Impact Assessment

The thirst step of LCA is life cycle impact assessment. This step assesses the impacts of activities under investigation. The LCI data is utilized to find out the affected areas. The LCI data is essentially analyzed in a two-step process:

Classification: The impact categories are established and LCI data is analyzed to mark the data and calculate the values of emissions corresponding to each category. The impact categories are based on the evaluation method utilized. As example, some of the categories are GWP, acidification, human toxicity, etc.

Characterization: It is the second step of assessment. Classification step groups the data in respective impact categories. Characterization step is used to evaluate the relative contribution of each type of emission to these impact categories.

Normalization and Weighting: This step is not mandatory according to the standards. The emissions are normalized corresponding to a standard and converted into a score system. The total score is utilized to identify the methods and processes of concern.

4.1.4 Interpretation of Results and Improvement

The last step is interpretation of results and feedback for improvement of system. The gray areas of system are identified and highly polluting processes can be eliminated with cleaner alternatives.

LCA can identify critical phases where process changes could significantly decrease impacts.

Performing an LCA brings some advantages as follows:

- Evaluating methodically the environmental consequences related with a given product or process
- Assessing the human and ecological effects of material and energy consumption and environmental emissions to the local community, region, and world
- Identifying impacts related to specific environmental areas of concern
- Assisting in identifying significant alterations in environmental impacts between life cycle stages and environmental media
- Comparing the health and ecological impacts of substitute products and processes
- Quantifying environmental releases to air, water, and land in relation to each life cycle stage and major contributing process

LCA is a methodology from cradle to grave. This tool helps to make effective decision by analyzing the system systematically. LCA analyzes the environmental impact of a product or process over the length of its entire life, beginning from raw material extraction to final disposal. LCA deliberates all the life periods of product or process to assess the overall environmental impact. There are a number of assessment methods progressed over the time to categorize and characterize the environmental flows of system. LCA can be performed using CML 2001 method which was proposed by a set of scientists under the principle of CML (Center of Environmental Science of Leiden University) including a group of impact classes and characterization procedures for the impact assessment phase in 2001. Some of the baseline indicators of CML method which are utilized in this study are explained as follows [13].

(a) *Depletion of Abiotic Resources*
The key concern of this category is the human and ecosystem health that is affected by the extraction of minerals and fossil as inputs to the system. For each extraction of minerals and fossil fuels, the abiotic depletion factor (ADF) is defined. This indicator has globe scale where it is related with concentration reserves and rate of de-accumulation.

(b) *Human Toxicity*
Toxic substances on the human environment are the core concerns for this category. In the working environment, the health risks are not included in this category. Characterization factors, human toxicity potentials (HTP), are determined with (the Uniform System for the Evaluation of Substances) USES-LCA, describing fate, exposure, and effects of toxic substances for an infinite time horizon. 1,4-dichlorobenzene equivalents/kg emissions is used to express each toxic substance. Depending on the substance, the geographical scale differs between local and global indicator.

(c) *Ozone Depletion*
Due to stratospheric ozone depletion, a bigger portion of UV-B radiation spreads the world surface. It may have damaging properties upon human health, animal health, terrestrial and aquatic ecosystems, biochemical cycles, and materials. The category is output related and it is at global scale. The model of characterization is advanced by the World Meteorological Organization (WMO) and describes ozone depletion potential of various gases in unit of kg CFC-11 equivalent/kg emission. The geographic scope of this indicator is at global scale and the span of time is infinity.

(d) *Global Warming*
The greenhouse gases to air are associated with the climate change. Adversative effects upon ecosystem health, human health, and material welfare can result from climate change. The Intergovernmental Panel on Climate Change (IPCC) developed the characterization model which is selected for the development of characterization factors. A kg carbon dioxide/kg emission is used to express the GWP for time horizon 500 years (GWP500). This indicator has a global scale.

The eco-indicator method specifies the environmental impact in terms of numbers or scores. It simplifies the interpretation of LCA by including a weighting method. After weighting, it supports to give single score for each of the product or process which is calculated based on the relative environmental impact. The score is represented on a point scale (Pt), where a point (Pt) means the annual environmental load (i.e., whole production/consumption undertakings in the economy) of an average citizen. Eco-Indicator 99 (E) uses load of average European [14, 15]. The Eco-Indicator 99 defines the environment damage in three broad categories:

(a) *Human Health*
It includes the number and duration of diseases and loss of life years due to permanent deaths caused by environmental degradation. The effects are included mainly by climate change, ozone layer depletion, carcinogenic effects, respiratory effects, and ionization.
(b) *Ecosystem Quality*
This category includes the impact of species diversity, acidification, ecotoxicity, eutrophication, and land use.
(c) *Resources*
This category corresponds to the depletion of raw materials and energy resources. It is measured in terms of the surplus energy required in future for the extraction of lower quality of energy and minerals. The agricultural resource depletion is studied under the category of land use.

5 Results and Discussion

SimaPro 7 software is used for life cycle assessment analyses. The two methods used for the current LCA analyses are CML 2001 and Eco-Indicator 99. Human toxicity can play an important role for decision of using alternative methods in ammonia production. Besides, GWP is the main characteristic to compare the total CO_2 equivalent emission from any source. Abiotic resources are natural resources including energy resources. Since fossil fuel resources are declining gradually, abiotic depletion potential is a significant category for LCA analysis.

Figure 6 shows the comparison of ozone layer depletion values for various transportation fuels. Ammonia has lowest ozone layer depletion even if it is produced from steam methane reforming and partial oxidation of heavy oil. In addition, production of fuel ammonia yields lower greenhouse gas emissions compared to petrol and propane production as shown in Fig. 7.

The production of various fuels is compared in terms of abiotic depletion of sources as shown in Fig. 8. Ammonia fuel has the lowest abiotic depletion value compared to others although the production process may be fossil fuel based. There are multiple pathways for ammonia production. Ammonia is cleaner when produced from renewable resources. Figures 9 and 10 compare the environmental impacts of various ammonia production pathways. Hence, ammonia from renewable resources has the least environmental impact. Furthermore, ammonia from hydrocarbon

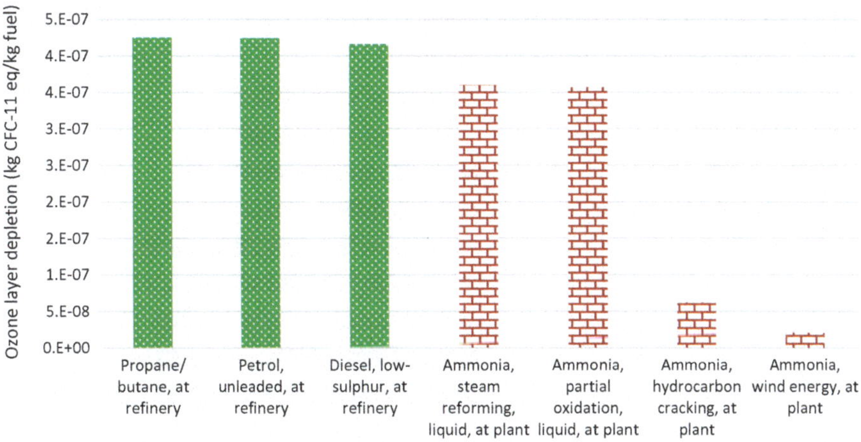

Fig. 6 Ozone layer depletion during production of various fuels

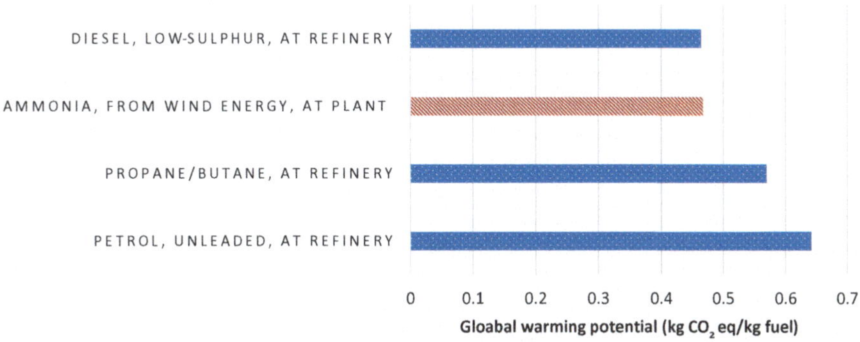

Fig. 7 Greenhouse gas emissions during production of various fuels

cracking and underground coal gasification is the most environmentally benign option among conventional methods.

When renewable source-based ammonia production options are compared as shown in Fig. 10, ammonia from PV and biomass-based electrolysis routes yields higher global warming values although they are quite lower than conventional methods.

Abiotic resources are natural resources including energy resources, such as iron ore and crude oil, which are considered as nonliving. The abiotic depletion is highest for coal electrolysis-based ammonia production methods followed by heavy oil- and natural gas-based methods as it is illustrated in Fig. 11. This is due to the fact that coal, heavy oil, and natural gas are primary sources of energy and feed source as well; it indicates the large consumption of fossil fuels for unit mass of ammonia produced.

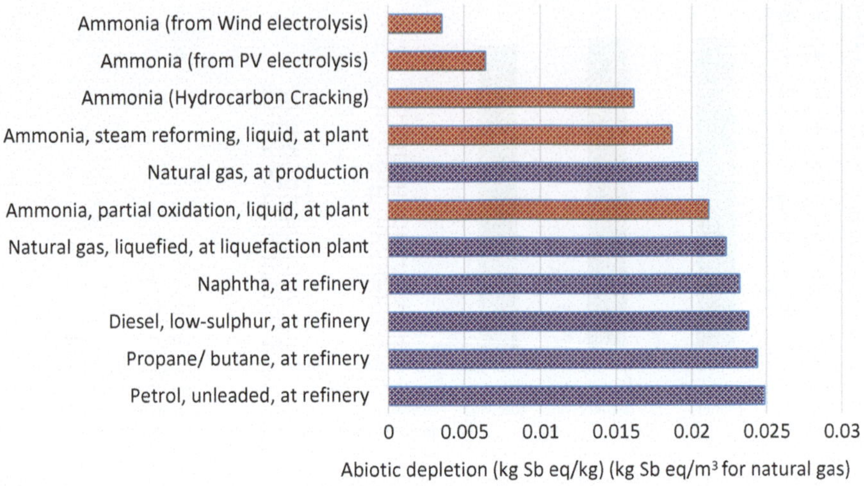

Fig. 8 Abiotic depletion values during production of various fuels

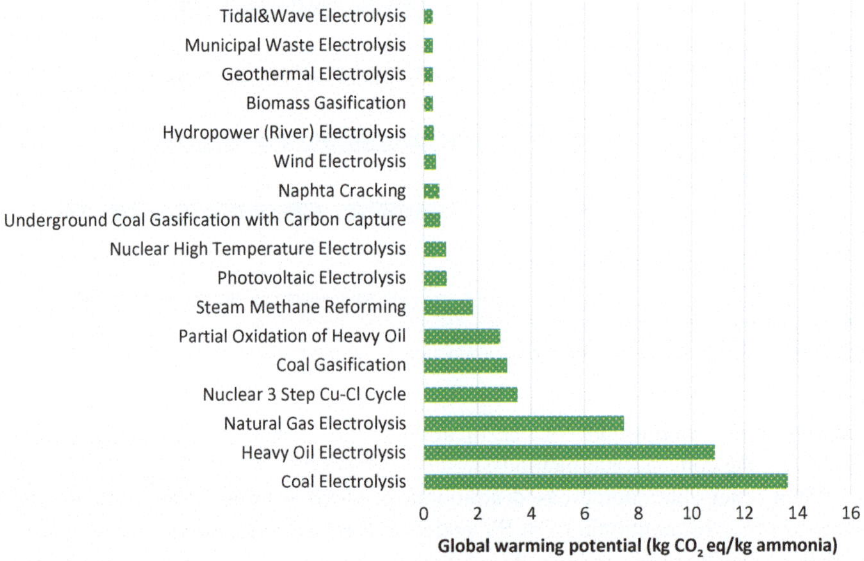

Fig. 9 Global warming values of various conventional and renewable ammonia production methods

The impact on human health due to human toxicity is maximum for the ammonia production from coal- and heavy oil-based electrolysis methods where the maximum is 2.92 kg 1,4-DB-eq per kg of ammonia for coal-based electrolysis method. Ammonia from both underground coal gasification and naphtha cracking-based methods yield lowest human toxicity values as seen in Fig. 12 among conventional

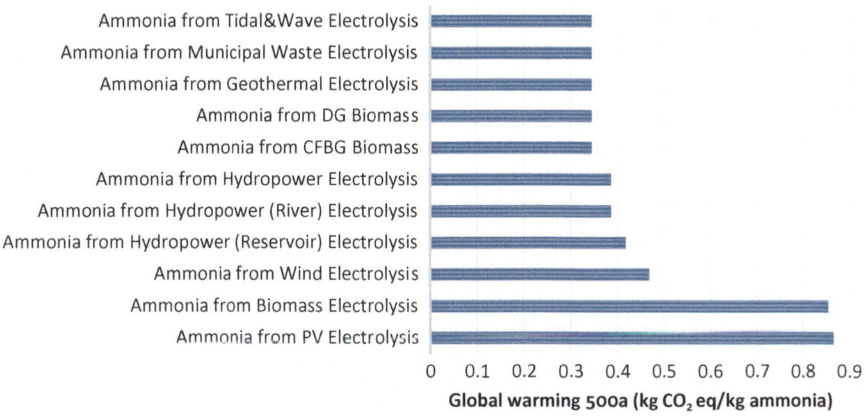

Fig. 10 Global warming values of renewable ammonia production methods

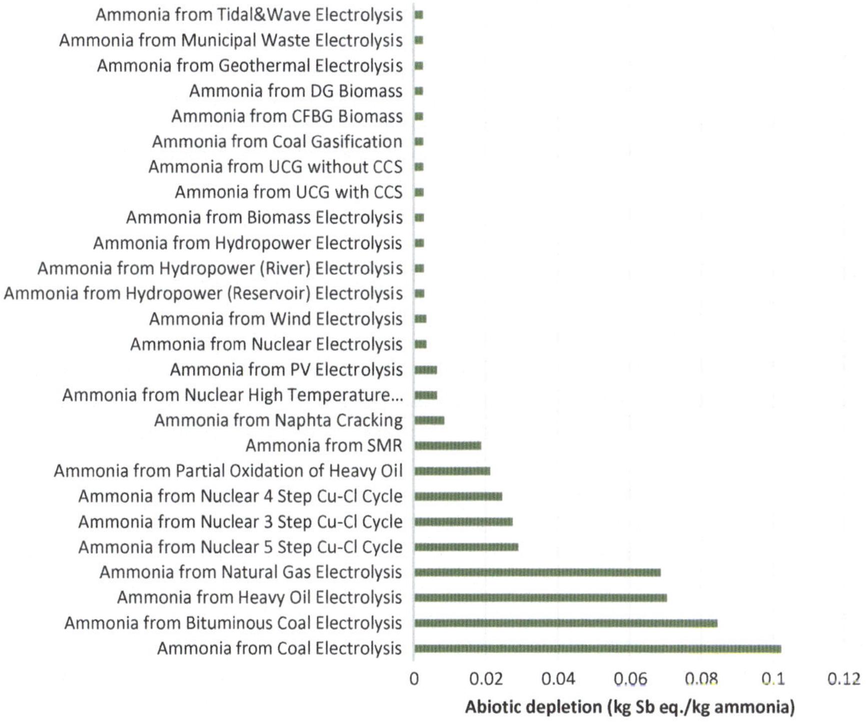

Fig. 11 Abiotic depletion values of various conventional and renewable ammonia production methods

option. On the other hand, renewable-based ammonia production methods such as Tidal&Waves, municipal waste, geothermal, and biomass have lower toxicity values.

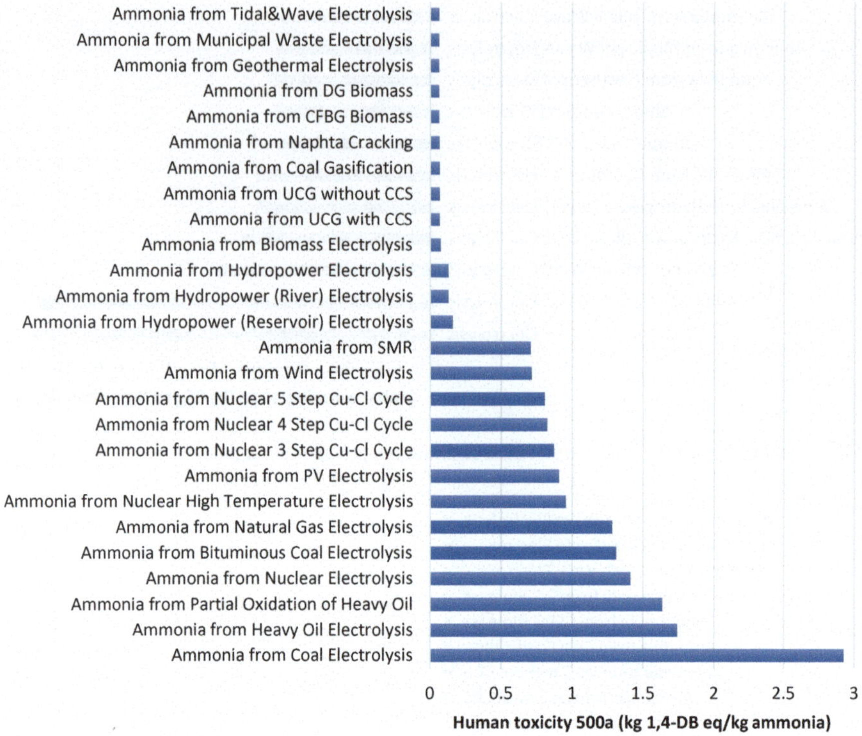

Fig. 12 Human toxicity values of various ammonia production methods

The Eco-Indicator 99 method combines emissions in a single score at highest level. Similar to other results, the single score of ammonia production from coal electrolysis-based methods has the maximum value while coal gasification methods have the minimum values among other methods. As illustrated in Fig. 13, heavy oil-based ammonia production method has the highest environmental impact on resources. The lowest single score is calculated for UCG with CCS method which can be evaluated as the most environmentally benign method among conventional methods.

6 Conclusions

Ammonia is the only carbon-free chemical energy carrier (other than hydrogen) suitable for use as a transportation fuel. Furthermore, ammonia has a high octane rating (110–130), and can be thermally cracked to produce hydrogen fuel using only ~12 % of the higher heating value. It has a well-established production and distribution infrastructure, and has zero GWP. In addition to its attractive qualities as a fuel, ammonia is widely used as a NO_x-reducing agent for combustion exhaust gases using

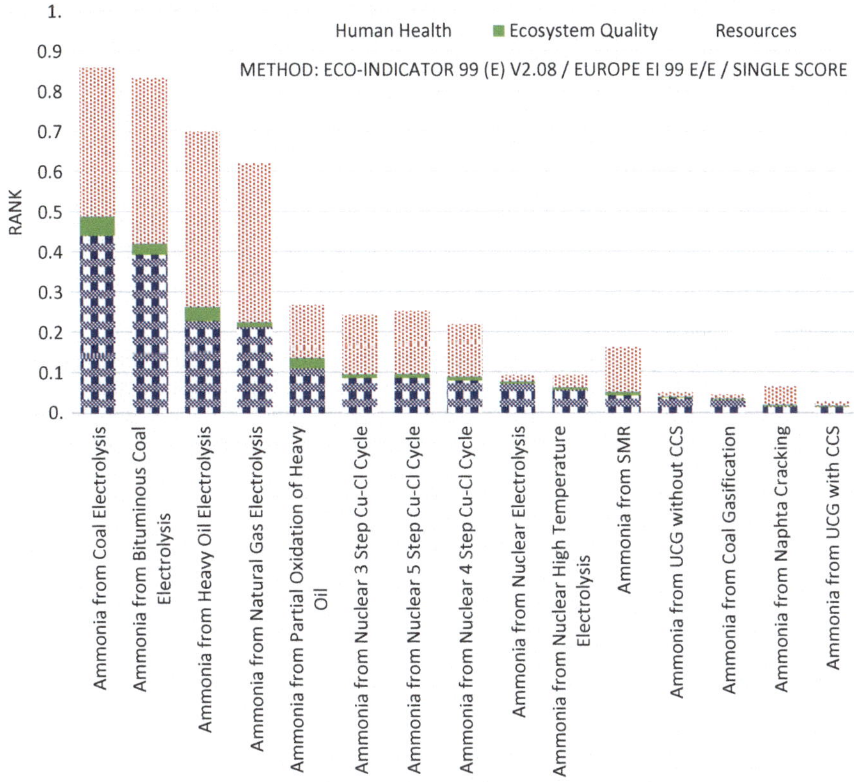

Fig. 13 Single score values of conventional ammonia production methods

selective catalytic reduction (SCR), and its capacity as a refrigerant can be applied to recover and further utilize engine heat that would otherwise be lost. In terms of environmental sustainability, ammonia can be produced using either fossil fuels, or any renewable energy source, using heat and/or electricity, which allows for evolution of ammonia production methods and technologies in parallel with sustainable development. The following concluding remarks are expressed based on the current study:

- LCA is a significant and reliable tool to study fuel production processes since it covers the period from cradle to grave.
- In terms of human toxicity, coal- and heavy oil-fired power plant-based electrolysis methods for ammonia production have highest values.
- Tidal&Waves-, municipal waste-, and geothermal-based ammonia production routes have lowest abiotic depletion, global warming, and human toxicity values, respectively, among all methods.
- Nuclear electrolysis- and naphtha cracking-based ammonia production methods have least effect on climate change among conventional methods while Tidal&Waves method is the most environmentally benign method in terms of climate change and global warming.

- The renewable sources with their improved efficiency can reduce the overall environmental footprint and can replace the current fossil fuel-based centralized ammonia production facilities.
- As the cost of renewable electricity catches the level of conventional electricity, renewable energy-based ammonia production systems will continue to gain practicality and popularity.

References

1. Industrial Efficiency Technology Database (IETD) (2015) The Institute for Industrial Productivity. http://ietd.iipnetwork.org/content/ammonia#key-data. Accessed Oct 2015
2. International Energy Agency (IEA) (2012) Energy technology perspectives 2012, pathways to a clean energy system. https://www.iea.org/publications/freepublications/publication/ETP2012_free.pdf. Accessed Oct 2015
3. Smith AR, Klosek J (2001) A review of air separation technologies and their integration with energy conversion processes. Fuel Process Technol 70(2):115–34
4. Bartels JR (2008) A feasibility study of implementing an ammonia economy [M.S.]. Iowa State University, Ann Arbor
5. Canadian Industry Program for Energy Conservation (2007) Canadian ammonia producers: benchmarking energy efficiency and carbon dioxide emissions. Canadian Industry Program for Energy Conservation (CIPEC), Ottawa
6. Olson N, Holbrook J (2012) NH_3—the other hydrogen. Iowa Energy Center. Ammonia Fuel Network. http://www.iowaenergycenter.org/wp-content/uploads/2012/03/NHA-2009v7-1-JHH-2.pdf. Accessed Oct 2015
7. Lan R, Tao S (2014) Ammonia as a suitable fuel for fuel cells. Front Energy Res 2:35
8. Hacker V, Kordesch K (2010) Ammonia crackers, Handbook of fuel cells. Wiley, Chichester
9. Thomas G, Parks G (2006) Potential roles of ammonia in a hydrogen economy: a study of issues related to the use ammonia for on-board vehicular hydrogen storage. US Department of Energy, p 23
10. Khemani H (2010) Properties of ammonia gas or R717 used as the refrigerant. http://www.brighthubengineering.com/hvac/64242-properties-of-ammonia-gas-or-r717-used-as-refrigerant/. Accessed Sept 2015
11. Kool A, Marinussen M, Blonk H (2012) LCI data for the calculation tool feedprint for greenhouse gas emissions of feed production and utilization, vol 34, GHG emissions of N, P and K fertilizer production. Blonk Consultants, Gouda, p 2805
12. International Organization for Standardization (ISO) ISO 14044 (2006) Environmental management—life cycle assessment—requirements and guidelines
13. Consultants P (2013) SimaPro Life Cycle Analysis Database version 7.3 (software)
14. Acero AP, Rodríguez C, Ciroth A (2014) Impact assessment methods in life cycle assessment and their impact categories, 12th edn. GreenDelta GmbH, Berlin, Germany
15. Baayen H (2000) Eco-indicator 99. A damage oriented method for life cycle impact assessment. Manual for designers. Dutch Ministry of Housing Spatial Planning and the Environment
16. Yara Fertilizer Industry (2014) Yara Fertilizer Industry handbook. http://yara.com/doc/124129_Fertilizer%20Industry%20Handbook%20slides%20only.pdf. Accessed Sept 2015
17. Wood S, Cowie A (2004) A review of greenhouse gas emission factors for fertilizer production, Research and Development Division, State Forests of New South Wales. Cooperative Research Centre for Greenhouse Accounting

Solar Trigeneration: Electricity, Cooling and Steam from the Sun

Lokurlu Ahmet and Saidi Karim

Nomenclature

CSP	Concentrating solar power
ORC	Organic Rankine cycle
PLC	Programmable logic controller
PTC	Parabolic trough collector
SOLTRIGEN	Solar trigeneration

1 Introduction

To supply the consumer with the required energy directly, in many cases both several energy conversion steps can be skipped, and the energy supply concept can be based on local energy plants.

Concentrating the solar energy, high-temperature applications such as multistage solar cooling, solar steam generation or direct generation of electricity can support the market using local solar thermal energy supply systems. Solar cooling avoids the consumption of electricity for compression cooling, and solar steam generation the firing of fuels. Additional to these new applications, the SOLITERM Group included the direct generation of electricity into the concept. For this, all kinds of energy that are required can be supplied with solar thermal, generated in one plant and at the same time, which is worldwide novelty. The SOLTRIGEN Solar Trigeneration has been set up successfully on Cyprus.

The state-of-the-art technologies—such as Photovoltaics, CSP or others—enable the conversion of solar energy for one purpose. For this, both the integration into the

L. Ahmet (✉) • S. Karim
SOLITERM Group, Süsterfeld Str. 83, 52072 Aachen, Germany
e-mail: info@solitermgroup.com; info@solitermgroup.com

T.S. Uyar (ed.), *Towards 100% Renewable Energy*, Springer Proceedings in Energy, DOI 10.1007/978-3-319-45659-1_8

existing structure and the challenges of energy storing, to buffer the differences between the solar energy offer and the demand, have to be solved in detail. In spite of this, the solar trigeneration can be used for the supply of that energy that is required at the given time. Additionally, the trigeneration can be operated in that way that solar energy is converted into that kind of energy which gives the best economical profits. This concept increases the potential and improves the economics of solar energy, and based on a supply structure with many local solar energy plants, an important step towards the "100 % renewable" aim can be taken.

2 Solar Energy in Turkey

All countries neighbouring the Mediterranean Sea show good boundary conditions for the economical operation of solar energy, as the irradiation values are good up to excellent, and the energy supply with the conventional energies such as electricity and fuel either is expensive, as in the south of Europe, or difficult, as in local regions with bad access to the common grids and infrastructure. For this, small up to huge photovoltaic systems and low-temperature solar thermal warm water supply systems are very well known already to be placed on the available areas. The technology of concentrating solar power (CSP) is realized or planned to be used in huge plants, with many parabolic trough collectors in long rows and many rows in one collector field. The used technology operates types of PTCs which are heavy and constructed for the ground installation.

Additionally to these opportunities, SOLITERM developed special lightweight PTCs that can be installed also onto flat rooftops or as a parking roof. As the concentration of solar energy enables higher usage temperature levels in comparison to conventional technologies, the operation of double-effect absorption cooling at high overall process efficiencies and the solar steam generation already proved the advantages. At this point, the direct generation of electricity using a solar-operated organic Rankine cycle (ORC) has been added.

With this novelty and the given boundary conditions, the integration of renewables into the Turkish energy supply structure undergoes a strong change towards the usage of a significantly higher technical potential. As the sun is well known as an energy source that can supply the whole energy demand about 10,000 times more than required: Why wait while the depletion of fossil fuel resources is going on, international aims for climate protection have been set already and the answer to these challenges is given already?

3 Integration of Solar Energy into the Given Structure

The SOLITERM solution takes account of the demand of a given customer. The examination of the collector installation possibilities, with the option of roof integration due to the lightweight design, the option to choose the suiting type of PTC

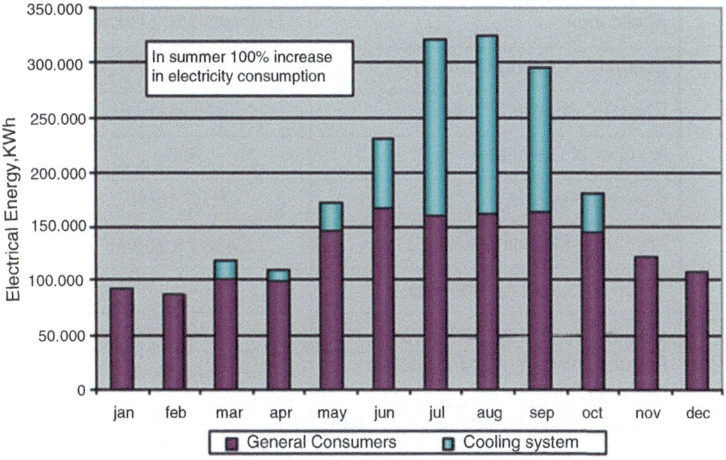

Fig. 1 Typical electricity consumption profile of a hotel

and a special arrangement regarding the number of collectors per row and the number of rows in the field, allows to deploy the best possibilities in the supply with solar energy.

Any single consumer may have a special energy consumption profile. A typical electricity consumption profile of a hotel is shown in Fig. 1.

The example shows that in the summer season, the highly efficient solar cooling can be used to cut the demand for electricity for state-of-the-art compression cooling. The operation of a solar energy plant then is split, e.g., into cooling in the summer and heating or warm water supply in the winter.

As a solar thermal plant that is able to supply cooling, heating and electricity may be used in different ways, the best economical solution should be chosen. Besides the advantages of zero emission and sustainable energy supply, at the given boundary conditions the economics of the plant will affect the operation mode.

3.1 Solar Cooling

Solar cooling uses several advantages. At first, the cooling demand, e.g., of a hotel in a Mediterranean country is influenced mainly by the solar irradiation. For this, the storing of solar energy is required with only small storages, as the offer of energy and the demand for cooling appear almost simultaneously. Figure 2 shows the data of a typical hotel.

The concentration of the solar energy using the parabolic trough collectors enables about 180 °C operation temperature of the solar hot water cycle. At this temperature level, which cannot be reached with conventional technologies, the operation of a double-stage absorption cooling process allows to generate about 1.4 units of cooling energy from 1 unit of solar thermal energy. Both the energy conversion with the solar

Application	Hotel with 300 Rooms
Estimated Cooling Demand	2.750 KW
DNI (Ref.Antalya / Turkey)	2.049 KWh/m^2
Number of Collectors	384
Collector Type	PTC 1800
Aperture / Mounting Area	3.456 / 8.200 m^2
Thermal Output (at 800 W/m^2)	1.700 KW
Cooling Capacity (at 800 W/m^2)	2.400 KW
Annual Chilled Water Production (summer mode)	4.400 MWh
Annual Electricity Savings by Cooling	1.900 MWh
Annual Heating (winter mode)	1.800 MWh
Annual Fuel Savings by Heating	1.500 MWh
Annual CO_2 Savings	1.700 t

Fig. 2 Typical data for solar cooling of a hotel

collector and that of the cooling process lead to an overall process efficiency that allows the economical operation; in best cases, such as calculated for a hotel planned to be installed on Cyprus, the amortisation time is less than 8 years already today.

Special applications occur if the electricity supplier offers energy at high tariff time in the evening, for example in Turkey. In this case, the usage of hot water storages to operate the absorption chiller after sunset can improve the economics additionally.

3.2 Solar Steam Generation

The high temperature level enables to generate steam at, e.g., 5 bars and about 150 °C, as required for several kinds of customers. Figure 3 shows the solar steam generation plant at the FritoLay PepsiCo plant in Tarsus, Turkey.

3.3 Electricity from the Sun

As it makes no sense first to convert conventional energy into electricity and then operating a compression cooling system, the efficient usage of solar energy as described above offers a better overall energy conversion efficiency. If electricity is

Fig. 3 Solar steam generation in Turkey

required additionally, all common technologies, such as photovoltaics or CSP, have the disadvantage that only one kind of energy is generated. For this, the challenges of storing electricity, improving the grid and setup of an intelligent consumer structure ("smart grid") that has the information when solar energy is offered and can be used have to be solved. This implies time and additional investment.

The way to combine the energy conversion units for cooling and electricity generation directly enables the customer to use either that kind of energy that is required, following the load curve, or that one that is most expensive at the given time. For this, the combination of solar electricity generation with the complete concept has to be evaluated different to the PV or CSP technologies.

ORC processes for electricity generation are operated at several temperature levels. The level of 180 °C that is used in the described applications is sufficient for the ORC operation. While several customers have demand for electrical power in the range of about 150 kW, the specific electricity generation costs can be decreased with increased power. However, to customize the component size to a small-scale application, the SOLTRIGEN plant is designed for only 15 up to 25 kW electrical power output. Figure 4 shows the first step of the SOLTRIGEN plant. To be able to operate tests of the solar-operated ORC at any weather conditions, additionally to the solar collector field a backup boiler is integrated into the solar hot water cycle. The ORC unit is driving the electric generator.

The SOLTRIGEN plant is the first step to combine the technologies. For any specified solar solution, the plant design and the sizing of the main components have to be adjusted to the demand structure; that is, the single functional groups for cooling or electricity generation have to be chosen according to the given requirements. In general, the higher the capacity (thermal capacity of the collector field, cooling capacity of the chiller, electrical power output of the ORC), the better the economics.

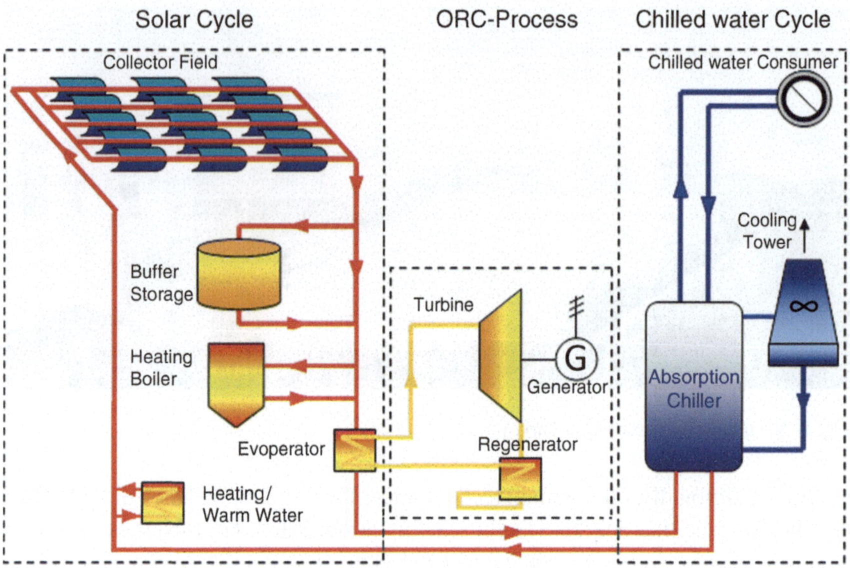

Fig. 4 Combination of solar cooling and solar electricity generation

Table 1 Data on the economics

Parameter	X	Unit
Solar cycle operation temperature	180	°C
Heat transfer medium	Water	–
Maximum operation pressure (design)	16	bar
Thermal efficiency of the PTC	60	%
COP absorption cooling process	1.3	–
COP compression cooling process	2.1	–
Electricity pricing	0.19	€/kWh
Gas pricing	0.10	€/kWh
Boiler efficiency	85	%
ORC efficiency	18	%

4 Economics

To give an example of the economics, Table 1 includes the main data for the described technologies.

The comparison of the economics gives the following results.

- *Solar Cooling*. 1 kWh solar irradiation is converted into 0.6 kWh thermal energy, which generates 0.78 kWh cooling. 0.37 kWh electricity for compression cooling is saved, i.e. 0.07€.

Fig. 5 SOLTRIGEN collector field on Cyprus

- *Solar Steam Generation*. 0.6 kWh solar thermal energy substitutes 0.71 kWh fuel, which leads to savings of 0.07€.
- *Solar Electricity Generation*. 0.108 kWh electricity is equivalent to 0.02€.

This comparison is based on the data given for Cyprus, which are based on the strong dependency on the imports of fossil fuels, i.e. high gas pricing. In other countries of the Mediterranean region, the economics may differ. Countries in the South of Europe (Spain, Italy, Turkey) have higher expenses on electricity in special ranges, and also high tariff times, so the highly effective solar cooling process can be the most economical solution (Fig. 5).

The electricity generation will be the more attractive, the worse electricity is available from the grid. Considering feed-in tariffs, the economics may differ for each application. To move towards the 100 % renewable aim, the electricity generation with the trigeneration technology will be a very important way to move away from the conventional or the carbon-based electricity generation structure.

The system is equipped with the SOLITERM online monitoring system, and the operation is done in a full automatic way. Using an overall energy management system, the energy distribution can be arranged in that way that the best economical operation mode is done automatically, i.e. the programmable logic controller PLC of the switching cabinet has the information on which kind of energy at which point of time should be used directly by the customer, or which one should be used considering the data given from the existing supply structure.

5 Conclusion

The described novelty to operate one solar system for cooling, heating and electricity generation offers several advantages that can be used in Turkey:

- Easy integration of renewable energy systems
- Increasing the technical potential of solar energy
- Offering all kinds of required energies with one solar energy plant and at the same time
- Using sustainable and zero-emission energy
- Enabling the economical usage of solar energy already today

Reference

1. The Republic of Turkey, Ministry of Energy and Natural Resources: Strategic Plan (2010–2014), 9 April 2010

Curriculum Development into Renewable Energies Through Coupled Research and Applied Projects

Drishtysingh Ramdenee, T.A. Poirier, R. Mohee, N. Barka, J. Chaumel, and A. Ilinca

1 Introduction

As fuel prices take roller-coaster ride over the past decades, renewable energies have known rising popularity to improve upon the concept of sustainable development and the idea of energy security. To support such development in a complex field as wind energy where highly complicated considerations as aerodynamics, aeroelasticity, ice accretion and control are required, well-structured and high-level research and training are important aspects. WERL-harboured ECO-UQAR aims at supporting research and training via applied projects. In this chapter, we wish to present the laboratory, its mission, its structure and undergoing projects. The chapter aims to provide an idea of undergoing projects and the outcomes of the projects both from a result-oriented point of view and from a pedagogical point of view.

D. Ramdenee (✉)
Wind Energy Research Laboratory, Université du Québec à Rimouski, 300 allée des Ursulines, Rimouski, QC, Canada

Institut Technologique de Maintenance Industrielle, Sept-Îles, Canada
e-mail: dreutch@hotmail.com

T.A. Poirier
Institut Technologique de Maintenance Industrielle, Sept-Îles, Canada

R. Mohee
University of Mauritius, Port Louis, Mauritius

N. Barka • J. Chaumel • A. Ilinca
Wind Energy Research Laboratory, Université du Québec à Rimouski, 300 allée des Ursulines, Rimouski, QC, Canada

T.S. Uyar (ed.), *Towards 100% Renewable Energy*, Springer Proceedings in Energy, DOI 10.1007/978-3-319-45659-1_9

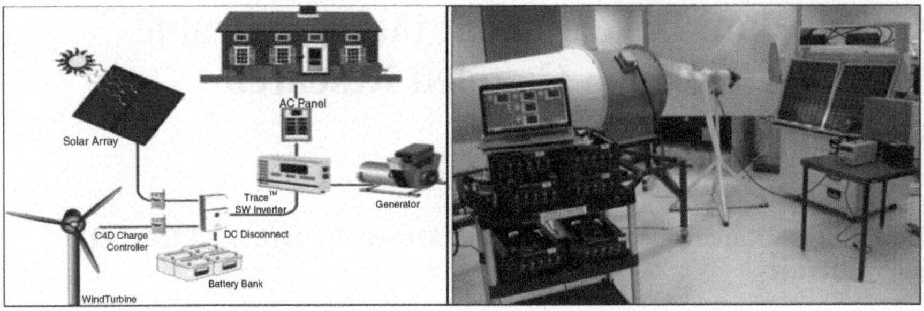

Fig. 1 Schematic representation and real setup of the laboratory

2 ECO-UQAR

ECO-UQAR proposes a real physical multi-renewable energy source bench test with an aim to allow students to learn the different aspects of renewable energy-related technologies through applied projects and setting up of bench tests. The laboratory contains a wind blower whose speed can be varied. The wind speed controller is connected via a data acquisition card to a computer. In a short term, we wish to simulate real wind in the blower from anemometry-collected data and analyse behaviour of wind turbines according to different wind regimes. This section of the laboratory is completely instrumented; that is, the parameters of the different sections are measured. For instance, the speed of the wind coming from the wind blower is measured and the data recorded in real time. The same applies to the instantaneously evolving turbine speed, voltage and current at the exit of the generator, at the exit of the batteries, etc. Similarly, the laboratory contains a number of solar panels and a variable insolation lamp. Like the wind section, the solar section is, also, completely instrumented. The light intensity and the output voltage and current are all measured. The two energy sources are of very different types but are coupled using a control panel. Figure 1 shows a schematic representation of the laboratory as well the real set-up of the laboratory.

Furthermore, development is in progress to render the laboratory accessible to a larger population via a Web-based virtual interface based on the real laboratory. The latter will be equipped with a high speed, high resolution and mobile Web camera that will allow real-time streaming video of the "in-operation" laboratory. The virtual laboratory (VL) user will, also, be able to see and manipulate, via a secured VPN service on a standalone machine, the variations of the different parameters via using LabView interface for the whole bench test. The VL users can, thus, observe the operation of the system, and assess the influence of different inputs (wind speed, luminosity and variability) on the output of the system (current, voltage, oscillations, etc.). Figure 2 illustrates part of the virtual laboratory.

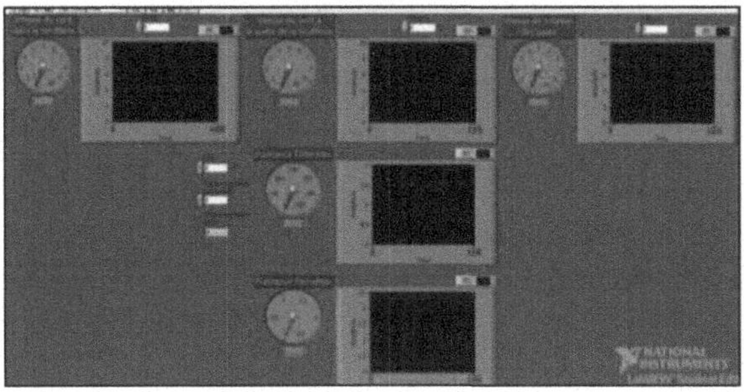

Fig. 2 Virtual laboratory interface

3 Mission and Projects of ECO-UQAR

ECO-UQAR has the mission to provide (1) a simple laboratory to popularise renewable energy technologies, (2) a bench test for calibration and characterisation of different wind turbines and solar panels, (3) a bench test to test new products to be applied for real projects, (4) a virtual laboratory to the general public and above all (5) research facilities and training via a project-oriented approach; that is, real-case projects related to renewable energy technologies are solved simultaneously with the learning of different principles and techniques. The projects cover a wide range of disciplines. In a given project, different students or teams are asked to work and specialise on a given problematic. The integration of the solutions of all the different projects makes up a single large and real-life project. The idea behind this is that renewable energy technology, mainly wind energy technology, is very vast and it is a utopia to think that a student can specialise in all the aspects of the technology. However, it is more convenient and realistic to provide a specialised training on a particular field of the technology and simultaneously provide a global training that equips the different students with sufficient aptitudes and skills to collaborate on a big project with optimised coordination. In this chapter we illustrate this mission of ECO-UQAR via an ongoing project: humanitarian project in Morocco (Safi region).

4 Installation of Coupled Wind-Solar Project in Morocco

This project aims at pumping water from an 85 m deep borehole in the region of Safi, Morocco, for a small village of around 500 people. The pump will be completely fuelled by a coupled wind-solar systems. This project integrates several aspects of renewable energy technology and can be defined as follows: to design the system, we need to evaluate equipment requirements. To do so, we need to predict pump energy needs as from water demand and consequently design the system according to the wind and solar potential of the region. High-level computational fluid dynamics (CFD) simulation has been run over the region and the wind potential evaluated. Similarly the insolation of the region was simulated. From these data

Fig. 3 Project set-up and location of site

and ECO-UQAR-designed wind turbine characteristics and bought solar panels, the number of the latter equipment was evaluated and the accessorial units (batteries, filters, etc.) bought. The control system that allows coupling of the two energy sources is also designed within ECO-UQAR. Similarly, the aerodynamics of the turbines, the aeroelastic and static effects on the blades as well the control of the overall system have also been performed by different teams of ECO-UQAR. Each team was led by an expert in the given field. Finally, in the light of all the equipment, a financial analysis was run to evaluate the cost and potential profitability of the project had the energy been sold. Weekly meetings regrouping the different specialised groups allowed an exchange of pertinent information via an M.S.-project-generated project management scheme and enabled all the different actors to have an idea of the different projects and clarify different matters related to the studied technologies. Figure 3 illustrates the project as a whole via a superposition of the equipment on the Safi region in GoogleEarth and a map of the region.

4.1 *Terrain Modelling, Energy Assessment and Wakes Modelling*

Bernoulli equations were used to evaluate the power requirements of the pump that will allow sufficient water availability to the village. The aim was that the wind turbines alone could cater for this demand. The use of the solar panels would be for safety measures in case of breakdown and for pedagogical reasons to study multi-source coupling technology. As we already had a choice for the wind turbine (ECO-UQAR characterised 300 W), the idea was to evaluate the number of such machines required to run the pump. CFD terrain aerodynamics was run to evaluate the wind potential of the region and the energy assessment was done via characterisation of the machine within ECO-UQAR laboratory. Furthermore, in order to avoid wake

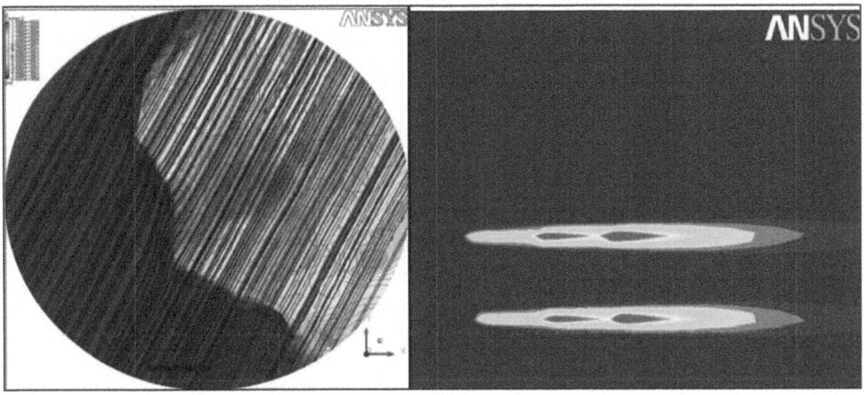

Fig. 4 Energy and wake assessment via CFD terrain aerodynamics to evaluate wind turbine requirements

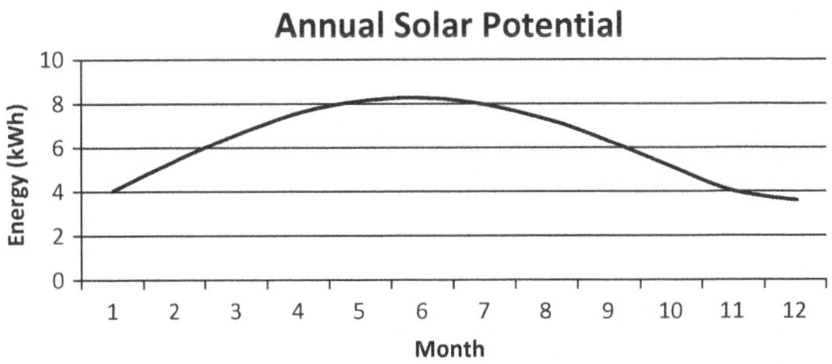

Fig. 5 Simulated annual solar potential for 2011 in the studied region

interaction between the turbines, wake modelling was done. Figure 4 shows the streamlines of the wind speed over the region and the wakes modelling using high-level CFD modelling.

4.2 Solar Panel Requirement

As for solar energy, as we mentioned in Sect. 4.1, the use is mostly for pedagogical reasons. We wished to develop an expertise in solar energy simulation and we anticipated power availability from some purchased solar panels via a Milankovitch function. This relates solar power availability to the position of the earth with reference to the sun as well as the tilt of the former. The purchased solar panels were, furthermore, tested in the laboratory. Figure 5 illustrates the annual solar potential simulated for the illustrated Safi region for the year 2010.

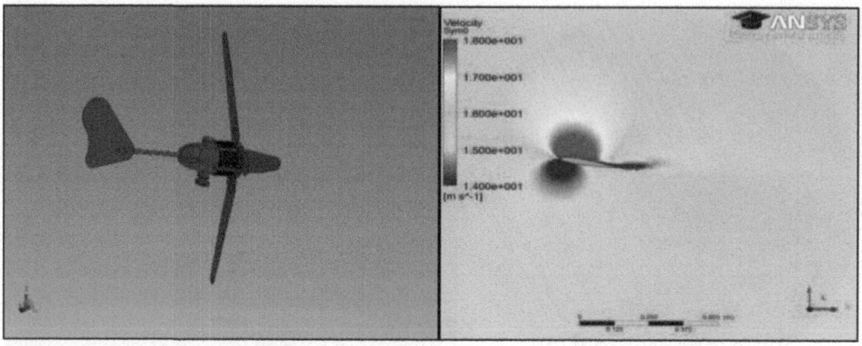

Fig. 6 CAD and aeroelastic analysis of the wind turbine

4.3 Wind Turbine Design, Static, Aerodynamic and Aeroelastic Analysis

ECO-UQAR, also, aspires to promote research in wind turbine engineering via optimisation. For Safi project, several actors worked on the wind turbine computer-aided design (CAD), the analysis of the latter to evaluate its resistance to static and aerodynamic forces. This part of the project is considered to be quite complex and focuses on graduate engineering students exclusively. Several articles dealing with this field have been published by the WERL [1, 2], and the acquired expertise was used for this particular project. Figure 6 illustrates the designed CAD model to be used for the project and run aeroelastic simulation on the blade airfoil.

4.4 Energy Integration and Control

Wind energy is variable, alternating while solar energy is more constant and direct. Coupling of these two energy types is a real challenge. Commercial control boxes exist to allow such coupling. However, to achieve this energy superposition fourth-year engineering students have been assigned as their end of bachelor project to design a new control module that allows coupling of different energy sources. This project is actually ongoing and is being supervised by three professors. In parallel, research is being conducted to devise control strategies to damp unwanted peaks and oscillations in the output energy signals. The students can observe these peaks in ECO-UQAR laboratory operation and learn about control strategies and tools on Matlab. The laboratory should offer, in short term, equipment that will allow testing of the proposed control strategies on ECO-UQAR installations.

5 Round-Up and Conclusion

The different projects have been and are being run in parallel under well-structured coordination. This is a very important part of wind energy projects and makes part and parcel of the integrative training scheme. A group of students and ECO-UQAR supervisors are presently working on HOMER software to produce a level financial feasibility study of the whole project. During weekly meetings, the different teams are invited to provide information about the cost, efficiency and output of their different modules to enter that are entered in the economic model of the software. The idea is to offer the students sufficient skills to evaluate the feasibility of a renewable energy project. HOMER integrates annuities, up-scaled operations and maintenance costs, machine costs and tax credits as well as probability distributions and other economic considerations to make a most refined and complete net actualised value analysis. This exercise marks the end of this very special training method offered by ECO-UQAR on this particular project. In the future, we will propose an appreciation scheme whereby students will be invited to compare their appreciation and a quantitative analysis of their acquired skills on renewable energy technologies as compared to classic courses on the subject.

References

1. Ramdenee D, Hussein H, Barka N, Ilinca A (2011) Modeling of aerodynamic flutter on a NACA 4412 airfoil, IMAACA conference, Italy
2. Ramdenee D, Minea S, d'Hamonville T, Ilinca A (2010) Numerical simulation of a wind turbine airfoil-part 1. Canadian Society for Mechanical Engineering, Victoria, BC, Canada

Part II
System Analysis, Modeling, Simulations

Barriers and Opportunities for Transformation of Conventional Energy System of Turkey to 100 % Renewable Community Power

Tanay Sıdkı Uyar

1 Introduction

Turkey has been a destination for the inefficient and polluting end-use technologies that are cast off, with the support of export credits, from more industrialized countries. The Turkish energy system is highly dependent on fossil fuels. This also places a heavy burden on the state budget of the Turkish Republic. The solution is to promote energy end-use efficiency by using the best available technologies and supply all of the energy needed with renewable energy, supported by renewable energy storage and smart grid technologies. We must come to a consensus regarding the sun's role in providing a living space for human beings on earth. The world is like a cell surrounded by a blanket of greenhouse gases that give it an atmosphere with an average temperature of 16 °C, in the middle of a space with an average temperature of −60 °C. Considering its natural solar, geothermal, wind, and biomass potential, Turkey definitely has more renewable energy resources than the world average. The official energy strategy in Turkey is constrained by the previous and current decisions of fossil fuel, nuclear, and hydropower investors. The externalities of energy consumption and production are not internalized in Turkey. Regarding the implementation of energy end-use efficiency and renewable energy technologies, Turkey is far behind the European Union directives. Turkish energy policies do not comply with European Union regulations on renewable energies and energy efficiency. Annual IRENEC (100 % Renewable Energy) Conferences organized by Renewable Energy Association of Turkey since 2011 aim to pursue improvements in energy end-use efficiency and renewable energies, and to create the infrastructure required to realize the 100 % renewable energy goal in industry, local communities, architecture, and transportation.

T.S. Uyar (✉)
Department of Mechanical Engineering, Marmara University, Istanbul, Turkey
e-mail: tanayuyar@marmara.edu.tr

T.S. Uyar (ed.), *Towards 100% Renewable Energy*, Springer Proceedings in Energy, DOI 10.1007/978-3-319-45659-1_10

2 The Global Transition of Humanity from the Fossil Age to the Solar Age

We know that up until the 1850s, human activities were carried out in substantial harmony with nature. The local damage caused by human activities was rarely irreparable and did not overstep the limits of the natural carrying capacity. Beginning in the 1850s, rising energy demand driven by the industrial revolution forced humanity to start consuming more fossil fuels. During the 1950s, thousands of people died from respiratory problems due to coal burning in the big cities of the world. As a result, "clean coal technologies" were devised to prevent such deaths. Burning coal with fewer externalities became one of the options for the future of the energy industry. After the Rio Conference in 1992, when it was acknowledged that the real threat to the atmosphere is the global warming caused by fossil fuel combustion, the arguments for clean coal combustion became obsolete. According to Sierra Club of the USA: More than *100,000 MW* of *coal* power retired since 2010 [1]. The 1970 petroleum crisis convinced the energy decision makers that dependence on petroleum is not a wise energy strategy. The first solution that human beings naturally adopted under those circumstances was using less energy than before. Countries that learned from this have initiated energy end-use efficiency programs, which ended up generating some of today's best available technologies. They consume ten times less energy than technologies of the 1970s, but provide the same transportation, industrial, and residential services. In 1973, the decision makers were convinced that the waste heat from nuclear weapon production facilities could be a way to supply the total energy requirements of the world. The majority of the existing nuclear power plants were built during this period. After the Three Mile Island nuclear disaster in 1978, the USA was the first country where plans to build new nuclear power plants were stopped. Approximately 100 nuclear power plants that had been ordered between 1973 and 1978 were also canceled. The main argument behind these actions was explained by Mark Halt: "No nuclear power plants have been ordered in the United States since 1978, and more than 100 reactors have been canceled, including all ordered after 1973. The most recent U.S. nuclear unit to be completed was TVA's Watts Bar 1 reactor, ordered in 1970 and licensed to operate in 1996. Reasons for the 30-year halt in U.S. nuclear plant orders include high capital costs, public concern about nuclear safety and waste disposal, and regulatory compliance costs" [2]. "Construction on Watts Bar Unit 2 originally began in 1973, but construction was halted in 1985 after the NRC identified weaknesses in TVA's nuclear program. The Tennessee Valley Authority's (TVA) Watts Bar Unit 2 was connected to the power grid on June 3, 2016 becoming the first nuclear power plant to come online since 1996, when Watts Bar Unit 1 started operations" [3].

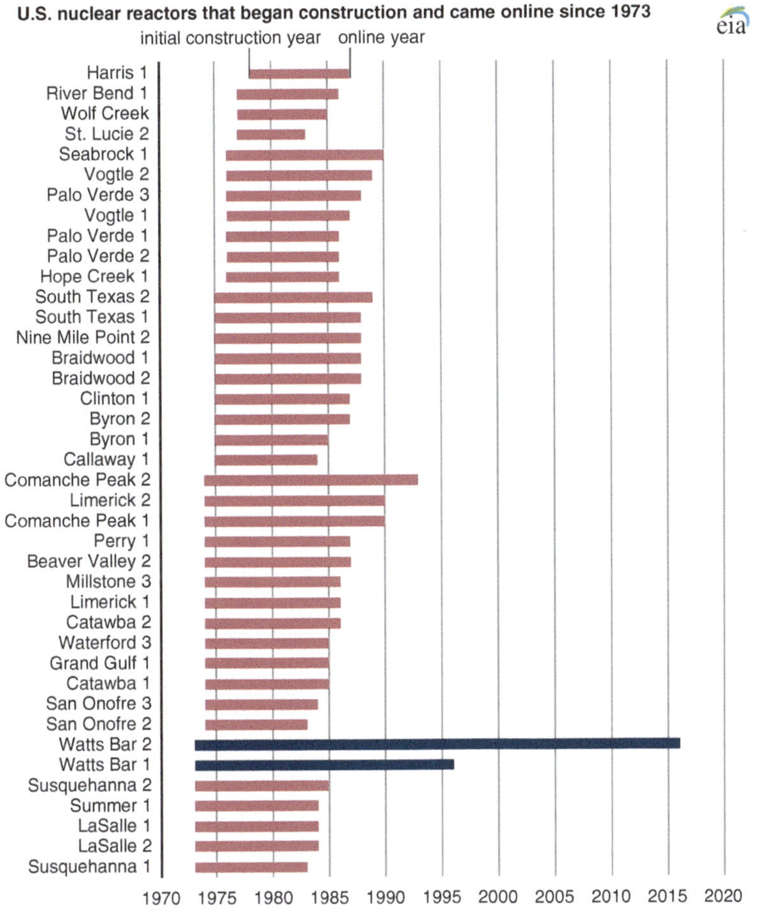

3 Energy Technology Initiatives: International Energy Agency, OECD

Beginning in the 1970s, the OECD countries started trying to define the energy problem and find feasible solutions to it. In 1974, to counter the Organization of the Petroleum Exporting Countries (OPEC), OECD countries established the International Energy Agency (IEA) as a political union of rich countries that would take measures such as ensuring a mandatory 3-month store of petroleum in each country. The IEA decided also to initiate 40 multilateral technology initiatives (Implementing Agreements) for facilitating energy technology cooperation between the scientists and researchers of OECD countries. (Implementing Agreements on technologies such as bioenergy, geothermal energy, photovoltaic power systems, renewable energy technology deployment, solar heating, and cooling and wind energy systems contributed a lot to the research, development, and commercialization of renewable energy technologies [4].) "Germany's plan: ramp up renewables, drive down energy consumption" can be seen as a realistic example which is being implemented by the decision makers of Germany as a result of dedicated research [5].

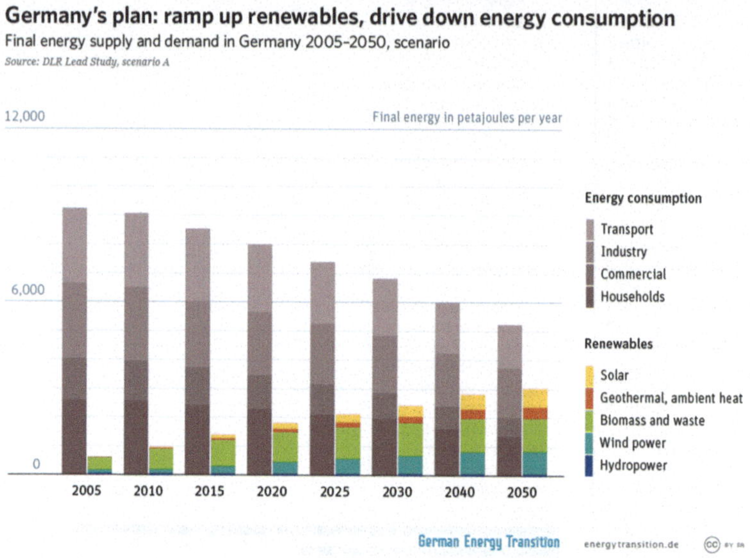

In March 2007, the EU's leaders endorsed an integrated climate and energy policy that aims to combat climate change and increase the EU's energy security while strengthening its competitiveness. They committed Europe to transforming itself into a highly energy-efficient, low-carbon economy. EU heads of state and government set a series of demanding climate and energy targets to be met by 2020 and 2030, respectively, as shown in the following figure [27].

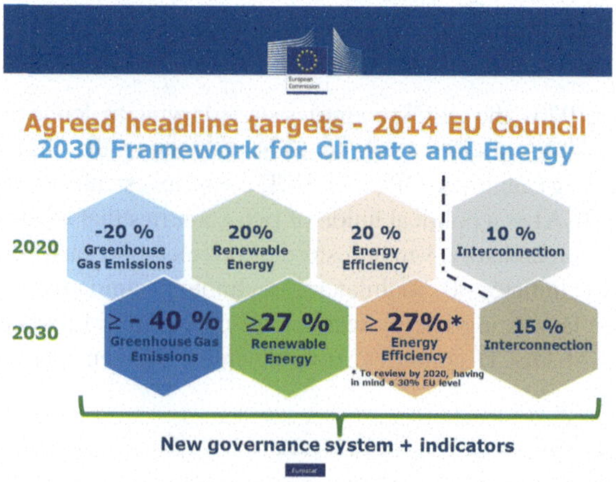

4 The Present Energy Situation in Turkey

To be able to analyze the present energy situation in Turkey, we need to keep in mind the above global historical development of energy problems and solutions. So far, Turkey has been a destination for the inefficient and polluting end-use technologies that are cast off, with the support of export credits, from more industrialized countries. Those more industrialized countries, meanwhile, are taking measures to be more efficient, healthy, and in harmony with nature. The Turkish energy system is highly dependent on fossil fuels. This also places a heavy burden on the state budget of the Turkish Republic. Turkey's substantial fossil fuel dependency is the main reason for the high external costs of energy production and consumption in Turkey. This also increases the cost of health services. The share of fossil fuels in Turkey's 2013 primary energy consumption was 88.9 %. Turkey's installed electricity capacity is dominated by hard coal, lignite, natural gas, fuel oil, diesel, and big hydropower plants. Natural gas is Turkey's main source of energy as shown below. Turkey imports 99 % of its gas, and Russia pipes in nearly 60 % of Turkey's total gas use [6].

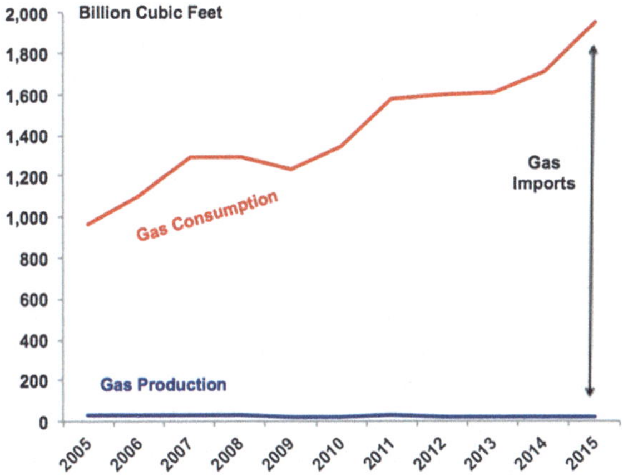

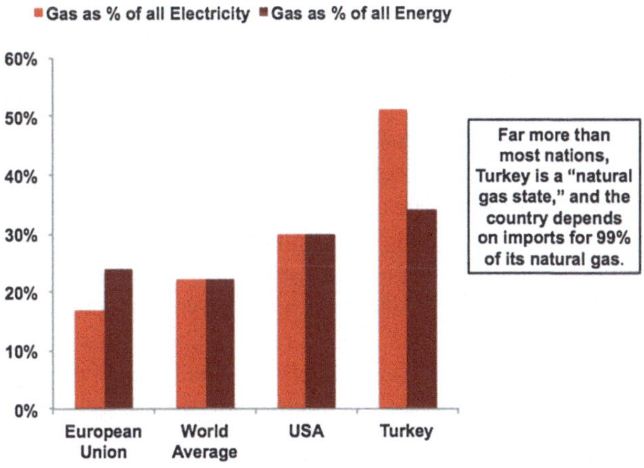

We have organized four workshops together with the energy committee of the Young Businessmen's Association of Turkey (TÜGİAD), inviting all actors in Turkey's energy sector. The final report, entitled "Energy Problems of Turkey and Suggested Solutions," included all the agreed-upon recommendations of the workshop participants [7]. The main factors contributing to Turkey's energy problem are:

1. Take-or-pay (TOP) arrangements based on long-term natural gas purchase agreements made with Russia, Azerbaijan, Iran, and other countries [8]
2. The willingness of the authorities to provide licenses to big hydropower, coal, natural gas, and nuclear power plants
3. Efforts to delay and slow renewable energy investments and energy end-use efficiency implementation

Turkey has terrific conditions for solar and wind power in Turkey. Exploiting these resources is already economically and technically possible. The problem is that the decision makers favor fossil fuels and nuclear energy. Turkey is locked into long-term agreements to purchase natural gas at fixed prices as well as nuclear energy technology, and these agreements are acting as a financial disincentive for the development of renewable energy.

Planned and proposed nuclear power reactors in Turkey are given below [9]:

	Type	MWe gross	Start construction	Start operation
Akkuyu 1	VVER-1200	1200	Late 2016	2023
Akkuyu 2	VVER-1200	1200	2017	2023
Akkuyu 3	VVER-1200	1200	2018	2024
Akkuyu 4	VVER-1200	1200	2019	2025
Sinop 1	Atmea1	1150	2017	2023
Sinop 2	Atmea1	1150	2018	2024
Sinop 3	Atmea1	1150		?
Sinop 4	Atmea1	1150		?
Igneada 1–4	AP1000 × 2	2 × 1250		
	CAP1400 × 2	2 × 1400		

More must also be done in Turkey to improve energy end-use efficiency. There is currently a huge amount of energy waste. Turkey can cut its electricity needs by 50 % if it uses more up-to-date, energy-efficient technology, and by doing so help keep down carbon emissions. According to official reports, electricity demand in Turkey is growing at an annual rate of 8 %. There are two issues with this. One is the lack of long-term energy-economy-environment decision support tool utilization for long-term energy demand estimates. Investments in Turkey's energy sector so far have not been decided using a long-term, strategic energy-economy-environment decision support tool. If there is an export credit available for a technology investment, the investment is accepted as "valuable" and included in the strategic plan of the Turkish Ministry of Energy and Natural Resources [10].

5 Global Dislocation of Obsolete Technologies from One Market to Another

Export credits are more available if the technology of the investment is obsolete and has no market value in the country where the export credit is coming from. When an obsolete technology is sold, however, its problems are also exported to the buyer country.

The second issue is the lack of regard for end-use efficiency by the decision makers, who are "business-as-usual energy sector" supporters. This means that they have promised and included in their program to allow and facilitate inefficient, dirty, and obsolete technologies to enter their energy market. To justify their obsolete technology investments, they always talk about the high growth rate of the energy demand of the countries. They never talk about energy end-use efficiency or try to reduce energy demand, because they do not want any barrier to their obsolete technology investments.

There are two types of decision makers: those who are part of the problem, and those who try to solve the problem. Since the major existing suppliers in the energy sector are based on coal, natural gas, petroleum, nuclear, and big hydropower, the existing decision-making mechanism is trying to slow the transition to 100 % renewable energy. The solution is to promote energy end-use efficiency by using the best available technologies and supply all of the energy needed with renewable energy, supported by smart grids and renewable storage technologies.

6 Free-of-Charge Service from the Sun

We must come to a global consensus regarding the sun's role in providing a living space for human beings on earth. The world is like a cell surrounded by a blanket of greenhouse gases that give it an atmosphere with an average temperature of 16 °C, in the middle of a space with an average temperature of −60 °C. Solar radiation makes light and heat available to all plants, animals, and human beings on earth, free of charge. Additionally, solar energy arriving in the atmosphere is stored as chemical energy in biomass, potential energy of water molecules, and kinetic energy of the wind. The scattered and high-entropy nature of these renewable energy sources makes them available all over the atmosphere, where they have the potential to boost peace, freedom, and employment in the human communities that make use of them. The renewable energy resources available in the world are more than we all need.

For example about 41 million tourists visited Turkey in 2014 to soak up its famous sun [11]. Turkey also has considerable geothermal resources, thanks to its position in a seismically active part of the world. Considering these, as well as its natural wind and biomass potential, Turkey definitely has more renewable energy resources than the world average.

Turkey has a huge potential for renewable electricity from wind, solar, and geothermal sources. The Turkish Wind Atlas Project was carried out between 1987 and 1988, supported by the State Planning Organization (DPT) of the Turkish Republic. The study covered 40 meteorological stations in Turkey. Twenty of these were taken as reference stations—using the European Wind Atlas Methodology [12–16]. The wind energy electricity generating potential of Turkey was estimated to be 83,000 MW by a team of scientists from Utrecht University in 1993 [17]. Today, with the available wind turbine technology, Turkey's potentially installable wind power-generating capacity is 150,000 MW.

Currently (end of March 2016), Turkey has a total installed generating capacity of approximately 74,039.4 MW (26,156.5 MW from hydro, 357.7 MW solar, 1001.3 MW renewable + waste + geothermal, 41,922.5 MW thermal, and 4601.4 MW wind) [18].

Installed Power of Electricity in Turkey (MW—End of March 2016)

TÜRKİYE'DE ELEKTRİK ENERJİSİ KURULU GÜCÜ (MW – 2016 MART SONU)

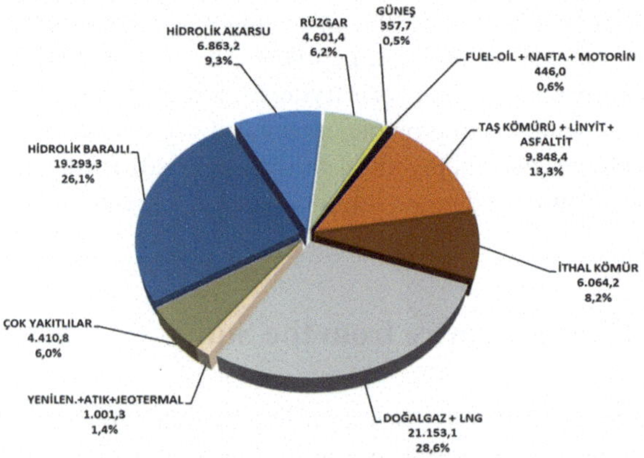

KURULU GÜÇ : 74.039,4 MW

Kaynak: TEİAŞ, 11.04.2016

Wind power could supply Turkey's future electricity needs twice over within 5–10 years if the decision makers have the political will to develop this sector. The share of electricity that comes from renewable energy sources in Turkey is tiny. Geothermal energy has the potential to supply five million households with heating. The potential for improved energy efficiency in Turkey is also vast. According to the Turkish Ministry of Foreign Affairs, "The primary aim of Turkey is to realize its own energy security. To this end, Turkey aims to

- Diversify its energy supply routes and source countries
- Increase the share of renewables and include the nuclear in its energy mix
- Take significant steps to increase energy efficiency
- Contribute to Europe's energy security" [19]

For the State Planning Organization's 5-year development plan, the New and Renewable Energy Resources and Technologies Unit of Kocaeli University (YEKAB) offered estimated wind energy targets in 1998. They are given in the figure below [20]:

Possible targets of wind energy for Turkey (YEKAB estimation)

Yıl	Kurulu Kapasite (MW)
2000	400
2003	1400
2005	5000
2010	10,000
2020	20,000

Turkey passed a new renewable energy law to bring it in line with European Union legislation. The law offered support to renewable sources, including wind power, by giving a government guarantee to purchase electricity at a set price for a period of 7 years. But the tariff given—approximately 5 Eurocents/kWh of electricity—was much lower than tariffs in most other European countries, and economic studies showed that it discouraged investment in the renewable energy sector. Turkey enacted its second renewable energy law, namely "Law No. 6094 Concerning the use of Renewable Energy Resources for the Generation of Electrical Energy," in 2010 [21]. Solar energy sources are covered by this law, which decrees that facilities generating electricity from renewable energy sources will be granted a renewable energy resources (RER) certificate, which will entitle such facilities to benefit from the incentives provided by the law. The Turkish Energy Market Regulatory Authority (EMRA) is the authority appointed to grant RER Certificates.

7 Global Renewable Energy Deployment

Technological research and development efforts accelerated after 1980 in industrialized countries, expediting the development of renewable energy technologies. Supporting wind power with tax credits contributed greatly to the establishment of a market for wind turbines in the USA. After 1996, we saw Germany and the European Union giving huge support to megawatt-scale wind turbine commercialization. Today, we know of 10 MW prototype production efforts in the USA.

Due to governmental support, including feed-in tariffs, the global installed capacity of solar PV reached 227,000 MW and global wind farms' installed capacity reached 433,000 MW by the end of 2015. According to recently published REN 21 Renewables 2016 Global Status Report 8.1 million people were working in the renewable energy industry at the end of 2015 [22].

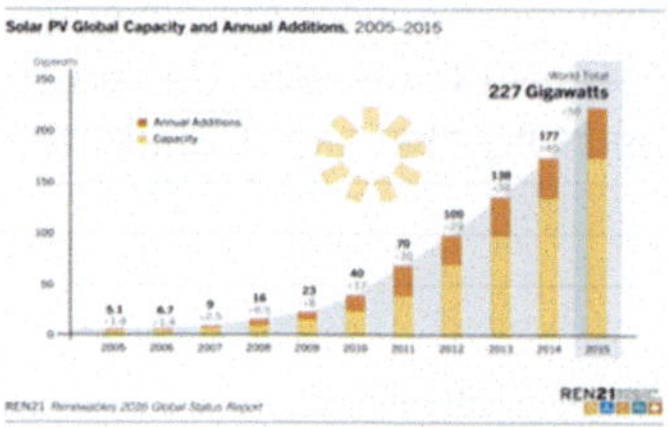

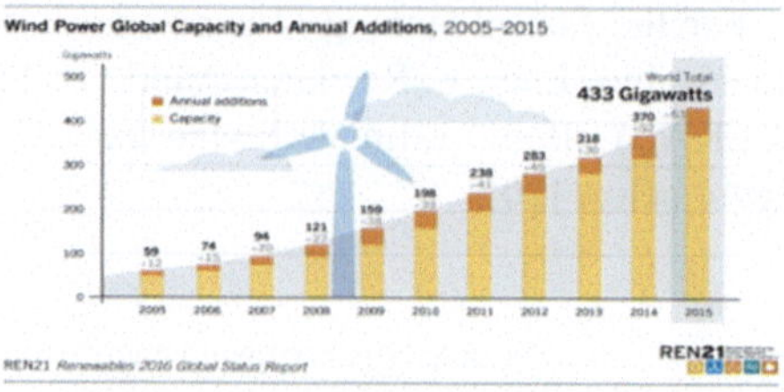

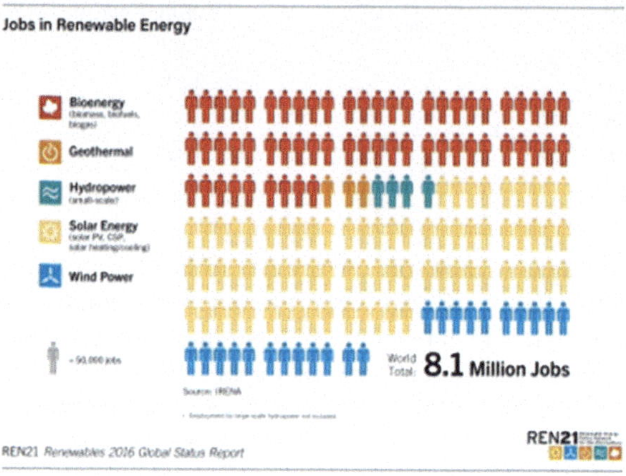

8 Turkey's Energy Strategy

The decision support tool developed under the Energy Technology Systems Analysis Program (ETSAP) implementing agreement of the IEA/OECD is not used in Turkey. The MARKet ALlocation (MARKAL) model for Turkey was devised by Uyar TS et al. at the Energy Section of Marmara University, and is operational [23].

The MARKAL model is a long-term strategic energy-economy-environment decision support tool that can perform the following:

- Analyze the effects of greenhouse gas emission-reduction strategies on Turkey's energy system and economy
- Establish energy-efficient utilization and cost-effective energy technology selection strategies for Turkey
- Establish mitigation strategies for energy-related emissions in Turkey

The official energy strategy of Turkey is constrained by the previous and current decisions of fossil fuel, nuclear, and hydropower investors. This tendency to build more conventional power plants is the main problem preventing Turkey from moving towards a solution. As the size of the problem increases, the damage from the problem also increases. The tax income of the national budget is proportional to the total fuel oil consumption in the country. Therefore, the decision-making bodies are unwilling to reduce the total amount of fuel oil used in the country. Coal is becoming more cheap and readily available in the market as the local and global externalities of coal burning in the atmosphere become more apparent. Countries like Turkey, with high dependency on expensive petroleum and gas imports, are trying to convince their citizens that in order to reduce this dependency, building coal-fired power plants is the only solution. The externalities of energy consumption and production are not internalized in Turkey. We know that if the externalities are not calculated and not included in project costs, they become social costs to be paid by citizens and the environment. When protective health services are not available in such societies, many people die as an effect, and even the resulting social security costs may strain the budget. The easily available obsolete energy technologies spreading around the world pose a global threat to humanity. As one country becomes cleaner and more efficient by increasing its standards and starting to use the best available technologies, the inefficient and dirtier technologies are snatched up by markets with lower standards, thanks to the availability of export credits. Investments in newer, renewable technologies tend to stop large-scale projects and promote decentralized alternatives. This contradicts the marketing bias towards large-scale projects. To justify the necessity of large-scale projects, the implementation of energy end-use efficiency and renewable energy technologies is consistently delayed. By taking per-capita energy consumption as its development criterion, Turkey's national climate strategy promotes the increase of carbon dioxide emissions. Regarding the implementation of energy end-use efficiency and renewable energy technologies, Turkey is far behind the European Union directives. The laws are available, but they ultimately slow down and possibly delay real implementation. Guaranteed prices of just \$0.073/kWh for wind and \$0.133/kWh for solar photovoltaics are two of the main obstacles. Renewable energy resources in Turkey and commercial renewable energy technologies available in global markets need only the support of decision makers for full implementation.

The existing distribution networks are designed for large-scale, centralized power plants. The main barrier to the full implementation of renewable energy technologies is the unwillingness of the energy transmission authority to redesign the grid so that it can accept more of the decentralized energy produced in the country. EMRA is limiting the number of renewable energy licenses it awards because of the arguments of the transmission company. Such constraints are not in place for the thousands of MW new nuclear and fossil-fuel power plants planned and licensed by EMRA.

9 External Obligations and Opportunities

Turkish energy policies do not comply with European Union regulations on renewable energies and energy efficiency, although they do comply with the "Privileged Partnership" statute envisioned for Turkey by some European decision makers. Even though accession negotiations have been initiated between the European Commission and the Turkish Republic, the energy chapter has not been opened due to the delay caused by country leaders who are trying to guarantee the privileged partnership statute for Turkey. The amount of licensed fossil fuel and nuclear power plant capacity in Turkey is increasing every day that passes without negotiations on the energy chapter opening. If Turkey is going to be a part of Europe in the future, the new investment decisions in Turkey should not be supporting an increase in the total cleaning cost of the Europe Union. Long-term estimates for the energy mix of the world demonstrate that renewable energy utilization is inevitable, as shown in the figure below.

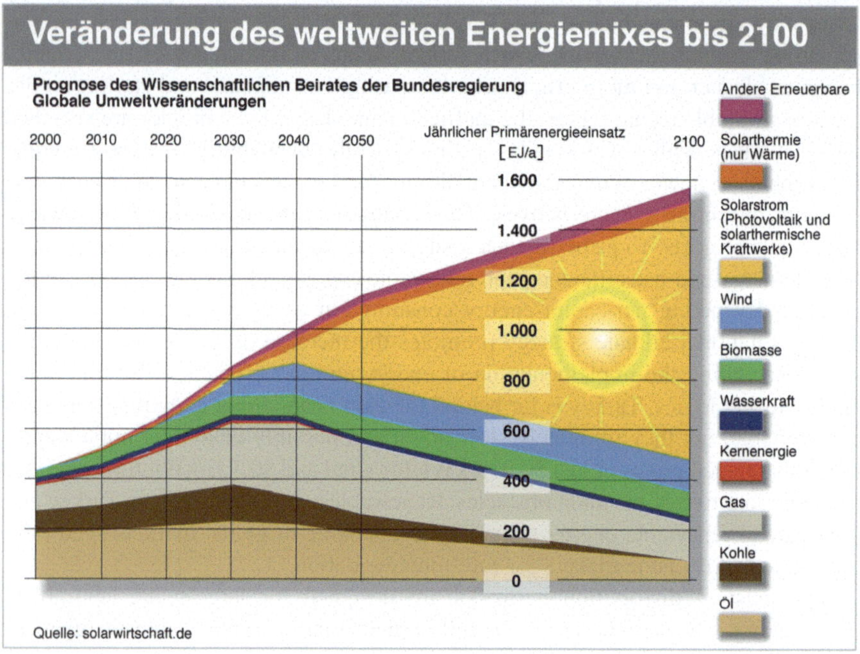

The question today is which countries will be in which part of the picture. In countries like Turkey, which insist on building fossil fuel and nuclear power plants despite their huge local renewable energy resources and energy end-use efficiency potential, the citizens will be the ones who suffer and pay the cost of the global transition from the fossil fuel age to the solar age.

10 The Efforts of Renewable Energy Association of Turkey (EUROSOLAR Turkey)

Since 2002, EUROSOLAR Turkey [24] has been:

- Laying the ground for solutions to Turkey's dependence on foreign fossil fuel sources and promoting the utilization of the Turkish renewable energy potential, which is exceptionally abundant and varied

- Bringing together all players in the field to fill in information gaps and try to put together a coalition with widespread public support that will force the politicians and bureaucrats to reverse their course on Turkey's energy policy
- Contributing to the development of legislation in the renewable energy field in Turkey, especially regarding harmonization with EU policy
- Mobilizing Turkey's civil society and internally dynamic groups to move sustainably towards a clean and independent energy policy

EUROSOLAR Turkey takes part in the Clean Energy Platform of Turkey (TEP), which includes several national and local institutions, and in the Environmental Platform of Turkey (TURÇEP), which comprises more than 200 grassroots environmental NGOs dealing with environmental issues.

EUROSOLAR Turkey has organized annual International 100 % Renewable Energy Conferences (IRENECs) since 2011, to set up an international platform to discuss the technical, economic, political, and administrative aspects of this monumental transition from fossil fuels to renewable energy sources, and to contribute to the 100 % renewable energy goal without using nuclear energy or carbon capture technology [25]. IRENEC conferences aim to pursue improvements in energy end-use efficiency and renewable energies, and to create the infrastructure required to realize the 100 % renewable goal in industry, local communities, architecture, and transportation.

The recent initiative from EUROSOLAR Turkey is a Civil Society Dialog Project, namely the Powering Communities Project, supported by the European Commission to develop roadmaps for four cities of Turkey for 100 % Renewable Community Power [26].

References

1. http://www.power-eng.com/articles/2016/05/sierra-club-more-than-100-000-mw-of-coal-power-retired-since-2010.html
2. RL33558 CRS report for congress, prepared for members and committees of congress. Nuclear energy policy, Mark Holt, specialist in energy policy, December 10, 2009, Congressional Research Service 7-5700. www.crs.gov
3. http://www.eia.gov/todayinenergy/detail.cfm?id=26652
4. https://www.iea.org/publications/freepublications/publication/technology_initiatives.pdf
5. http://energytransition.de/wp-content/themes/boell/pdf/en/German-Energy-Transition_en.pdf
6. http://www.forbes.com/sites/judeclemente/2016/02/07/turkeys-rising-natural-gas-demand-needs-u-s-lng/#20701f2a2ac7
7. Uyar TS (January 2003) Türkiye'nin Enerji Sorunları ve Çözüm Önerileri. English edition: Energy problems and solution suggestions for Turkey. Ajans-Türk Basın ve Basım A.Ş., Batıkent, Ankara. TÜGİAD 2004.
8. http://www.todayszaman.com/news-268794-turkey-prepares-to-sue-iran-over-too-high-gas--prices.html
9. http://www.world-nuclear.org/information-library/country-profiles/countries-t-z/turkey.aspx
10. The Republic of Turkey Ministry of energy and natural resources strategic plan (2015–2019). http://sp.enerji.gov.tr/ETKB_2015_2019_Stratejik_Plani.pdf

11. https://en.wikipedia.org/wiki/Tourism_in_Turkey#cite_note-csmonitor-4
12. Uyar TS et al (1988, February) Wind atlas statistics calculated for Çorlu, Uzunköprü and Kırklareli. TÜBİTAK project report no: 2, MAE, Gebze, Kocaeli, pp 190
13. Uyar TS et al (1988, June) Wind atlas statistics calculated for Afyon, Seydişehir, Elmalı, Yumurtalık and Anamur. TÜBİTAK project report no: 3, MAE, Gebze, Kocaeli, pp 301
14. Uyar TS et al (1988, August) Wind atlas statistics calculated for Bozcaada, Çeşme, Fethiye and Sultanhisar. TÜBİTAK project report no: 4, MAE, Gebze, Kocaeli, pp 148
15. Uyar TS et al (1988, November) Wind atlas statistics calculated for Hopa, Samsun, Trabzon and Sinop. TÜBİTAK project report no: 5, MAE, Gebze, Kocaeli, pp 148
16. Uyar TS et al (1988, December) Wind atlas statistics calculated for Ağrı, Gemerek, Çanakkale and Akçakoca. TÜBİTAK project report no: 6, MAE, Gebze, Kocaeli, pp 148
17. van Wijk AJM, Coelingh JP (1993) Wind power potential in the OECD Countries. Utrecht University, Department of Science, Technology and Society, pp 35
18. http://www.teias.gov.tr/
19. Turkey's energy profile and strategy. http://www.mfa.gov.tr/turkeys-energy-strategy.en.mfa
20. Uyar TS (2013) Emerging wind energy sector in Turkey. In: Preben Maegaard, Anna Krenz, Wolfgang Palz (eds) Wind power for the world: international reviews and developments. Pan Stanford series on renewable energy, vol 3, pp 534–556. Pan Stanford Series on Renewable Energy, USA
21. Renewable energy law of Turkey. https://www.tbmm.gov.tr/develop/owa/kanunlar_sd.durumu?kanun_no=6094
22. REN 21 Renewables 2016 Global status report. http://www.ren21.net/status-of-renewables/global-status-report/
23. Uyar TS, Kırlıdoğ M, Sulukan E, Sağlam M (2009) Determining optimum energy strategies for Turkey by MARKAL model. Research project in Marmara University, Turkey. (Project # FEN-KPS-080805-0177)
24. www.eurosolar.org.tr
25. www.irenec.org
26. www.poweringcommunities.org
27. http://ec.europa.eu/clima/policies/strategies/2030_en

From Planning to Operation: Wind Power Forecasting Model for New Offshore Wind Farms

Melih Kurt, Jan Dobschinski, Bernhard Lange, and Arne Wessel

Nomenclature

ANN	Artificial neural network
C_t	Thrust coefficient
DWD	German weather service
EWEA	European Wind Energy Association
IPCC	Intergovernmental Panel on Climate Change
ME	Mixture of experts
MOS	Model output statistics
MW	Megawatt
NNS	Nearest-neighbor search
nRMSE	Normalized root mean square error
NWP	Numerical weather prediction
RMSE	Root mean square error
SVM	Support vector machines
TSO	Transmission system operator
WCMS	Wind farm cluster management system
WPMS	Wind power management system

M. Kurt (✉) • J. Dobschinski • B. Lange • A. Wessel
Energy Economy and Grid Operation, Fraunhofer IWES,
Koenigstor 59, Kassel 34127, Germany
e-mail: melih.kurt@gmail.com

T.S. Uyar (ed.), *Towards 100% Renewable Energy*, Springer Proceedings
in Energy, DOI 10.1007/978-3-319-45659-1_11

1 Introduction

The IPCC indicated in Assessment Report of the Intergovernmental Panel on Climate Change that human activities are directly related to the increased atmospheric levels of greenhouse gases, i.e., carbon dioxide, methane, chlorofluorocarbons, and carbon monoxide. Additionally, according to NASA Earth Observatory a correlation also exists between global warming involving greenhouse gases and environmental problems. The main artificial source of carbon dioxide discharge is derived from fossil fuels (conventional power plants). Therefore, many recent researches have focused on reducing the consumption of fossil fuels and replacing those with renewable, environment-friendly energy sources. Currently, the wind energy is considered as one of the most promising renewable energy sources. Especially the offshore wind power with its higher production capacity, approximately twice more full-load hours than onshore [1], is significant to achieve energy targets such as reduction of greenhouse gases and improvement of energy independency. In addition to these, many countries have revised their energy policy after Fukushima nuclear disaster in March 2011 in Japan; for example the German Government has decided to shut down all nuclear reactors in Germany by 2022 and to accelerate installation of renewable energies.

As a result of the situation above, the penetration of the offshore wind energy is progressing evidently in Europe in the last few years. As it is reported by the EWEA, with the grid connection of 101 new offshore wind turbines with a capacity of 348.1 MW in the first half year of 2011, offshore wind power reached 3294 MW in total in 49 wind farms in 9 European countries. Figure 1 illustrates the development of installed wind power capacity in Europe until the end of the year 2010 [2].

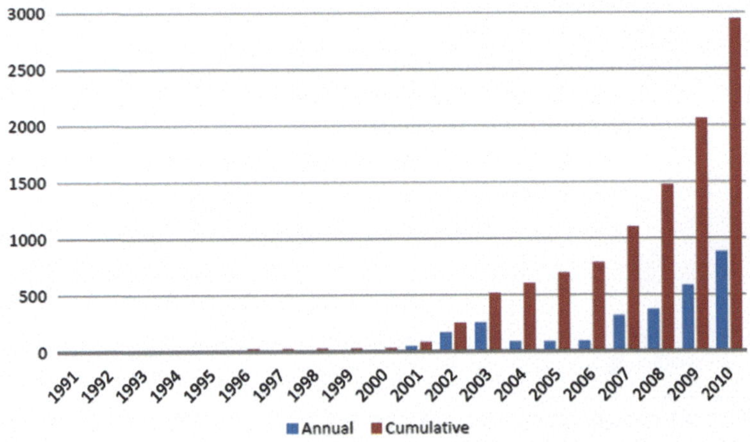

Fig. 1 Development of installed offshore wind power capacity in MW—Europe. *Source* EWEA

Fig. 2 Offshore wind farm "alpha-ventus" © alpha ventus

In Germany, the first offshore wind farm "alpha ventus" with 12 wind turbines and 60 MW capacity was erected in 2009 in the North Sea, from which this study takes its roots mainly (see Fig. 2). And it is aimed by the German Government that the installed capacity of electricity generation from offshore wind power in Germany shall increase to 25 GW installed capacity in 2030.

Besides the advantages of the high potential, there are however several challenges as to safe and consistent grid integration of offshore wind farms. An important amount of wind energy will be produced in a relatively small area; hence smoothing effects are limited compared to onshore wind power. Therefore fluctuating offshore wind power generation will become a critical issue concerning required balancing and reserve power with increasing offshore wind power deployment.

2 Methods of Wind Power Forecasting

There are basically three different approaches to develop wind power forecast: statistical, physical, and learning approach. Some studies handle learning method under statistical approach. There are also models existing, which combine these different forecasting models [3, 4].

The aim of physical approach is to describe the physical process of conversion of wind into energy and to model each of the related steps. The physical approach uses physical considerations concerning the terrain information such as roughness and orography and models the shadowing effects of the wind farm [5].

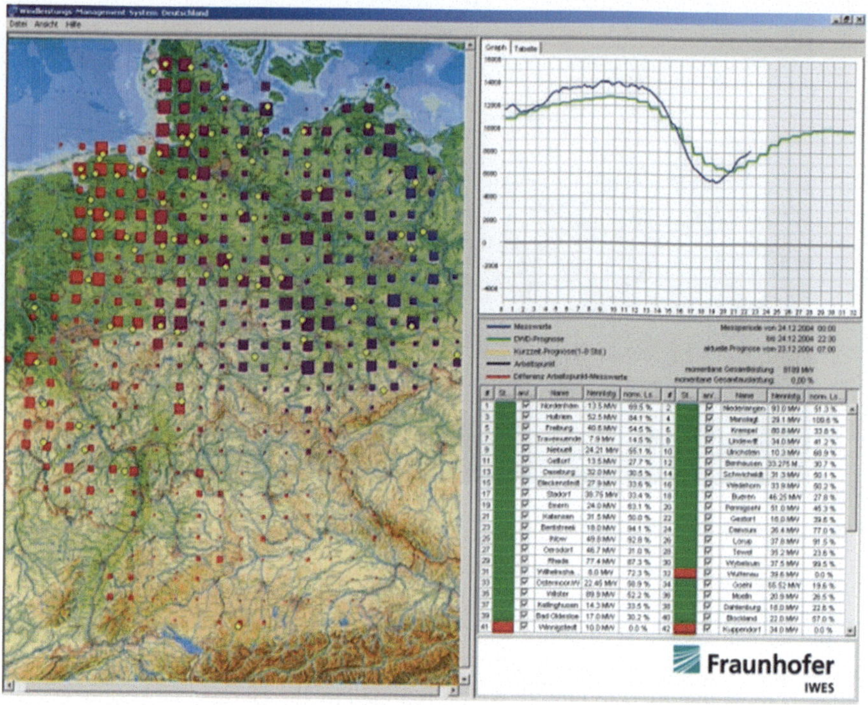

Fig. 3 Screenshot of WPMS. *Source* IWES

The statistical approach aims to describe the connection between the predicted wind and the wind power directly through the statistical analysis of time series. Statistical approaches analyze the relationship between weather prediction and wind energy via time series from the past and describe this relationship in a way that enables to use the assumptions for the future [6, 7].

The artificial intelligence methods are the other widely used approaches to define the relation between NWP and power generation of the wind farms. In the studies, these methods are considered as a part of statistical approach but these methods use, instead of a detailed statistical analysis, the algorithms that are able to describe non-linear and highly complex relationships between these data implicitly [3]. Different methods are used for this approach: ANN, ANFIS, ME, SVM, and NNS. Wind power management system (WPMS) see Fig. 3 is a prominent example of this approach [8].

2.1 *Wind Power Management System*

The WPMS has been developed by Fraunhofer IWES and it is an approved and flexible tool to forecast power feed-in by wind energy into electrical supply systems. The WPMS is used by four German transmission system operators (TSOs) for several years, as well as by the TSOs of Austria, GB, Romania, Egypt, China, and Italy

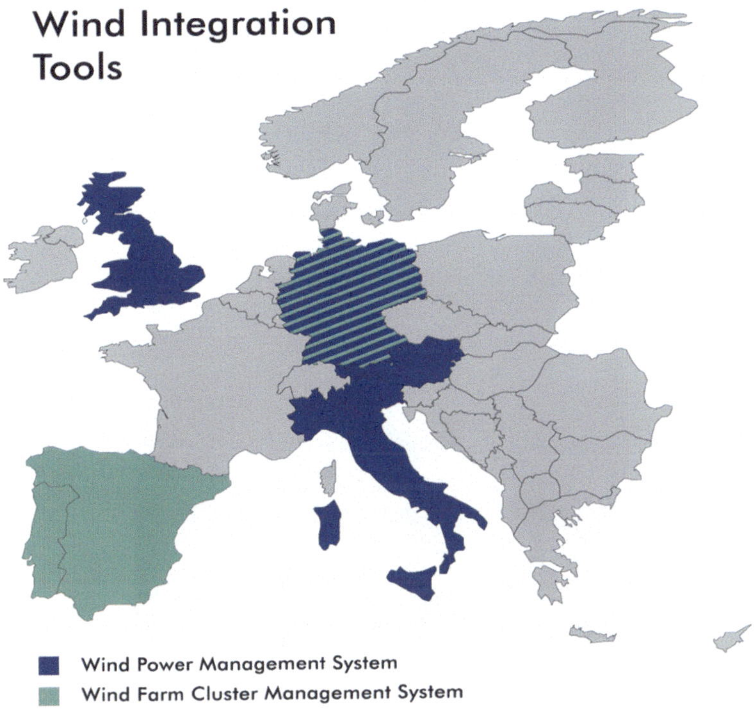

Fig. 4 Usage of WPMS in Europe. *Source* IWES

(see Fig. 4 for Europe). In order to forecast the wind power to be expected from representative wind farms on the basis of the predicted meteorological data and measurements, the WPMS uses ANN. This software can forecast wind power prediction of wind farms in time range from 1 to 96 h [9].

3 Data

This work is carried out with measured meteorological data from measurement mast Fino1, measured power from offshore wind farm "alpha-ventus," and NWP from DWD.

The research platform Fino1 is located in North Sea approximately 45 km off the island of Borkum and it is very close to "alpha-ventus" wind farm (see Fig. 5). Wind speed, wind direction, and other meteorological parameters such as temperature and pressure, at different heights, were available from the Fino1. After excluding implausible data, the wind speed data has been corrected based on atmospheric stability in boundary layer. The bulk Richardson number method has been used for the determining of atmospheric stability. The NWP data has been delivered by DWD twice in a day with a forecast schedule of 76 h in an hourly resolution. The measured power data was provided by TenneT TSO GmbH, a TSO in Germany. The

Fig. 5 Offshore research platform Fino1

main problem on power data was the information lack about states of the wind turbines, namely the count of turbines in operation was unknown. This information deficit restricts the determination of the wind farm power capacity at a selected time point and therefore forecasting of the wind power.

4 Methodology and Results

In this work, wind power forecast models for different phases of a new offshore wind farm for day-ahead wind power forecast were developed.

4.1 Planning Phase: Development of a Physical Model

In this phase a physical model has been developed, which allows modeling a wind farm without any historical data. The physical model is built in three levels. The first level is the data acquisition, pre-processing, and data conversion. In this first step after validating and correcting wind speed data, the wind speed data will be calculated on hub height via different methods, such as interpolation-extrapolation and

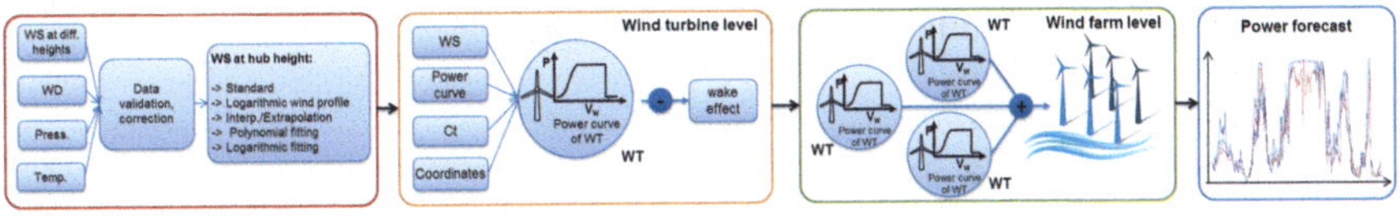

Fig. 6 Physical model based on power curve of wind turbines

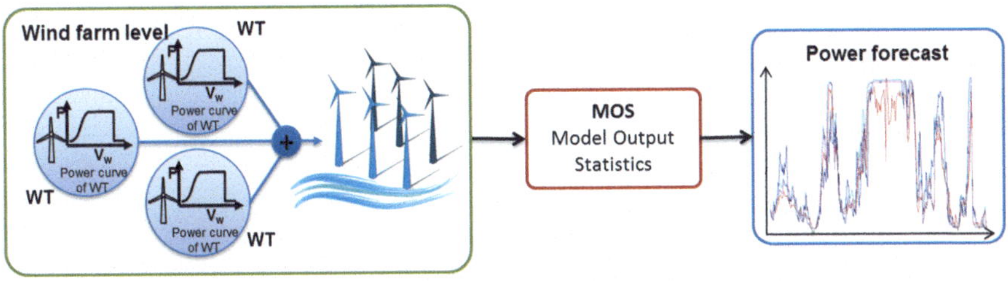

Fig. 7 Physical model extended with MOS

logarithmic fitting. In the second level, wind turbine level, the power production of each wind tribune will be calculated. This step considers wake effects via using N.O. Jensen-Park model approach. Inputs of this level are wind speed at hub height, power curve of each turbine, c_t curves, and position of the turbines. The last level, wind farm level, is calculation of production from wind farm. In this step, production of each wind turbine from second level will be aggregated to calculate power production of whole wind farm via considering wake effects. The output of this model is a lookup table of wind power generation dependent on wind speed and wind direction. Figure 6 illustrates the workflow of the physical model, including defined three levels.

At the planning phase of wind farm technical data, positions and power curves of wind turbines and NWP were available. This step uses the corrected wind speed data from Fino1 to extend and to improve the physical model.

4.2 *First Months After Installation: Optimization of the Physical Model*

The physical model is optimized with an MOS, based on historical measured power; see Fig. 7. The target of this approach is reducing systematical forecast error via using measured wind power. MOS correction can be applied via different methods such as linear regression and double-bias correction. For this work the linear regression method was selected, which is shown in Eq. (1):

$$Y = \beta_0 + \beta_1 \cdot X_1 + \beta_2 \cdot X_2 + \cdots + \beta_n \cdot X_n \tag{1}$$

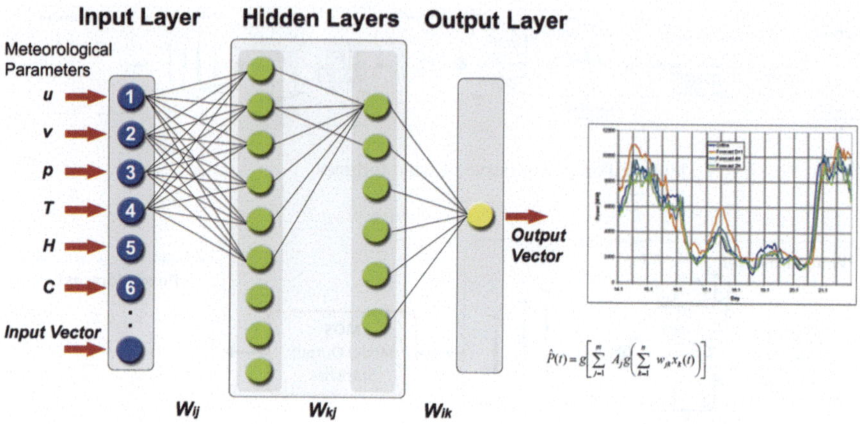

Fig. 8 ANN used to predict wind power

Table 1 Results of day-ahead wind power prediction

Method	Correlation (%)	RMSE	nRMSE (%)
Physical model	86	11.69	19.45
Physical model + MOS	86	11.06	18.18

where Y is predicted (dependent) variable; $X_1, X_2, \ldots, X_n$ are predictor (independent) variables; β_0 is regression constant; and $\beta_1, \beta_2, \ldots, \beta_n$ are regression coefficients (Fig. 8).

Table 1 shows the results of the physical model and the physical model extended with MOS. The statistical correction could improve the prediction quality approximately 7 %. The results are explained in conclusion section.

4.3 Half a Year After Installation: Development of an ANN Model

One of the main advantages of ANN compared to other prediction methods is that they learn from historical data and interpolate results, even when their inputs are contradictory or incomplete. ANN has a great performance if there is enough data available [10]. For the purpose of learning the relationship between meteorological data and wind farm power output, the artificial ANN needs to be trained with numerical weather forecasts and measured power values from the past.

The first 3-month data from 2011 has been used for the training of ANN and the next 3-month data has been used for testing. Cross-validation method has been used in order to avoid overfitting.

The wind power was forecasted with using same dataset via three different methods: the ANN model, the physical model, and the physical model extended with MOS. This phase can be seen as validation of physical model. The results of the models are listed in Table 2. The results are explained in conclusion section.

Table 2 Results of day-ahead wind power prediction

Method	Correlation (%)	RMSE	nRMSE (%)
Physical model	86	11.69	19.49
Physical model + MOS	86	11.06	18.44
ANN (WS and WD)	86	11.05	18.43

A further experiment has been performed to analyze the effects of other meteorological parameters additional to wind speed and wind direction on wind power forecasting. For this purpose two different ANN have been trained with longer training dataset, one of them is only with wind speed and direction and the other one with all available meteorological parameters. While first model has an nRMSE of 16.18 % the second model has an nRMSE of 14.62 %.

5 Conclusion

For the different phases of new offshore wind farms it is possible to use different approaches to predict wind farm energy production. The physical model is developed based on power curve of wind turbines. This model forecasts the wind power with considering wake effects. This model doesn't need any historical data for the purpose of learning the relationship between meteorological data and wind farm power output. The physical model allows simulating a wind farm and the results show that if there is not enough data available, this model can be used to predict the wind power. The physical model is extended with a statistical correction method MOS that gives out better results than pure physical model. The half-a-year results show that the extended physical model can be used in the first months of offshore wind farms after installation. This model can be improved by including other available parameters to the model.

Based on half-a-year data the extended physical model and ANN give out similar results. It is a known fact that if there is enough data, containing different weather conditions, the ANN will give better results than limited data. The ANN gives better results with usage of other meteorological parameters such as pressure, temperature, addition to wind speed, and wind direction.

References

1. Danish Energy Authority (2005) Offshore wind power, Danish experiences and solutions, Danish energy authority. p 18. http://ec.europa.eu/ourcoast/download.cfm?fileID=984
2. EWEA (2011) The European offshore wind industry—key trends and statistics: 1st half 2011. http://www.ewea.org/fileadmin/ewea_documents/documents/00_POLICY_document/Offshore_Statistics/20112707OffshoreStats.pdf

3. Cali Ü (2010) Grid and market integration of large-scale onshore and offshore wind farms using advanced wind power forecasting techniques: technical and energy economical aspects. PhD thesis, University of Kassel
4. Boyle G (2007) Renewable electricity and the grid. Earthscan, London, pp 95–117
5. Landberg L (1999) Short-term prediction of the power production from wind farms. J Wind Eng Ind Aerodyn 80(1):207–220
6. Monteiro C, Keko H, Bessa R, Miranda V, Botterud A, Wang J, Conzelmann G (2009) A quick guide to wind power forecasting: state-of-the-art 2009. Argonne National Laboratory. p 3. http://ceeesa.es.anl.gov/pubs/65614.pdf
7. Schwartz M, Ela E (2009) Utility sector wind power forecasting: status and measurement needs, National Renewable Energy Laboratory, Golden Colorado. p 2. https://ams.confex.com/ams/pdfpapers/154155.pdf
8. Lange B, Rohrig K, Schlögl F, Cali Ü, Jursa R (2007) Wind power prediction in Germany—recent advances and future challenges, Institut für Solare Energieversorgungstechnik e.V., Kassel, Germany. pp 3–6. https://www.researchgate.net/publication/228893382_Wind_power_prediction_in_Germany-Recent_advances_and_future_challenges
9. Mackensen R, Slaby W, Rohrig K, Gesino A, Sanit-Drenan Y-M (2010) Strategies and tools for the grid integration of renewables. In: Renewable energy forum—sector forum: research meets industry, Hannover Messe 2010, Hannover. pp 8–12. http://publica.fraunhofer.de/eprints/urn_nbn_de_0011-n-1342698.pdf
10. Cali Ue, Lange B, Dobschinski J, Kurt M, Moehrlen C, Ernst B (2008) Artificial neural network based wind power forecasting using a multi-model approach, Kassel, Germany. p 5. https://www.researchgate.net/publication/242601478_Artificial_neural_network_based_wind_power_forecasting_using_a_multi-model_approach

Melih Kurt is working as a research assistant at the R&D Division Energy Economy and Grid Operation at Fraunhofer IWES in Kassel. He is a B.Sc. information scientist from University of Kassel (Germany) and from Trakya University (Turkey). He received an M.Sc. degree in informatics from the University of Kassel in 2009. His main research interests are artificial neural networks, wind power forecasting, and assessment of meteorological data.

Jan Dobschinski is the head of Energy Meteorology and Wind Power Management of the R&D Division Energy Economy and Grid Operation at Fraunhofer IWES in Kassel. He has studied Physics at the University of Göttingen with main focus on space weather and elementary particle physics. Within the scope of his PhD he analyzes the benefit of using a multi-model ensemble weather prediction system to improve deterministic and probabilistic wind power forecasts.

Bernhard Lange is the head of Information and Prediction Systems of the R&D Division Information and Energy Economy at ISET. He is a physicist with M.Sc. from the University of Oldenburg. After graduation, he worked in Denmark with Risø National Laboratory and Wind World A/S. From 1998 to 2002 he prepared his Ph.D. about offshore wind power meteorology at Risø National Laboratory and University of Oldenburg. His main research interests for the last 10 years are wind power meteorology, wind farm modeling, and wind power forecasting.

Technical Efficiency Improvement Scenario Analysis for Conversion Technologies in Turkey

Egemen Sulukan, Mustafa Sağlam, and Tanay Sıdkı Uyar

1 Introduction

Energy is getting used to satisfy every demand of mankind in different sectors. Lately, the main perspective about energy is getting focused on more efficient and cost-effective applications throughout the world. Shifting to more efficient technologies naturally creates additional costs, in general terms. Various energy-saving solutions are applied on all end-user technologies, with the effect of increasing public situational awareness about the environmental issues.

Although it sounds pretty difficult to figure out to both *least-cost* and *environmentally acceptable* solution while the energy system meets all the demands, most of the recent efforts are given in this direction.

2 Model Structure

In this analysis, energy carriers, energy technologies, and demands are determined at the first step, their interactions and positions in MARKAL hierarchy is identified then, and a general network of energy from source to end use which is named "Reference Energy System (RES)" is constructed.

Energy supply commodities, sectors and sectoral demands, conversion and process technologies, and consumption and demand technologies are determined for Turkey, based on energy balance statistics published by the Ministry of Energy and

E. Sulukan (✉)
Department of Naval Sciences, Turkish Naval Academy, Istanbul, Turkey
e-mail: egemensulukan@gmail.com

M. Sağlam • T.S. Uyar
Department of Mechanical Engineering, Marmara University, Istanbul, Turkey
e-mail: fulbars@yahoo.com; tanayuyar@marmara.edu.tr

T.S. Uyar (ed.), *Towards 100% Renewable Energy*, Springer Proceedings in Energy, DOI 10.1007/978-3-319-45659-1_12

Natural Resources. Timeframe of the model is set on the year 2005, projected along the period of 2005–2025. User-defined variables are identified according to MARKAL hierarchy and naming conventions with relevant data by using ANSWER user interface.

This database is called "RES," which includes parameters characterizing each of the technologies and resources used to obtain the energy equilibrium, including fixed and variable costs, technology availability, performance, and pollutant emissions.

After developing alternative scenarios to achieve cost-effective technology selection options and running each alternative against the Base Scenario, model responses and scenario results are analyzed to provide technical recommendations. Finally, an operational "technology-economy-environment" integrated model calculating final energy consumption from primary energy supply is created giving "optimal solutions" including both current energy technologies and future candidates. This study brought out a native energy-economy-environment model by using MARKAL model generator for our energy system.

3 Base Scenario and Conversion Technologies in RES

Base Scenario, namely, the RES, is the generalized representation of Turkish energy system's current situation with all components as the supplies including domestic resources and imports, to be used in the various energy generation and process technologies used in conversion to the final energy carriers as electricity or heat and finally to satisfy all demands of all the requirement and services to the consumers. There have been some studies concerning the fundamentals and about various further Turkish MARKAL model implementations [1–6].

As mentioned also in previously published official reports, the optimistic assumption of migrating to more efficient technologies has been the main motivation of this study.

Conversion technologies are utilized for converting primary energy carriers to final energy carriers as electricity or heat. Parameters of annual availability, bound on capacity, bound on activity, annual delivery cost, emission coefficient per activity, ratio of electricity/heat produced, transmission efficiency, annual fixed and variable O&M cost, total cost of investment in new capacity, fraction of unavailability which is forced outage, units of activity/unit of capacity, lifetime of new capacity, and start year are entered into model to characterize each of conversion technologies.

Conversion technologies are as follows: Asphaltite-Based Power Plant (ASP-E), Biogas-Based Power Plant (BIO-E), Biomass Tree Power Plant (BIO-PLT), Hard Coal-Based Power Plant (COA-E), Nuclear Power Plant (E-21), Wave Farm (E-36), Geothermal Electricity (GEO-E), Geothermal Heating-Heat Exchanger Center (GEO-H), Small-Hydro Power Plant (<=10 MW) (HYD1-E), Big-Hydro Power Plant (>10 MW) (HYD2-E), Industrial Cogeneration-Gas Turbine (IN-COGEN), Lignite-Based Power Plant (LIG-E), Natural Gas-Based Power Plant (NGA-E), Oil

Products-Based Power Plant (OILP-E), Solar Power Plant (SOL-PNT), Solar Photovoltaic (SOL-PV), Waste Incineration Plant (WAS-IP), Wind Decentralized (WIN-DECENT), Wind Onshore (WIN-E), and Wind Offshore (WIN-E2).

Some of these technologies are coupled (electric and district heat), some of them are only generating electricity or heat, and some of them are identified as stand-alone power systems, which are not connected to the national grid (decentralized).

4 Improving Efficiency Scenario (TECH-1)

Technical Efficiency Improvement Scenario is focused to analyze the results of technical efficiency improvement option specifically in conversion technologies. Case of 1 % increase in technical efficiencies of currently active power generation technologies is analyzed. In case of 1 % increase in technical efficiencies, effects of the scenario take place in efficiencies and capital costs of conversion technologies, starting from 2015. Additional capital costs of conversion technologies were estimated as $2/kW for 1 % improvement in efficiencies.

Technical efficiencies of conversion technologies identified in Base Scenario and 1 % increase for power generation technologies are given in Table 1 below. Investment costs of conversion technologies are given for each technology in Table 2 below. These costs are Base Scenario values, and investment costs of conversion technologies for TECH-1 Scenario are also given in the same table, simply calculated with additional $2/kW for each technology.

After giving the main inputs into the model with the fundamental assumptions, the model runs with this TECH1 Scenario against BAU. Basic results given in Table 3 indicate that electricity export levels remained the same as the Base Scenario trends. System imports and exports fossil fuels at the same levels in Base and TECH1 Scenarios.

TECH1 Scenario ceases fossil fuel exportation in 2010–2025 periods. Final energy and use of final energy levels are identical in these two scenarios. Mining activities increased 55 % in TECH1. Total primary energy decreased 6 % in TECH1 Scenario.

Table 1 Technical efficiencies for Base and TECH1 Scenarios

Technology	EFF (base)	EFF + 1 %	Technology	EFF (base)	EFF + 1 %
ASP-E	0.40	0.41	LIG-E	0.45	0.46
BIO-E	0.40	0.41	NGA-E	0.45	0.46
BIO-PLT	0.25	0.26	OILP-E	0.40	0.41
COA-E	0.47	0.48	SOL-PNT	0.30	0.31
E-21	0.35	0.36	SOL-PV	0.15	0.16
GEO-E	0.15	0.16	WAS-IP	0.15	0.16
GEO-H	0.75	0.76	WIN-DECENT	0.40	0.41
HYD1-E	0.65	0.66	WIN-E	0.40	0.41
HYD2-E	0.85	0.86	WIN-E2	0.40	0.41
IN-COGEN	0.85	0.86			

Table 2 Investment costs of conversion technologies for Base Scenario (2000$US M)

	Base Scenario					TECH-1		
Technology	2005	2010	2015	2020	2025	2015	2020	2025
ASP-E			1841	1841	1841	**1843**	**1843**	**1843**
BIO-E		700	700	700	700	**702**	**702**	**702**
COA-E	1000.5	1000.5	1000.5	1000.5	1000.5	**1002.5**	**1002.5**	**1002.5**
E-21				1613.1	1613.1		**1615.1**	**1615.1**
E-36				3790	3790		**3792**	**3792**
GEO-E	831.8	831.8	831.8	831.8	831.8	**833.8**	**833.8**	**833.8**
GEO-H	550	550	550	550	550	**552**	**552**	**552**
HYD1-E	1080	1080	1080	1080	1080	**1082**	**1082**	**1082**
HYD2-E	1090	1090	1090	1090	1090	**1092**	**1092**	**1092**
LIG-E	750	750	750	750	750	**752**	**752**	**752**
NGA-E	950	950	950	950	950	**952**	**952**	**952**
OILP-E	670	670	670	670	670	**672**	**672**	**672**
SOL-PNT		2935.4	2935.4	2935.4	2935.4	**2937.4**	**2937.4**	**2937.4**
SOL-PV	2935.4	2935.4	2935.4	2935.4	2935.4	**2937.4**	**2937.4**	**2937.4**
WAS-IP	1815	1815	1815	1815	1815	**1817**	**1817**	**1817**
WIN-DECENT		930	930	930	930	**932**	**932**	**932**
WIN-E	1000	1000	1000	1000	1000	**1002**	**1002**	**1002**
WIN-E2					1500			**1502**

Table 3 Basic results summary of Base and TECH1 Scenarios (PJ)

Case	Parameter	2005	2010	2015	2020	2025
Base-TECH1	TOT.EXP.ELC	20.3	39.1	39.1	39.1	39.1
TECH1	TOT.EXP.ENC	210	0	0	0	0
Base	TOT.EXP.ENC	210	315	370.5	341.8	405
Base-TECH1	TOT.FINALNRG	3022.5	3611.7	4248.3	5005.1	5885.9
Base-TECH1	TOT.FINALUSE	2928.8	3523.5	4144.6	4883.1	5742.4
TECH1	TOT.IMP.ENC	3020.3	3061.2	3500.6	3952.3	4320.2
Base	TOT.IMP.ENC	3031	3593	4231.6	5076.6	5864.3
TECH1	TOT.MIN	573.5	886	1116.5	1457.7	2068.4
Base	TOT.MIN	573.5	826.1	973	1303.6	1335.6
TECH1	TOT.PRIMNRG	3850.5	4456.1	5218.1	6112.3	7237.9
Base	TOT.PRIMNRG	3861.6	4620.4	5447.6	6800.8	7706.9

Fuel consumption by sectors is given in Fig. 1. Fuel consumption increases annually in commercial (6.28 %), industrial (5.60 %), and residential (5.59 %) sectors in higher levels among others. Average annual increase rate of fuel consumption for all sectors is 5.30 % during the analysis period.

Changes in electricity outputs of conversion technologies are given in Table 4. All conversion technologies give electricity output at the same levels as in Base Scenario. Heat outputs are identical with the Base Scenario results. Technology capacities are not changing due to the efficiency and investment cost changes by this improvement scenario.

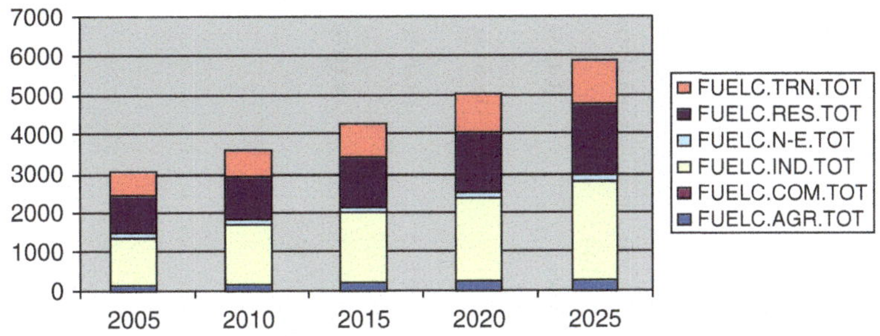

Fig. 1 Fuel consumption by sectors (PJ)

Table 4 Electricity output from conversion technologies (PJ)

Case	Parameter	Technology	2000	2005	2010	2015	2020	2025
Base-TECH1	OUTPUT.ELC	ASP-E	0	0	0	3.18	3.74	4.39
Base-TECH1	OUTPUT.ELC	COA-E	13.74	47.68	115	136	238	189
Base-TECH1	OUTPUT.ELC	E-21	0	0	0	0	45.5	45.5
Base-TECH1	OUTPUT.ELC	GEO-E	0.27	0.34	0.4	0.47	0.48	0.65
Base-TECH1	OUTPUT.ELC	HYD1-E	9.49	12.18	14.33	16.86	17.41	23.32
Base-TECH1	OUTPUT.ELC	HYD2-E	101.67	130.24	153.2	180.2	211.96	249.32
Base-TECH1	OUTPUT.ELC	LIG-E	123.72	107.79	126.79	149.14	154.06	206.35
Base-TECH1	OUTPUT.ELC	NGA-E	166.38	264.4	311	365.82	377.89	506.14
Base-TECH1	OUTPUT.ELC	OILP-E	33.52	19.74	23.22	27.31	28.21	37.79
Base-TECH1	OUTPUT.ELC	SOL-PV	0	0	0	0	0	0
Base-TECH1	OUTPUT.ELC	WAS-IP	0.44	0.44	0.52	0.61	0.63	0.84
Base-TECH1	OUTPUT.ELC	WIN-E	0.12	0.21	0.25	0.29	0.3	0.38

5 Conclusion

Annual total costs of TECH1 decrease 3.77 % in average, while the domestic fuel cost increases 3.43 % in average. Net expenditures on fuel and net import cost decreased with an average of 15.61 % and 18.09 %, undiscounted total system cost decreased 8.10 %, but discounted total system cost has decreased 1.6 % in TECH1 Scenario as can be seen in the cost summary of this scenario, given in Table 5.

Table 5 Summary of costs ($US M)

Case	Parameter	2005	2010	2015	2020	2025	Change
TECH1	U.ANN.ADJ.TOT.COS	78,158.9	93,132.4	109,454.1	118,517.5	152,500	**–3.77 %**
Base	U.ANN.ADJ.TOT.COS	78,200.6	94,526.1	111,463.6	127,235.5	161,144.5	
TECH1	U.DOM.FUEL.COST	2008.5	2408.2	3091.8	3491	4296.1	**3.43 %**
Base	U.DOM.FUEL.COST	2008.5	2425	2861.1	3528.9	3947.6	
TECH1	U.NET.EXP.FUEL	19,099.2	22,682.3	26,349.6	29,699.5	34,917.6	**–15.61 %**
Base	U.NET.EXP.FUEL	19,141	24,076	28,361.2	38,370.9	43,520.3	
TECH1	U.NET.IMP.COST	17,090.7	20,274.1	23,257.8	26,208.5	30,621.5	**–18.09 %**
Base	U.NET.IMP.COST	17,132.5	21,651.1	25,500.1	34,842	39,572.7	
TECH1	U.TOT.COST	29,387.3	43,463.1	47,496.8	39,083.7	97,394.6	**–8.10 %**
Base	U.TOT.COST	29,429	44,856.8	49,505.1	47,802.9	106,046.8	
Change							
TECH1	D.TOT.COST	896,415		**–1.603 %**			
Base	D.TOT.COST	911,015					

Domestic fuel costs increased, because the system prefers more domestic fuel instead of imports. Therefore, import costs also slightly decrease. A tiny improvement gives a chain effect in the whole system and this turns out in a positive way to the country's economy.

References

1. Sulukan E, Saglam M, Uyar TS, Kırlıdoğ M (2010) A preliminary study for post-Kyoto period for Turkey by MARKAL model. In: Proceedings of 5th ege energy symposium and exhibition (IEESE-5), Denizli, 27–30 June 2010
2. Sulukan E, Saglam M, Uyar TS, Kırlıdoğ M (2010) Determining optimal energy strategies for Turkey by Markal model. J Nav Sci Eng 6(1):27–38
3. Sulukan E, Sağlam M, Çelebi UB, Uyar TS (2011) Greenhouse gas emission assessment from maritime transportation for Turkey by MARKAL model. In: INT-NAM 2011, first international symposium on naval architecture and maritime, Istanbul, 24–25 Oct 2011
4. Sulukan E, Saglam M, Uyar TS (2011) Analysis of demand side management option with cogeneration implementations in Turkish energy system by MARKAL model. In: IRENEC 2011, first international 100% renewable energy conference and exhibition, Istanbul, 6–8 Oct 2011
5. Saglam M, Sulukan E, Uyar TS (2011) Analysing cost-effective energy contribution options for Turkey. In: IRENEC 2011, first international 100% renewable energy conference and exhibition, Istanbul, 6–8 Oct 2011
6. Saglam M, Sulukan E, Uyar TS (2011) An alternative carbon dioxide emission estimation for Turkey. In: IRENEC 2011, first international 100% renewable energy conference and exhibition, Istanbul, 6–8 Oct 2011

Balancing of Fluctuating Power to Obtain 100 % Supply with Renewable Energy

Preben Maegaard

1 Introduction

Renewable energies offer the opportunity for a sufficient and secure global energy supply. The amount of solar energy (including biomass, wind, waves) reaching the earth is equal to 12,000 times the present world energy consumption so there are no natural limitations to its use for the benefit of mankind. Due to the limited fossil and nuclear resources, it is obvious that the steady growth of renewable energy utilization will be continued in the future.

More and more countries see the need to use local renewable resources instead of relying on scarce fossil and nuclear resources with their high political and economical risks. Renewable energy in comparison is a clean source of energy with the potential to contribute efficiently to a more ecologically, socially and therefore economically sustainable future for the planet [12].

Especially the successful wind energy development worldwide clearly reflects the political priority that a few leading countries are giving to wind energy by creating favourable legal frameworks. Wind energy use will be one cornerstone of the energy supply of the future. It will be part of a future integrated energy mix including solar energy, hydropower, biomass, geothermal energy, etc [9].

Once the share of wind energy exceeds 20 % or more, however, initiatives have to be taken. With 20 % of wind power at the annual basis, there will be hours and even days when wind power production can fully cover the actual need for power. In regions with a high concentration of wind power, they can deliver most of the base load for power, but also periodically when wind speeds are high, they can even deliver 300 % or more of the actual power needs. The structural and technological challenges involved in a transition to renewable energy are obvious when taking

P. Maegaard (✉)
Nordic Folkecenter for Renewable Energy,
Kammersgaardsvej 16, 7760 Hurup Thy, Denmark
e-mail: pm@folkecenter.dk; Preben@Maegaard.net

T.S. Uyar (ed.), *Towards 100% Renewable Energy*, Springer Proceedings
in Energy, DOI 10.1007/978-3-319-45659-1_13

into account that in the transition away from fossil fuels and atomic energy, by 2016 some countries have plans of 100 % wind and solar in their supply of power.

Especially the emerging countries with their high need for additional energy sources to cover their economic growth like China, India and Brazil are now more and more focussing on the renewable energies [14]. Some of these countries also develop their own renewable energy industrial sectors that will create thousands of new jobs and expand the supply of renewable energy equipment. This has also demonstrated the need for appropriate forms of international transfer of technology.

2 Future Energy Strategies

Conversion to 100 % renewable energy requires the mobilization of all forms of renewable energy resources and both large and small installations. The technological building bricks for the transition to a renewable energy future already exist in the form of wind turbines, solar energy and many types and sizes of local biomass and biogas cogeneration plants [6].

The global attachment to the fossil fuel-based energy system has limited the development of combining the fluctuating solar and wind energies into coherent, autonomous systems. One consequence of this is that renewable energies when generated in excess remain unutilized or even wasted. Wind turbines in regions with high shares of wind energy may periodically be shut down when they produce too much power. Similarly, when the energy-efficient combined heat and power (CHP) production coincides with excess wind energy, an excess power capacity may occur. These problems will become increasingly frequent as more wind turbines feed power into the grid and more CHP systems are utilized. Electric boilers and heat pumps have proved to be a low-cost solution to capture excess energy, by using excess wind power in fuel-efficient combined heat and power systems [16].

Besides an electricity grid, the future energy structure with extensive pipe networks for district heating and cooling will have ancillary functions. Due to their low efficiencies of 45 % or less, conventional condensation power stations will not have an important role to play.

With wind and solar as the primary sources of energy, biomass will be for backup. Because biomass functions as an ideal seasonal storage solution, and due to its limited availability, it is necessary that it be reserved for combustion in combined heat and power stations with efficiencies of 90 % or more. Their primary function is for balancing by upregulation when solar and wind energy cannot cover the demand loads [15].

On the other side, biomass should not be combusted when wind and solar is sufficient. Environmentally and economically, the conversion of excess wind power for the local district heating supply and in their hot water reservoirs will have the per kWh value of the substituted combustible fuel. Thus, it becomes an optimal solution instead of exporting the excess power to neighbouring countries, sometimes at low or even negative spot market prices.

It is often claimed that with the expansion of wind energy, periods of a surplus of wind energy will become more and more frequent and contribute to instability in the power system. By combining and integrating CHPs with electric boilers, heat pumps and hot water storage in combination with fluctuating energy forms of wind and solar, it will be possible to have much higher shares of solar and wind in the energy structure as a whole, without causing instability in the power system. With further integration of the mobility sector using cars with batteries, additional possibilities of handling fluctuating power will become available.

Export of power to neighbouring countries involves heavy investments in long-distance transmission lines and is not a long-term realistic solution; periods with high wind speeds are a transborder phenomenon, and neighbouring countries are often expanding their wind power capacity as well. Power balancing by the use of interregional compensation with strong connections to neighbouring countries still plays an important role for upregulation and downregulation, however, as the present importers of excess power most likely in the future will be less interested in buying power as the deployment of fluctuating forms of renewable energy will only increase in neighbouring countries as well. The reality is that outlets for periodical overcapacities will be required locally using integrated systems that see power supply, mobility, heating and cooling as a whole together with existing possibilities of demand-side management.

3 Wind and Solar Energy for Base Load Supply

They are, however, fluctuating which causes a need for adaptation and backup from other supply solutions or storage. Flexibility and response time are important requirements to match satisfactorily with the integrated supply of power, heat and cooling.

Some regions and even countries already have relatively high shares of fluctuating power supply. In 2015, Denmark saw 42 % of its demand for electricity from wind turbines, which by 2020 will grow to approximately 50 %. During periods of low peak power demand and high wind speeds, wind power can currently fully cover the consumption of electricity; at the local level, the share of wind power may even be 400 % of actual consumption.

With 20 % of wind power at the annual basis, there will be hours and even days when wind power production can fully cover the actual need for power and initiatives have to be taken. The structural and technological challenges involved in a transition to renewable energy are obvious when taking into account that in the transition away from fossil fuels and atomic energy, some countries have plans of 50 % wind and solar in their supply of power.

Power peaks coming from wind and solar can be balanced in the local CHP stations with gas engines that can respond within seconds or minutes to the variations in the productions of wind and power. In the long term, solar and wind can often cover power base load needs and at periods even cover both the need for electricity and the need for heating. This saves natural gas or biomass that will be available as backup fuel when solar and wind are not sufficient.

In the case of Denmark, with its many hundreds of CHP stations, it is possible to stop gas engines at periods when the wind and solar power used in electric boilers and heat pumps is sufficient for both the actual supply of power, heat and hot water-based cooling. As the CHP stations already have large hot water storage tanks, additional excess power can be stored for some days of hot water supply with no electricity to water conversion loss.

Building of new district heating structures will at the same time create the needed flexibility for the management of a 100 % renewable energy supply and is an affordable, highly efficient and well-developed solution with a mature technology [10].

Initial investment in the electric boilers for downregulation combined with decentralized natural gas CHP for upregulation (900 per kW) is less than investment costs of conventional fossil fuel power stations and significantly lower than atomic power where costs of long-term storage of radioactive waste have to be added. Based on experiences of numerous installations in Denmark, the investment in a new district heating network is often similar to the costs of the CHP station that it is connected to.

4 CHP as the Key to Efficiency

District heating and cooling, CHP and fluctuating energy forms from solar and wind energy can be combined and integrated to create truly autonomous systems. Pioneering projects demonstrate that a decentralized heat and power system using biomass and wind energy can be cost-effective and pave the way for a sustainable energy supply. It has been demonstrated as well that conventional fossil fuel-based power production from big centralized power plants can be phased out with improved safety of supply and prevention of climate changes.

Because it is impractical to transport heat over long distances, cogeneration equipment must be located physically close to its heat application. A number of environmental benefits flow from this fact:

- *Power is generated close to its consumer.*
- *Significantly reducing transmission losses.*
- *The need for distribution equipment.*

Cogeneration plants are usually smaller and owned and operated by smaller and more localized consumer-owned companies than conventional power plants. Cogeneration is at the heart of district heating and cooling systems.

Operators that construct new district heating schemes often consider application of CHP a rational and viable solution, or CHP may be a requirement by the local or national energy authorities as it improves energy efficiency significantly and makes it easier to meet CO^2 emission reductions [2]. The same policies that have created a high penetration of district heating in Scandinavia have also created a high share of

CHP with hot water storage connected to the district heating systems. This as well paves the way for increased use of the fluctuating energy from solar and wind leading to 100 % renewable energy.

Since the early 1980s in Denmark, no new power plants have got permission for commissioning unless provided with the ability to use CHP and to supply heat to the district heating networks. This was motivated by security of supply and environmental concerns encouraging energy efficiency [13]. Construction of new electricity generating capacity must be justified by the need for new heat-production capacity.

Advantages of community-based CHP units are significant, the main benefits being:

Reliability: Gas engines can be used where reliability is of the utmost importance. Typically, these gas engines are installed in transcontinental gas compressor stations, drilling rigs, offshore oil platforms and villages not served by the national power grid.

Community autonomy: Having local CHP provides the local community with autonomy giving the "power to the people". This enables the community to ensure that the power is developed in an appropriate manner.

The ability to incorporate renewable energy in the future: Having CHP with district heating opens opportunities to incorporate large shares of renewable energy in the form of biogas, solar thermal heating, wind for heat, biomass gasification, plant oil-based fuels and combustion of locally grown biomass.

Scalability and flexibility: Local CHP is scalable and flexile to operate. This makes it easy to increase capacity in the future and matches well with the incorporation of wind and solar power in the supply system.

High efficiency: Medium-sized stationary gas CHP units boast an electrical efficiency of around 42 %, and with heat recovery of the jacket water, exhaust, lube oil and turbo charger, they can achieve an overall efficiency of over 95 % (with the condensation of exhaust gas even over 100 % efficiency).

Cost-effective heat and power: With high total efficiencies and two energy products from the same fuel source, cost of power and heat can be reduced. As an example, Denmark according to Eurostat periodically had the fifth lowest power prices (without taxes) for GWh size consumers in Europe with Sweden, Norway, France and Finland being lower [3].

The challenge for the system operator in some countries is already and will in the future be even more to optimize the available resources. While Europe as a whole has 11 % of the efficient combined heat and power (CHP) in the energy system, the share in Denmark is 60 %, the world's highest, which at the same time represents not only the problem mentioned but also its solution [4].

Compared to other uses and storage of excess power, electric boilers and heat pumps cause no conversion loss. Heat pumps even boost the excess production of power by a factor of 2–6 depending on the energy source available. As they require substantial investments, the operation of heat pumps using excess wind and solar power should result in yearly operation of 3,000 h or more. This amount of excess power will in a more distant future be available at medium peak periods.

As fluctuating solar and wind will always cause short-term high peaks, electric boilers will, due to their low initial investment, be the preferred solution to utilize the high peaks of solar and wind energy. Therefore, it is not a relevant discussion whether electric boilers or heat pumps should be preferred. In the early stage of using fluctuating solar and wind in the heating system, electric boilers will be optimal, while increased amounts of fluctuating power will pave the way for heat pumps to convert the medium peak electricity into district heating and cooling, while the electric boilers will take the high peaks.

5 The 100 % Renewable Energy in Thisted Municipality

In the Thy peninsular with its 46,000 inhabitants, the 225 windmills and other renewables since the 1990s cover 100 % of the annual need for electricity. The local energy production has become an important local source of income. On days with strong wind, the wind turbines may even produce four times more than the actual consumption, and the power quality still lives up to the highest standards [11].

The local utility, Thy-Mors Energi, has demonstrated real-time management of such big quantities of wind energy to visitors from all parts of the world. In the towns and villages in Thy, people get their space heating from hot water pipelines in the streets. It is environmentally and economically the best solution to use the excess wind power for the actual supply and storage in the big hot water reservoirs of the local district heating suppliers instead of exporting the surplus power to neighbouring countries sometimes at low-spot market prices.

(Data Thisted Municipality (2008)):

- 225 windmills
- 124,600 kW installed wind power capacity
- 35,800 kWel installed CHP capacity
- Power from wind energy 265 GWh
- Power consumption of 340 GWh
- 80 % from wind
- 20 % from biogas and CHP waste
- A small amount of PV [8]

The further development of wind power will make it the primary source of electricity and heating, WHP. With an estimated 80 MW, new additional wind power, already at 5–6 m/s wind power, will cover the demand for electricity, while peaks will be more frequent and available for other energy purposes with heating as the major outlet [7].

In Thisted all the CHP plants and the district heating are not-for-profit consumer owned. Local surplus of wind electricity will in the future be used in combined heating and power plants and periodically replace natural gas and biomass. Biomass will function as backup storage when wind and solar energy is not sufficient to cover the need for electricity and heating. Solid and liquid biomass is easy and cheap to store.

The local community owns, instals and operates the green energy infrastructure; therefore, the benefits from the infrastructure are reaped by the community. The strong local involvement is crucial and is obtained by local ownership. Local CO2 reduction, job creation, business development, economic diversification and skill building are of general interest for the community. Once a community has experience with community power, that skill can be transferred to other communities [1].

6 Folkecenter Autonomous Energy System

At the Nordic Folkecenter for Renewable Energy located in the Thisted municipality, a prototype autonomous renewable energy system is installed. The energy system supplies heat and electricity to 2,000 m^2 of offices, meeting rooms, laboratories, workshops and residential facilities. Sources of energy supply are wind turbines of 75 kWel and 6 kWel, 42 kWth electric boiler, 40 kWth wood pellet stoker with automatic start-up and stop, 8 kWel/20 kWth plant oil CHP unit with automatic start-up and stop, 35 kWel PV and 30 m^2 solar thermal panels. Wind and solar energy are the primary sources for heat and electricity. Biomass (wood pellets and plant oil, PPO) is used for backup.

The overall operational strategy is:

1. *The power flow control (PFC) directs excess wind electricity through the thyristors to the electric boiler.*
2. *When the electric boiler does not supply sufficient wind-generated heat, the wood pellet stoker is automatically activated.*
3. *In case of no power from solar and wind, the CHP unit using plant oil is activated.*
4. *Excess production of solar and wind-generated heat is stored in a 10 m^3 hot water storage. The system is connected to the public grid. The overall principle is not-to-sell-not-to-purchase from the grid. The CHP unit, however, can operate in island mode in case of a power blackout.*

In practice, excess wind power covers up to 60 % of the annual demand for heat, with the balance coming from solar and biomass. The technology and strategy of the autonomous system was pioneered, developed and implemented by the Folkecenter in 2007.

In 2013 internet activation of excess power was implemented for the supply of individual consumers as well as larger groups of population depending on the amount of excess power to be deployed. For a 100 % supply of power and heat/cooling from renewable energies, the same principles and strategy can be applied at the regional and national level as well. The system delivers a realistic solution to questions often made about alternative energy sources.

7 The Need for New Political Priorities

Renewable energy resources exist everywhere. Using renewable energy will create a more balanced relationship with nature. A new culture of energy efficiency can lead to concerned, socially responsible use of all natural resources. The use of renewable energy in a local resource can contribute to the preservation of local cultures and also promote new lifestyles, new prosperity and security of supply [5].

Energy savings and energy efficiency are equally important in all sectors of the society that will lead to recycling of scarce resources, ambitious building codes in favour of passive house principles, urban planning that reduces the need for mobility, smart energy management and district heating pipe network for the distribution of low-grade energy in the form of hot water for space heating and cooling. In combination with the most successful incentives such as feed-in tariffs that are used in more than 50 countries, mankind will be able to meet the challenges of the twenty-first century.

Policies with appeal and encouragement of the population will have to replace unpopular and unjust quota systems, green certificates and carbon credits. As they wrongly have been seen as a sacrifice and burden for the national economies, there is a fundamental lack of rational of the policies applied when for a decade or more governments during annual COP negotiations have competed to avoid the needed transition to the renewable energy forms.

References

1. Abramsky K (2010) Sparking a worldwide energy revolution, social struggles in the transition to a post-petrol world. AK Press, Oakland
2. Bassam NE, Maegaard P (2004) Integrated renewable energy for rural communities, planning guidelines, technologies and applications. Elsevier, The Netherlands
3. Pieper C, Rubel H (2010) Electricity storage. In: Making large-scale adoption of wind and solar energies a reality. Boston Consulting Group. https://www.bcg.com/documents/file41973.pdf
4. Danish Energy Agency (2010) Heat supply: goals and means over the years, http://dbdh.dk/
5. Etcheverry J (2008) Challenges and opportunities for implementing sustainable energy strategies in coastal communities of Baja California Sur. University of Toronto, Mexico
6. Lund H (2010) Renewable energy systems: the choice and modeling of 100% renewable solutions. Academic, Burlington
7. Maegaard P (2008) Denmark: wind leader in standby. Int Sustainable Energy Rev
8. Maegaard P (2009/2010) Thisted, 100% renewable energy municipality. PPT, presented 2009/2010 at conferences/events in 34 cities in 18 countries
9. Maegaard P (2009) Danish renewable energy policy. http://WCRE.org homepage
10. Maegaard P (2010) Wind energy development and application prospects of non-grid-connected wind power. In: Proceedings of 2009 world non-grid-connected wind power and energy conference, IEEE Press
11. Ruby JU (2006) Continuous headwind pioneering the transition from fossil fuels and atomic energy to the renewable energies. Hovedland (in Danish language)
12. Scheer H (2006) Energy autonomy, the economic, social and technological case for renewable energy. Earthscan, London

13. The Poul la Cour Foundation (2009) Wind power, the Danish way, from Poul la Cour to modern wind turbines. Self-Published, Denmark
14. Maegaard P, Palz W, Krenz A (2013) Power for the world. The emergence of wind energy, vol I–II. Pan Stanford, Singapore
15. Maegaard P (2012) Integrated systems to reduce global warming. In: Handbook of climate mitigation. Springer Science, New York
16. El Bassam N, Maegaard P, Schlichting ML (2012) Distributed renewable energies for off-grid communities. Elsevier Science, New York

Analysis of Demand-Side Management Option with Cogeneration Implementations in Turkish Energy System by MARKAL Model

Egemen Sulukan, Mustafa Sağlam, and Tanay Sıdkı Uyar

1 Introduction

Base Scenario, namely, the Reference Energy System (RES), is the generalized representation of Turkish Energy System's current situation with all components as the supplies including domestic resources, imports, to be used in various energy generation and process technologies used in conversion to the final energy carriers as electricity or heat, finally to satisfy all demands of all the requirement and services of the consumers. There have been some studies concerning the fundamentals and about various further Turkish MARKAL Model implementations [1–10].

As also mentioned in previously published official reports, the optimistic assumption of managing the demand side to a more efficient level by activating the potential of cogeneration has been the main motivation of this study.

In order to indicate the possible effects of promoting the expanded use of cogeneration in the industrial sector, it is assumed that demand levels will drop by 15 % in industry and 10 % in residential sector. Besides the demand reduction measures in mentioned sectors, cogeneration type plants are supposed to be used to increase the cogeneration share in the total electricity production, which is 4 % by end of the year 2005 and projected to reach 16 % in 2025. Therefore, 16 % of total electricity generation would come from cogeneration in 2025.

E. Sulukan (✉)
Department of Naval Sciences, Turkish Naval Academy, İstanbul, Turkey
e-mail: egemensulukan@gmail.com

M. Sağlam • T.S. Uyar
Department of Mechanical Engineering, Marmara University, Istanbul, Turkey
e-mail: fulbars@yahoo.com; tanayuyar@marmara.edu.tr

T.S. Uyar (ed.), *Towards 100% Renewable Energy*, Springer Proceedings in Energy, DOI 10.1007/978-3-319-45659-1_14

2 Demand-Side Management in Power and Industry

Emission reductions and cost efficiencies of alternative scenario were assessed while comparing the main indicators are estimated of alternative DSM Scenario versus Base Scenario, i.e., total economic cost, cost increase, net energy import cost, net energy import increase, total CO_2 emission, emission increase, and cost reductions (%).

For each technology used in power generation, industry and residential sectors, bounds on activity and capacity have been fixed for 2000 and 2005, set free for other periods. Annual new investment in all these technologies have been identified as 100 Million Turkish Liras (66.2 M USD) starting from the year 2010 by using energy efficient technologies [11, 12].

3 Demand-Side Management Scenario

In order to indicate the possible effects of promoting the expanded use of cogeneration in the industrial sector, it is assumed that demand levels will drop by 15 % in industry, 10 % in residential sector. Reduced demands of industrial and residential subsectors are given in Table 1. Demand level is reduced by 13.9 % in 2020 and by 25.9 % in 2025 compared with Base Scenario demand levels.

Besides the demand reduction measures in mentioned sectors, cogeneration type plants are supposed to be used to increase the cogeneration share in the total electricity production, which is 4 % by end of the year 2005 and reach to 16 % in 2025. The total efficiency of cogeneration type plants is taken as 85 %, whose 55 % is for heat and 30 % is for electricity generation. It is also assumed that all new cogeneration capacities will be natural gas fired.

We can have a look at the basic results summary of the DSM Scenario in Table 2. Electricity export levels remained same as the Base Scenario trends. System exports no fossil fuels during 2015–2020 periods. Imports of fossil fuels decreased 2141.8 PJ in 2025 from 5864 PJ. Mining activities are also decreased from 1335.6 to 372.7 PJ in 2025, but the total primary energy is reduced to 3763 from 7706.9 PJ. Total final energy and total final use of energy decreased from 5885.9 PJ to 3607.6 PJ and from 5742.4 PJ to 3494 PJ, respectively.

Fuel consumption by sectors is shown in Fig. 1. Fuel consumption annually increases in commercial (6.28 %); transport (4.74 %) and agriculture (4.61 %) sectors are in higher levels among other sectors. Annual average increase rate of fuel consumption for all sectors is 1.70 % during the analysis period. Despite the demand reductions in industrial and residential sectors, industry holds the biggest share among other sectors with 1295.05 PJ in 2025 and average annual increase rate of 1 %. Residential sector consumed fuels of 786.25 PJ with average annual increase rate of 0.09 %.

Table 1 Change in demand levels of DSM scenario (PJ)

DSM	15 % DROP in IND. and 10 % DROP in RES.						
Demand	2005	2010	2015	2020 (Base)	2020 (DSM)	2025 (Base)	2025 (DSM)
ICEM	148.8572	152.7285	156.5998	160.471	136.40035	164.3423	136.40035
ICPCM	32.3242	38.0215	44.723	52.6057	44.714845	61.8777	44.714845
IH	92.665	108.9977	128.2091	150.8067	128.185695	177.3872	128.185695
IISM	86.3057	101.5175	119.4105	140.4573	119.388705	165.2136	119.388705
IMSC	473.0145	556.3858	654.4518	769.8024	654.33204	905.4842	654.33204
INFM	33.7068	39.6478	46.6359	54.8558	46.62743	64.5244	46.62743
ISGR	16.4242	19.3191	22.7241	26.7294	22.71999	31.4406	22.71999
RCD	2.254	2.6513	3.1186	3.6682	3.30138	4.3148	3.30138
RCO	52.944	62.2757	73.2521	86.1632	77.54688	101.3499	77.54688
RCW	7.2652	8.5457	10.052	11.8237	10.64133	13.9077	10.64133
RDW	3.75	4.411	5.1884	6.1029	5.49261	7.1786	5.49261
RLI	5.3671	6.3131	7.4258	8.7346	7.86114	10.2742	7.86114
RME	9.6125	11.3068	13.2996	15.6438	14.07942	18.4011	14.07942
ROE	10.9537	12.8843	15.1553	17.8265	16.04385	20.9685	16.04385
RRF	10.5658	12.4281	14.6186	17.1952	15.47568	20.2259	15.47568
RSC	3.1207	3.6707	4.3177	5.0788	4.57092	5.9739	4.57092
RSH	100.8751	118.6549	139.5684	164.1681	147.75129	193.1036	147.75129
RWH	41	48.2265	56.7266	66.725	60.0525	78.4857	60.0525
Total IND. and RES. demand					1515.186055		1515.186055
Demand reduction					13.9 %		25.9 %

Table 2 Basic results summary of DSM (PJ)

Case	Parameter	2005	2010	2015	2020	2025
BASE-DSM	TOT.EXP.ELC	20.3	39.1	39.1	39.1	39.1
DSM	TOT.EXP.ENC	210	77.5	0	0	0
BASE	TOT.EXP.ENC	210	315	370.5	341.8	405
DSM	TOT.FINALNRG	3022.5	2792.5	3068.4	3380.1	3607.6
BASE	TOT.FINALNRG	3022.5	3611.7	4248.3	5005.1	5885.9
DSM	TOT.FINALUSE	2928.8	2722.7	2986.4	3283.6	3494
BASE	TOT.FINALUSE	2928.8	3523.5	4144.6	4883.1	5742.4
DSM	TOT.IMP.ENC	3031	1701.3	1729.9	1924.7	2141.8
BASE	TOT.IMP.ENC	3031	3593	4231.6	5076.6	5864.3
DSM	TOT.MIN	573.5	306.2	362.6	364.9	372.7
BASE	TOT.MIN	573.5	826.1	973	1303.6	1335.6
DSM	TOT.PRIMNRG	3861.2	2925.4	3212	3536	3763
BASE	TOT.PRIMNRG	3861.6	4620.4	5447.6	6800.8	7706.9

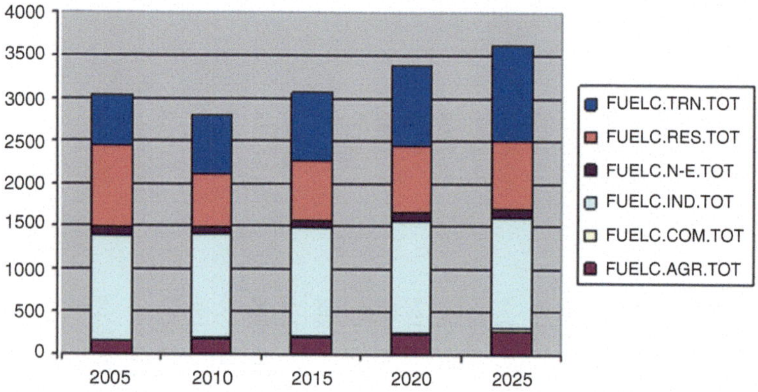

Fig. 1 Fuel consumption by sectors (PJ)

Electricity outputs of conversion technologies are given in Table 3. Biogas-based power plants are activated in 2015, as planned. Hard coal, lignite, natural gas, oil products, and waste incineration-based power plants generate no electricity in 2010–2025 periods. Asphaltite and nuclear power plants operate at the same levels as in Base Scenario. Geothermal power plants operate about 0.27 PJ electricity generation level. Solar power plants generate 13.05 PJ in 2010, 12.4 PJ in 2015, 11.65 PJ in 2020, and 11.31 PJ in 2025. Wind power plants generate 0.12 PJ in 2005 and 0.21 PJ in the 2005–2025 periods.

Small hydroelectric power plants generated 4 PJ in 2010 and 67.82 PJ in 2025; big hydroelectric power plants also generated 303.96 PJ in 2025.

No heat output is taken from coal, lignite, natural gas, and waste-fired power plants from 2010 to 2025. Geothermal heating reduced from 38.77 PJ in 2010 to 6-7 PJ levels. Both changes in electricity and heat outputs, which are directly related with the capacity levels of technologies, are given in Table 4.

Table 3 Output of electricity from conversion technologies (PJ)

Case	Parameter	Region	Technology	2005	2010	2015	2020	2025
DSM	OUTPUT.ELC	REGION1	BIO-PLT	0	0	3	3	3
DSM	OUTPUT.ELC	REGION1	COA-E	47.68	0	0	0	0
DSM	OUTPUT.ELC	REGION1	GEO-E	0.34	0.28	0.27	0.27	0.27
DSM	OUTPUT.ELC	REGION1	HYD1-E	12.18	4	31.45	67.82	67.82
DSM	OUTPUT.ELC	REGION1	HYD2-E	130.24	303.96	303.96	300.27	303.96
DSM	OUTPUT.ELC	REGION1	LIG-E	107.79	0	0	0	0
DSM	OUTPUT.ELC	REGION1	NGA-E	264.4	0	0	0	0
DSM	OUTPUT.ELC	REGION1	OILP-E	19.74	0	0	0	0
DSM	OUTPUT.ELC	REGION1	SOL-PV	0	13.05	12.4	11.65	11.31
DSM	OUTPUT.ELC	REGION1	WAS-IP	0.44	0	0	0	0
DSM	OUTPUT.ELC	REGION1	WIN-E	0.21	0.21	0.21	0.21	0.21

Table 4 Output of heat from conversion technologies (PJ)

Case	Parameter	Region	Technology	2005	2010	2015	2020	2025
DSM	OUTPUT.LTH	REGION1	COA-E	14.45	0	0	0	0
DSM	OUTPUT.LTH	REGION1	GEO-H	38.77	6.65	6.62	7.15	7.15
DSM	OUTPUT.LTH	REGION1	LIG-E	20.34	0	0	0	0
DSM	OUTPUT.LTH	REGION1	NGA-E	58.76	0	0	0	0
DSM	OUTPUT.LTH	REGION1	WAS-IP	0.04	0	0	0	0

Changes in technology capacity levels are given in Table 5 for the analysis period. We can see asphaltite-based power plants are not operated until 2015 period as in Base Scenario. Capacity of briquette factories stayed below under Base Scenario values. Biogas power plants had an installed capacity of 0.17 GW in 2015 and 2020, then increases to 0.2 GW in 2025 although no capacity was assumed in Base Scenario. Hard coal-fired power plants are fixed at 2.03 GW between 2005 and 2025 periods.

Nuclear power plants and wave energy converters had the same installed capacities as in Base Scenario. Wave energy power plants will have installed capacities of 0.01 GW in 2020 and 0.04 GW in 2025. Geothermal power plant stay tuned constantly in 0.02 GW until 2025. Geothermal heating capacities remained constant level of 1.52 GW from 2005 to 2025.

Hydroelectric power plants of two types, HYD1-E and HYD2-E as identified into model, were 52 % levels compared with of Base Scenario capacities.

Lignite-based power plants remained at 7.13 GW capacity levels, natural gas-fired power plants capacity decreased to 6.75 GW, and oil product-based power plants to 0.97 GW, as great residual capacities in the system.

Waste incinerator capacities stayed constant at 0.04 GW level during the entire period. Refineries' capacity is narrowed 40 % in 2025. Decentralized wind power plant capacity levels went to 0.1 GW in 2025; onshore wind capacities shifted to 0.02 GW in 2025; but no offshore capacity came out throughout the entire period.

Table 5 Changes in technology capacity levels

Case	Units	Technology	2005	2010	2015	2020	2025
BASE	GW	ASP-E	0	0	0.13	0.15	0.18
DSM	PJ/a	BF	3.64	3.64	3.64	3.64	1.99
BASE	PJ/a	BF	3.64	4.28	5.03	5.92	6.96
DSM	GW	BIO-PLT	0	0	0.17	0.17	0.2
BASE	GW	BIO-PLT	0	0	0	0	0
DSM	GW	COA-E	2.03	2.03	2.03	2.03	2.03
BASE	GW	COA-E	2.03	5	6	10.5	11
BASE	GW	E-21	0	0	0	4.5	4.5
BASE	GW	E-36	0	0	0	0.01	0.04
BASE-DSM	GW	GEO-E	0.02	0.02	0.02	0.02	0.05
DSM	GW	GEO-H	1.52	1.52	1.52	1.52	1.52
BASE	GW	GEO-H	1.52	1.66	1.95	2.2	2.45
DSM	GW	HYD1-E	2.36	2.36	2.36	2.36	2.36
BASE	GW	HYD1-E	2.36	2.77	3.26	3.83	4.51
DSM	GW	HYD2-E	10.55	10.55	10.55	10.55	10.55
BASE	GW	HYD2-E	10.55	12.41	14.6	17.18	20.2
DSM	GW	LIG-E	7.13	7.13	7.13	7.13	7.13
BASE	GW	LIG-E	7.13	8.39	9.87	11.61	13.65
DSM	GW	NGA-E	13.79	13.79	13.79	13.79	6.75
BASE	GW	NGA-E	13.79	16.22	19.08	22.44	26.4
DSM	GW	OILP-E	2.96	2.96	2.96	2.96	0.97
BASE	GW	OILP-E	2.96	3.48	4.1	4.82	5.67
DSM	kt/a	PRAF	1296.94	1297	1368.77	1508.07	1508.07
BASE	kt/a	PRAF	1296.94	1526	1794.42	1853.64	2482.72
BASE	GW	SOL-PNT	0	1.18	1.38	1.63	1.91
DSM	GW	SOL-PV	2	2	2	2	2
BASE	GW	SOL-PV	2	2.35	2.77	3.25	3.83
DSM	GW	WAS-CL	0	0	0	0	0
BASE	GW	WAS-CL	0	0.01	0.01	0.01	0.01
DSM	GW	WAS-IP	0.04	0.04	0.04	0.04	0.04
BASE	GW	WAS-IP	0.04	0.04	0.05	0.06	0.07
BASE	GW	WIN-DECENT	0	0.02	0.07	0.09	0.1
DSM	GW	WIN-E	0.02	0.02	0.02	0.02	0.02
BASE	GW	WIN-E	0.02	0.1	0.41	1.75	2
BASE	GW	WIN-E2	0	0	0	0	0

General cost summary of the DSM Scenario is given in Table 6. Annual investment costs decrease 35.4 %, domestic fuel cost is decreased 62.9 %, and system gives 85 % excess of the investments compared with Base Scenario. Net expenditures on fuel and net import cost decreased with an average of 73.1 % and 74.4 %, undiscounted total system cost also decreased 89.3 %. Discounted total system cost has

Table 6 Summary of costs (2000$US M) and total emission (kt)

Case	Parameter	2005	2010	2015	2020	2025	Average
DSM	U.ANNADJTOTCOS	78,200.60	71,556.50	75,936.20	83,768.70	8,864,320	
BASE	U.ANNADJTOTCOS	78,200.60	94,526.10	111,463.60	127,235.50	161,144.50	
DIFFERENCE		0.0 %	−32.1 %	−46.8 %	−51.9 %	−81.8 %	**−35.4 %**
DSM	U.DOMFUELCOST	2008.50	1317.70	1734.80	1738.60	1749.20	
BASE	U.DOMFUELCOST	2008.50	2425.00	2861.10	3528.90	3947.60	
DIFFERENCE		0.0 %	−84.0 %	−64.9 %	−103.0 %	−125.7 %	**−62.9 %**
DSM	U.INV.TOT	3657.60	7556.00	5695.80	11,035.50	96,105.50	
BASE	U.INV.TOT	3657.60	20,047.60	22,395.60	9816.50	141,558.40	
DIFFERENCE		0.0 %	−165.3 %	−293.2 %	11.0 %	−47.3 %	**−82.5 %**
DSM	U.NETEXPFUEL	19,141.00	13,639.80	14,847.20	16,561.10	18,178.80	
BASE	U.NETEXPFUEL	19,141.00	24,076.00	28,361.20	38,370.90	43,520.30	
DIFFERENCE		0.0 %	−76.5 %	−91.0 %	−131.7 %	−139.4 %	**−73.1 %**
DSM	U.NETIMPCOST	17,132.50	12,322.10	13,112.40	14,322.50	16,429.60	
BASE	U.NETEMPCOST	17,132.50	21,651.10	25,500.10	34,842.00	39,572.70	
DIFFERENCE		0.0 %	−75.7 %	−94.5 %	−135.1 %	−140.9 %	**−74.4 %**
DSM	U.TOTCOST	29,429.00	14,173.10	20,188.10	24,639.50	58,922.50	
BASE	U.TOTCOST	29,429.00	44,856.80	49,505.10	47,802.90	106,046.80	
DIFFERENCE		0.0 %	−216.5 %	−145.2 %	−94.0 %	−80.0 %	**−89.3 %**
DSM	**D.TOTCOST**						
BASE	**D.TOTCOST**						
DIFFERENCE		**−15.0 %**					
DSM	**EMISSION.TOT**	5,630,289.00					
BASE	**EMISSION.TOT**	9,882,005.00					
DIFFERENCE		**−43.0 %**					

Table 7 Comparison of base and alternative scenarios

Scenarios	Total economic cost (2000 M USD)	Cost increase (2000 M USD)	Net energy import cost (2000 M USD)	Net energy import increase (2000 M USD)	Total CO_2 emissions (kt)	Emission increase (kt)	% CO_2 reduction
BASE	911,015	0	149,149.4	0	9,882,005	0	0
DSM	774,103	−136,912	84,270.1	−64,879.3	5,630,289	−1,468,229	−43 %

decreased 15 % and total CO_2 emissions decreased 43 %, respectively, throughout analysis period. All costs are reduced in a serious manner, but this can only be possible with a narrowing economy profile, which contradicts with the under an assumed strategic long-term development goals of the country.

4 Conclusion

In comparing the scenarios, emission reductions and cost efficiencies of alternative scenarios were assessed. Total economic cost, cost increase, net energy import cost, net energy import increase, total CO_2 emission, emission increase, and cost reduction (%) were calculated by taking the values of alternative scenarios and the Base Scenario.

While obtained minus (−) values in results given in Table 7 suggest reduction compared to the reference scenario, plus (+) values point at an increase. As the results are given in the table, the DSM scenario is evaluated as a cost-effective and climate friendly implementation during our modeling studies. This scenario provides CO_2 reduction in 43 % level. Naturally, such measures will cost an additional burden to country's economy at the first level of this implementation but will bring out both economical and environmental advantages in the long run. This view will strengthen our country's profile in international community in terms of indicating the determination of sustainable development goal together with the IPCC process.

References

1. Sulukan E, Saglam M, Uyar TS, Kırlıdoğ M (2010) A preliminary study for post-Kyoto period for Turkey by MARKAL model. In: Proceedings of 5th ege energy symposium and exhibition (IEESE-5), Denizli, 27–30 June 2010
2. Sulukan E, Saglam M, Uyar TS (2010) An ındigenous and new energy model for Turkey: wind power ıntegration options by 2025. In: Proceedings of 9th world wind energy conference and exhibition, Istanbul, 15–17 June 2010
3. Uyar TS, Kırlıdoğ M, Sulukan E, Saglam M (2010) Analysing the carbon mitigation alternatives on transport in the MARKAL-Turkey model. In: International energy workshop 2008, International Energy Agency, Paris, 30 June–2 July 2008

4. Sulukan E, Saglam M, Uyar TS, Kırlıdoğ M (2010) Determining optimal energy strategies for Turkey by Markal model. J Nav Sci Eng 6(1):27–38
5. Sağlam M (2010) Establishing mitigation strategies for energy related emissions for Turkey using the MARKAL family of models. Ph.D. Thesis 2010, Marmara University, Istanbul
6. Sulukan E (2010) Establishing energy efficient utilization and cost-effective energy technologies selection strategies for Turkey using MARKAL family of models. Ph.D. Thesis 2010, Marmara University, Istanbul
7. Uyar TS, Sulukan E, Saglam M (2010) Emission mitigation scenarios in Turkish MARKAL model. In: Proceedings of the ETSAP workshop, New Delhi (India), YTAM, New Technologies Research and Application Center, Marmara University, Istanbul, 18–23 Jan 2010
8. Uyar TS, Sulukan E, Saglam M (2008) About the new version of Turkish MARKAL model. In: Proceedings of the ETSAP workshop, Nice (France), New Technologies Research and Application Center, Marmara University, Istanbul, 15–17 Dec 2008
9. Sulukan E, Saglam M, Uyar TS (2009) Establishing energy efficiency and mitigation strategies in Turkey. In: Proceedings of the ETSAP workshop, Venice (Italy), Marmara University, Istanbul, 16 June 2009
10. Uyar TS, Kırlıdoğ M, Sulukan E, Sağlam M (2005–2009) Determining optimum energy strategies for Turkey by MARKAL model. Project FEN-KPS-080805-0177, BAPKO Research Project in Marmara University, Istanbul, 2005–2009
11. Çalıkoğlu E (2007) Energy efficiency in Turkey. In: TAIEX workshop on demand side management in energy efficiency. http://www.eie.gov.tr/duyurular/EV/TAIEX/ErdalCalikoglu_TAIEXWorkshop25625onDSM_221107.pdf
12. Turkish Cogeneration Association Website (2010) http://www.kojenerasyon.com

Analyzing Cost-Effective Renewable Energy Contribution Options for Turkey

Mustafa Sağlam, Egemen Sulukan, and Tanay Sıdkı Uyar

1 Introduction

Scientists believe that environmental corrosion and climate change are caused by increase in the amount of greenhouse gases in the atmosphere. Although as the most important one of GHGs, CO_2, is naturally emitted much more than that amount released during the mankind activities, a minor change affecting the equilibrium then causes big changes in the nature [1].

Limited fossil fuel resources and technologies using the products from them have caused the global warming up to this century. But now, we have renewable energy resources scattered equally and abundantly all over the world where people can use the innovative RES technologies. From now on, only the results of poorly planning of energy usage can keep the current worsening situation.

The standard MARKAL model combines energy demands, capital requirements, subsidies, investment costs, greenhouse gas emissions, conversion, resource (import–export) and process technologies, energy carriers, and end-use technologies. The model determines the optimal solution satisfying the energy demands with the least cost by choosing the energy-supply chain and respective technologies and facilitates to reach the accurate economic and environmental decisions [2].

In this context, the whole Turkish energy system has been represented numerically in the Turkish MARKAL model as Base Scenario (Business-as-Usual Scenario, BAU), and various implementations have been executed on mitigation strategies for energy-related emissions for Turkey with different scenarios within the periods 2005–2025 [3–12].

M. Sağlam (✉) • E. Sulukan • T.S. Uyar
Department of Mechanical Engineering, Marmara University, 34722 Istanbul, Turkey
e-mail: fulbars@yahoo.com; egemensulukan@gmail.com; tanayuyar@marmara.edu.tr

T.S. Uyar (ed.), *Towards 100% Renewable Energy*, Springer Proceedings in Energy, DOI 10.1007/978-3-319-45659-1_15

2 Energy Usage and Potentials in Turkey

Base Scenario comprises all the resources of Turkey: geothermal, solar, wind, biomass, hydro and renewable waste resources, etc. as renewables, domestic/import hard coal, asphaltite, oil, natural gas, and lignite as fossils. People use energy in the transportation, power generation, agriculture, residential, and industrial sectors. In the reference year, 2005, the supply rates of hydro, animal-biomass waste, biomass, solar, geothermal heat, and geothermal electricity in the energy system are 3.5 %, 1.2 %, 4.3 %, 0.4 %, 1 %, and 0.1 %, respectively. That is the total share of renewable energy as 10.5 % as illustrated in Fig. 1.

Turkey is a country with an abundant solar energy potential. Possible geothermal electricity, heat, and total solar energy reserves are 412 MWh/year, 28.2 GWh/year, and 380 TWh/year, respectively. Despite different estimations on wind energy potential, the potential of 120 TWh/year has been the input to our model [13]. Turkish Statistical Institute says that Turkey has a potential of 21.2 TWh/year biogas productions, big and small hydro energy reserve of Turkey is about 127.4 TWh/year together [14]. While each country classifies hydroelectric power plants separately, UNIDO (United Nations Industrial Development Organization) hydropower classified the range of 1–10 MW installed capacity as small hydropower [15].

3 Renewable Scenario (REN100)

This scenario aims to use all renewable energy sources in a great extent to satisfy the energy demand in Turkey. Possible potentials of renewables have become the upper bound on activities in the model. All prices are given in US dollars in 2005. That especially the projection of the demands and the bound capacities are proportional to the economic growth of Turkey forecasted by the World Bank has been one of the assumptions [16]. Total hydro energy potential in the country is assumed as renewable, disregarding the size of the dams or their capacities by applying an upper limit in the scenario.

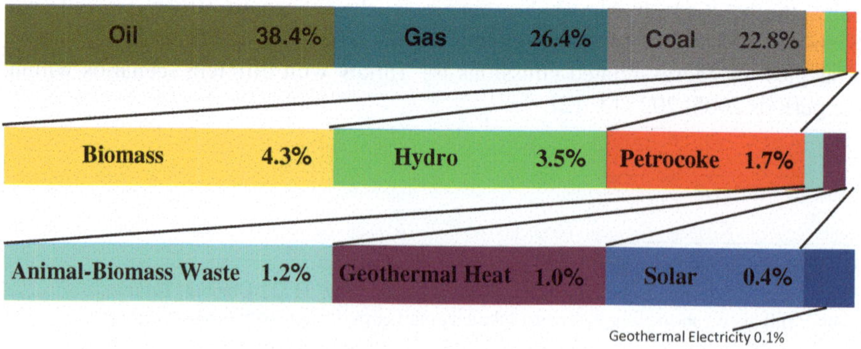

Fig. 1 Shares of energy types in Turkish energy usage (2005)

To get an idea about the situation of the possible highest rate of the renewable in the energy system, first, "fix and upper bounds on activity and in the capacity" of all the renewable conversion technologies have been set free in a separate scenario. Second, renewable process technologies such as landfill, hydrogen production, and forestry have also been set unlimited in order to determine the maximum usage level of renewables with the current technologies.

Thus, in letting the model use more than even the country's reserves and setting the fixed bounds of fossil fuel based on technologies' unlimited turns toward a favorable result as the model output. The model results in "optimal" eventually in this case. It is possible to see much increase in renewable usage; on the contrary there is a decrease in fossil fuel usage. Overall demand in the system is optimized, and the energy total system cost gets minimized severely caused by excess oil products export and abundant renewable electricity produced.

After this step, when it is restricted to use one of the fossil fuels (e.g., coal) completely in the energy system by setting the bounds "zero" on fossil-based and the end-use (demand) technologies, the energy carriers used in the model preferred by the end-use technologies cause to give "infeasible" result. This means the marketing system does not change direction without subsidies, taxes, some limitations, or innovation in end-use technologies.

To prepare a more factual scenario to give a feasible result with favorable input, each assumption should be selected carefully and should be applied one by one indicating the model's responses obviously for each trial.

For example, putting a limit on the burden of import fossil energy technologies by calibrating to zero level in the periods other than reference year as long as the model gives the optimum result would be a good start point for mitigating options. Here the reserves are kept as upper bounds set on the resource level. The bounds on activities and capacities of renewable technologies are set free to permit the system to use them independently. Bounds on capacities of demand technologies using electricity are set free in order them to use renewable electricity instead of diesel, for example, in railway transportation.

In this case in the whole system, while the imports of fossil fuels including crude oil and NGA stop, the increase of domestic lignite and coal usage increases. No briquette is produced in the system. Model concludes hard coal and crude oil as the necessary components of the energy mix of the country until the market has the transportation technologies using hydrogen and an alternative fuel production as hydrogen. So it is possible to decrease the fossil fuel import approximately 100 percent if the market has a brand new fuel produced by renewable electricity. Meanwhile, it is possible for the system to satisfy the electricity demand without producing electricity in the planned nuclear power plants.

It is admissible that 100 % renewable energy utilization would not be possibly applied till 2025 as long as the current energy demand technologies are kept to meet the end-use demands. That means you have diesel cars and you need to combust the diesel fuel in the diesel motor to meet your riding need. Unless you swap this end-use technology with a hydrogen or biodiesel car, you have to provide fossil diesel. Coal is necessary for the stoves used for residential space heating. But, the main question will arise: "How much renewable energy can be used at the same situation in the overall energy system as long as the grid is smart?"

To set the renewable scenario (REN100) giving an optimum result, assumptions have been selected in a way that least import and export resources and energy carriers, fossil fuel, and uranium are consumed. As a breakthrough, hydrogen fuel produced from renewable electricity and hydrogen using end-use technologies is introduced. Wave energy power plant is suggested to produce electricity from 2020. All the potential energy resources are available. Good security points are given for mining resources domestically. Model does not permit to enter values for denoting the security of wind and solar energy. The system works and gives optimum result in these circumstances. Fossil resource usage decreased remarkably, as given in Table 1.

The model says no nuclear energy is compulsory to sustain the energy system, concerning the conversion technologies.

Without caring whether it is expensive or cheap or other properties of the energy type, when it is permitted model to use all renewable electricity potential, it is apparent that renewable utilization is extremely high.

Thus, no electricity from asphaltite, natural gas, oil products, lignite, and hard coal power plants is possible when the renewable potential of country is used in REN100 Scenario. The model avoids using fossil energy for electricity production. In this condition, total electricity production is obtained from renewable energy sources (Table 2).

As the main indicator of dependency on foreign resources, total imports of fossil energy carriers are decreasing 100 % in REN100 Scenario, and the decrease of the total domestic supply of fossil energy carriers also becomes 94.4 %.

There is much less payment for fossil fuel each year until the end of the periods. In REN100 Scenario, the less energy supply can meet the demand identically suggested in REN100 and BAU Scenarios, and, the most important one, investment cost for the demand technologies is not included in the calculation so discounted total system cost decreases 17.3 % in the alternative scenario. It should be noted that different variations in investment cost for the demand technologies could be analyzed under different scenarios to indicate the effects of end-use technologies over all energy systems.

According to BAU results, total primary energy of the country is 3861.6 PJ in 2005 and is going up to 7705.5 PJ in 2025, increasing approximately 128.9 % in 20 years of analysis period under the main economical assumptions. This increase is 90.1 % in REN100 Scenario. Renewable power use changes this view with an increase rate of 97.2 % in BAU Scenario with 892.3 % in REN100 Scenario resulting in 100 % renewable electricity production throughout the scenario period (Table 3). All these parameters could be modified, or applied on model as other scenarios, by the user depending on the foremost indicators indicators of economy.

CO_2 emission levels are also influenced by the increasing share of renewable energy and decreased combustions due to less consumption of fossil energy carriers. BAU projects that CO_2 emissions will rise from 236.8 Mt in 2005 level to 479.9 Mt

Table 1 Resource levels in REN100 Scenario

Case	Parameter	Units	Technology	2005	2010	2015	2020	2025
BASE	RESOURCE.L	PJ	Animal-biomass waste	49.36	58.06	68.3	80.33	94.49
TRIAL3	RESOURCE.L	PJ	Animal-biomass waste	49.36	95	95	95	95
BASE	RESOURCE.L	PJ	Export oil products	210.01	314.96	370.47	341.78	405
TRIAL3	RESOURCE.L	PJ	Export oil products	210.01	0	0	0	0
BASE	RESOURCE.L	PJ	Export electricity	6.49	12.5	12.5	12.5	12.5
TRIAL3	RESOURCE.L	PJ	Export electricity	6.49	0	0	0	0
BASE	RESOURCE.L	PJ	Import briquette	2.14	2.52	2.96	4.75	4.75
TRIAL3	RESOURCE.L	PJ	Import briquette	2.14	0	0	0	0
BASE	RESOURCE.L	PJ	Import hard coal	478.64	563	662.23	778.95	916.24
TRIAL3	RESOURCE.L	PJ	Import hard coal	478.64	0	0	0	0
BASE	RESOURCE.L	PJ	Import electricity	2.3	2.71	3.19	3.75	4.41
TRIAL3	RESOURCE.L	PJ	Import electricity	2.3	0	0	0	0
BASE	RESOURCE.L	PJ	Import coke	10	21.37	29.47	39.98	49.7
TRIAL3	RESOURCE.L	PJ	Import coke	10	0	0	0	0
BASE	RESOURCE.L	PJ	Import lignite	0	0	0	0	0
TRIAL3	RESOURCE.L	PJ	Import lignite	0	0	0	0	0
BASE	RESOURCE.L	PJ	Import natural gas	1026.69	1207.65	1420.5	1670.87	1965.37
TRIAL3	RESOURCE.L	PJ	Import natural gas	1026.69	0	0	0	0
BASE	RESOURCE.L	PJ	Import crude oil	1015.16	1199.66	1411.1	1659.82	1952.37
TRIAL3	RESOURCE.L	PJ	Import crude oil	1015.16	0	0	0	0
BASE	RESOURCE.L	PJ	Import petroleum products	428.78	498.79	586.71	782.74	811.75
TRIAL3	RESOURCE.L	PJ	Import petroleum products	428.78	0	0	0	0
BASE	RESOURCE.L	PJ	Import petrocoke	69.58	100	117.63	138.36	162.74
TRIAL3	RESOURCE.L	PJ	Import petrocoke	69.58	0	0	0	0

(continued)

Table 1 (continued)

Case	Parameter	Units	Technology	2005	2010	2015	2020	2025
BASE	RESOURCE.L	PJ	Import tree	0	0.75	1.5	1.76	2.08
TRIAL3	RESOURCE.L	PJ	Import tree	0	0	0	0	0
BASE	RESOURCE.L	PJ	Import of uranium	0	0	0	130	130
TRIAL3	RESOURCE.L	PJ	Import of uranium	0	0	0	0	0
BASE	RESOURCE.L	PJ	Asphaltite mine	15.99	17	26.3	30.93	36.38
TRIAL3	RESOURCE.L	PJ	Asphaltite mine	15.99	0	0	0	0
BASE	RESOURCE.L	PJ	Hard coal mine	49.57	211.71	244.02	446.08	326.94
TRIAL3	RESOURCE.L	PJ	Hard coal mine	49.57	136.04	136.57	138.3	139.03
BASE	RESOURCE.L	PJ	Lignite mine	373.46	439.29	516.71	607.79	714.91
TRIAL3	RESOURCE.L	PJ	Lignite mine	373.46	0	0	0	0
BASE	RESOURCE.L	PJ	Gas at wellhead	34.16	40.19	47.27	55.6	65.4
TRIAL3	RESOURCE.L	PJ	Gas at wellhead	34.16	34.16	34.16	34.16	34.16
BASE	RESOURCE.L	PJ	Crude oil extraction	100.27	117.95	138.74	163.19	191.95
TRIAL3	RESOURCE.L	PJ	Crude oil extraction	100.27	121.26	121.26	121.26	121.03

Table 2 Total electricity production

Case	Parameter	Units	Technology	2005	2010	2015	2020	2025
BASE	Output of electricity	PJ	Asphaltite - power plant	0	0	3.18	3.74	4.39
REN100	Output of electricity	PJ	Biogas power plant	0	72	72	72	72
BASE	Output of electricity	PJ	Nuclear power plant	0	0	0	45.5	45.5
REN100	Output of electricity	PJ	Nuclear power plant	0	0	0	0	0
REN100	Output of electricity	PJ	Wave farm electricity	0	0	0	36	36
BASE	Output of electricity	PJ	Geothermal electricity	0.34	0.4	0.47	0.48	0.65
REN100	Output of electricity	PJ	Geothermal electricity	0.34	57.6	57.6	57.6	57.6
BASE	Output of electricity	PJ	Hydro power plant	142.42	167.53	197.06	229.37	272.64
REN100	Output of electricity	PJ	Hydro power plant	142.43	600	600	600	600
BASE	Output of electricity	PJ	Oil products power plant	19.74	23.22	27.31	28.21	37.79
REN100	Output of electricity	PJ	Oil products power plant	19.74	0	0	0	0
BASE	Output of electricity	PJ	Solar photovoltaic	0	0	0	0	0
REN100	Output of electricity	PJ	Solar photovoltaic	0	198.53	314.58	417.83	565.43
BASE	Output of electricity	PJ	Wind onshore	0.21	0.25	0.29	0.3	0.38
REN100	Output of electricity	PJ	Wind onshore	0.21	29	29	29	29
REN100	Output of electricity	PJ	Wind offshore	0	0	0	0	19
BASE	Output of electricity	PJ	Hard coal power plant	47.68	115	136	238	189
REN100	Output of electricity	PJ	Hard coal power plant	47.68	0	0	0	0
BASE	Output of electricity	PJ	Lignite power plant	107.79	126.79	149.14	154.06	206.35
REN100	Output of electricity	PJ	Lignite power plant	107.79	0	0	0	0
BASE	Output of electricity	PJ	Natural gas power plant	264.4	311	365.82	377.89	506.14
REN100	Output of electricity	PJ	Natural gas power plant	264.4	0	0	0	0

(continued)

Table 2 (continued)

Case	Parameter	Units	Technology	2005	2010	2015	2020	2025
REN100	Output of electricity	PJ	Solar power plant	0	9.19	11.91	15.16	18.92
BASE	Output of electricity	PJ	Waste incineration plant	0.44	0.52	0.61	0.63	0.84
REN100	Output of electricity	PJ	Waste incineration plant	0.44	14.25	14.25	14.25	14.25
REN100	Output of electricity	PJ	Biomass tree power plant	0	0	3	3	3

in 2025, which means 131.6 % increase during the total analysis period. The change in CO_2 emission goes down 64.4 % in REN100 Scenario. Giving up the import fossil fuels caused security level to increase by 6.3 %.

4 Conclusion

Conventional energy resources still dominate energy mix due to current energy system composition, while relatively lower costs for electricity per kWh is mainly caused by excluded externalities. Some technologies like coal furnaces used in steel industry are invented in a way to consume energy carriers coming from the mentioned resources, and so the end-use technologies like trucks are developed in line with this type of combustion fuel.

Slipping this habit out and lead to innovate the conversion, process, and all end-use technologies may bring out right effects on the best and fastest way of introducing renewable usage in every phase of our lives with higher levels.

However, political guidance is the most important part of the right path into diminishing contaminating hazardous energy resources consumption for mitigating global warming effects and sustain society life. Reaching consensus for decision-makers, engineers, local authorities, and nongovernmental organizations on energy and environmental planning can be improved by efforts of the experts with energy optimization modeling systems. Modeling systems should determine the respective features of energy and energy usage, which involves energy economy, environmental effects, energy production, and ethical responsibility to reduce energy consumption through energy efficiency and conservation considering also security issues in order to get a sustainable system fulfilling the national requirements.

Table 3 Comparison of BAU and the REN100 Scenarios

Scenario	Units	2005	2010	2015	2020	2025
BAU	CO_2 emission (Mt)	236.8	294.0	346.8	411.6	480.0
REN100		236.842	53.1	58.0	63.2	73.9
Diff. BAU-REN100		0.00 %	−81.9 %	−83.3 %	−84.6 %	−84.6 %
BAU	Total primary energy (PJ)	3861.6	4620.3	5446.6	6799.6	7705.5
REN100		3861.6	3497.0	4368.8	5200.9	6393.3
Diff. BAU-REN100		0.00 %	−24.3 %	−19.8 %	−23.5 %	−17.0 %
BAU	Total imports of fossil energy carriers (PJ)	3031	3593.0	4230.6	5075.5	5862.9
REN100		3031	0.0	0.0	0.0	0.0
Diff. BAU-REN100		0.00 %	−100.0 %	−100.0 %	−100.0 %	−100.0 %
BAU	Total domestic supply of fossil energy carriers (PJ)	3394.4	4104.2	4833.2	6037.3	6793.5
REN100		3394.4	291.5	292.0	293.7	294.2
Diff. BAU-REN100		0.00 %	−92.9 %	−94.0 %	−95.1 %	−95.7 %
BAU	Total usage of renewable energy carriers (PJ)	639.1	751.7	884.2	994.0	1226.2
REN100		639.1	3258.3	4123.0	4946.0	6137.1
Diff. BAU-REN100		0.00 %	200.20 %	240.50 %	260.80 %	258.20 %
BAU	Total system cost (million USD)	78,476.5	94,435.9	111,327.1	127,582.0	161,524.1
REN100		78,476.5	68,647.2	72,211.6	75,943.3	80,174.0
Diff. BAU-REN100		0.00 %	−27.3 %	−35.1 %	−40.5 %	−50.4 %
BAU	New investment cost (million USD)	57,542.5	68,133.7	80,349.1	86,096.9	114,324.0
REN100		57,592.1	64,679.7	67,809.7	71,169.8	75,109.7
Diff. BAU-REN100		0.10 %	−5.1 %	−15.6 %	−17.3 %	−34.3 %
BAU	Total system cost, net of taxes/subsidies (million USD)	**913,096**				
REN100		**754,866**				
Diff. BAU-REN100		**−17.3 %**				

References

1. Growth and CO2 emissions: how do different countries fare? (2007) Environment Department The World Bank, pp 10–13. http://www3.imperial.ac.uk/pls/portallive/docs/1/34721711. PDF
2. Loulou R, Goldstein G, Noble K (2004) Standard MARKAL user manual, vol I. http://www.etsap.org/documentation.asp
3. Sulukan E, Saglam M, Uyar TS, Kırlıdoğ M (2010) A preliminary study for post-Kyoto period for Turkey by MARKAL model. In: Proceedings of 5th ege energy symposium and exhibition (IEESE-5), Denizli, 27–30 June 2010
4. Sulukan E, Saglam M, Uyar TS (2010) An indigenous and new energy model for turkey: wind power integration options by 2025. In: Proceedings of 9th world wind energy conference and exhibition, Istanbul, 15–17 June 2010
5. Uyar TS, Kırlıdoğ M, Sulukan E, Saglam M (2010) Analysing the carbon mitigation alternatives on transport in the MARKAL-Turkey model. In: International energy workshop 2008, International Energy Agency, Paris, 30 June–2 July 2008
6. Sulukan E, Saglam M, Uyar TS, Kırlıdoğ M (2010) Determining optimal energy strategies for Turkey by MARKAL model. J Nav Sci Eng 6(1):27–38
7. Sağlam M (2010) Establishing mitigation strategies for energy related emissions for Turkey using the MARKAL family of models. Ph.D. Thesis 2010, Marmara University, Istanbul
8. Sulukan E (2010) Establishing energy efficient utilization and cost-effective energy technologies selection strategies for Turkey using MARKAL family of models. Ph.D. Thesis 2010, Marmara University, Istanbul
9. Uyar TS, Sulukan E, Saglam M (2010) Emission mitigation scenarios in Turkish MARKAL model. In: Proceedings of the ETSAP workshop, New Delhi (India), YTAM, New Technologies Research and Application Center, Marmara University, Istanbul, 18–23 Jan 2010
10. Uyar TS, Sulukan E, Saglam M (2008) About the new version of Turkish MARKAL model. In: Proceedings of the ETSAP workshop, Nice (France), New Technologies Research and Application Center, Marmara University, Istanbul, 15–17 Dec 2008
11. Sulukan E, Saglam M, Uyar TS (2009) Establishing energy efficiency and mitigation strategies in Turkey. In: Proceedings of the ETSAP workshop, Venice (Italy), Marmara University, Istanbul, 16 June 2009
12. Uyar TS, Kırlıdoğ M, Sulukan E, Sağlam M (2005–2009) Determining optimum energy strategies for Turkey by MARKAL model. Project FEN-KPS-080805-0177, BAPKO Research Project in Marmara University, Istanbul, 2005–2009
13. Ministry of Energy and Natural Resources of Turkey Energy Statistics (2010) http://www.enerji.gov.tr/index.php?dil=tr&sf=webpages&b=y_istatistik&bn=244&hn=244&id=398
14. Energy Resources Website (2010) http://www.enerjikaynaklari.net/keyf/turkiye%E2%80%99_de_biyokutle_biyogaz_enerjisi_alanindaki_calismalar-246.html
15. Gençoğlu MT, Cebeci M (2001) Büyük Hidroelektrik Santrallar ile Küçük Hidroelektrik Santrallarin Karşılaştırılması. Fırat Üniversitesi Mühendislik Fakültesi Elektrik-Elektronik Mühendisliği Bölümü, 18–20 Jan 2001
16. Global Economic Prospects 2010 (2010) International Bank for Reconstruction and Development (World Bank)

A Native Energy Decision Model for Turkey

Egemen Sulukan, Mustafa Sağlam, and Tanay Sıdkı Uyar

1 Introduction

There are different energy forms mankind extract or explore from various reserves such as mines or wells; types of coal, petroleum, natural gas; and renewables such as wood, animal or plant wastes, hydraulic, wind, geothermal, or solar power. Even at the first phase of obtaining these forms of energy as primary energy carriers, people should convert them to electricity or heat to use or meet their demand. Mankind's needs could be classified in basic or higher levels. Air, water, and food are metabolic requirements for survival in all animals, including human beings. Basic physiological needs such as housing, cooking and eating, heating, cooling, lighting, washing, drying at the places we live must be satisfied at least to survive or to ensure the quality of life according to a society's life standards. But in evolution process throughout the history, some other increasing standards occurred especially starting from the industrial revolution.

The industrial revolution was a period from the eighteenth to the nineteenth century where major changes occurred in agriculture, manufacturing, mining, and transport and reformed the socioeconomic and cultural conditions starting in the Europe and then spreading to North America and eventually the world. The start of the industrial revolution marked a critical point in human history; almost every aspect of daily life was eventually affected.

Industrial subsectors began shaping as conversion and energy involving power plants generating electricity and heat with coke or briquette ovens at smokeless fuel plants; petroleum refineries, iron and steel, chemical-petro chemical, cement, sugar,

E. Sulukan (✉)
Department of Naval Sciences, Turkish Naval Academy, İstanbul, Turkey
e-mail: egemensulukan@gmail.com

M. Sağlam • T.S. Uyar
Department of Mechanical Engineering, Marmara University, İstanbul, Turkey
e-mail: fulbars@yahoo.com; tanayuyar@marmara.edu.tr

T.S. Uyar (ed.), *Towards 100% Renewable Energy*, Springer Proceedings in Energy, DOI 10.1007/978-3-319-45659-1_16

and nonferrous metals producing factories; transportation by maritime, land, rail and airways, agriculture, and some nonenergy segments.

While using final energy carriers of electricity and heat generated by various energy carriers, these industrial subsectors became the driving force of the countries' economies. The gross domestic product or gross domestic income has been a basic measure of a country's overall economic output by the time, which means the market value of all final goods and services made within the borders of a country in a year as it is often positively correlated with the standard of living.

Innovation in energy technology may change the social life and even shape it. The steam engine triggered the industrial revolution. The internal combustion engine made widespread use of public transport. Gas turbines in aviation made the world a smaller globe. However, each of these options has a price.

Society aims to increase its life standards for each individual while keeping this is an economical growth progress for the scale of a country or a union like formed in Europe. Is it possible to keep this change positively in the long term? History denotes some conflictions throughout time so far because of the societies' or nations' benefits intersection especially in sharing the resources on the world. The story of "sharing the resources" forced the nations to discover untouched areas at first, but now different types of races surrounded this old globe and we came to a point that, there is no way other than sustainable development for each nation while diminishing the resources. Sustainable development is a pattern of resource use that aims to meet human needs while preserving the environment so that these needs can be met not only in the present, but also for future generations. Sustainable development ties together concern for the carrying capacity of natural systems with the social challenges facing humanity. As early as the 1970s "sustainability" was employed to describe an economy "in equilibrium with basic ecological support systems." Ecologists have pointed to "The Limits to Growth" and presented the alternative of a "steady-state economy" in order to address environmental concerns [1, 2].

Energy prices affect investments and have an important role in the development of energy systems and economic growth, which depends on the following main factors: demand, reserves, and imports if needed or exports as income, with total costs.

2 Scope

A country should determine the domestic resources, forecast the total demands, and then make a complete strategic plan balancing the supply side and the demands within the energy system to minimize its total system cost for a sustainable economy.

Responsible authorities in Turkey for managing the process on economic development while protecting the environment are the Ministry of Environment (MoE), the Ministry of Energy and Natural Resources (MENR), and the Ministry of Health, as well as the Turkish Electricity Generation & Transmission Company (TEAS).

The World Bank also had cooperative works with these agencies for planning to estimate the costs and benefits of various energy policy alternatives under an Energy and Environment Review (EER). A number of individual studies have been conducted under these activities to analyze energy technology issues [3].

Energy modeling studies have been carried out by Argonne National Laboratory's Centre for Energy, Environmental and Economic Systems Analysis (CEEESA) with a team by the Turkish Ministry of Energy and Natural Resources (MENR) and the Turkish Electricity Generation and Transmission Corporation (TEAS). Energy and Power Evaluation Program (ENPEP), an integrated energy modeling system developed by Argonne has been used in these studies. Different modules under this program are used such as MAED, WASP, BALANCE, and VALORAGUA. MAED has been used for energy demand levels projection, WASP for electricity generation expansion planning, BALANCE for projection of future, fossil and nonfossil energy flows in all sectors, and VALORAGUA for evaluation for operation of the hydro-portion of the electric system [4].

Following these activities, different reports were published about energy demand projections, greenhouse gases abatement measurements or energy generation capacity expansion planning issues by governmental or nongovernmental organizations such as Ministry of Energy and Natural Resources (MENR), Energy Market Regulatory Authority (EMRA), Turkish Electricity Transmission Corporation (TEIAS), and World Energy Council-Turkish National Committee (WEC-TNC), which are based on the same results obtained from the studies carried out by Argonne National Laboratory's CEEESA [5–9].

Moving from the point of necessity to have an indigenous energy model developed in our own hands, we have started searching the energy modeling activities around the world and seen that there are different types of modeling research programs or activities in many countries and institutions. A list of energy models currently being used is given under the next title.

3 Current Applications in Energy Modeling

Regarding the main objectives of a country's energy policy; various energy modeling tools have been developed so far and some of them are listed as follows.

- Cities for Climate Protection Software
- Comparative Techno-Economic Project Evaluation of Energy Supply and Demand Options
- Carbon Dioxide Technology Database
- EnergyPLAN
- Energy Costing Tool
- Energy and Power Evaluation Program (ENPEP)
- Global Emission Model for Integrated Systems (GEMIS)
- HOMER

- Long Range Energy Alternatives Planning System (LEAP)
- MAED
- MESSAGE
- REAP
- RETScreen
- SUPER

4 Model Approach and Methodology

4.1 Overview and RES Concept

MARKAL means market allocation. The model's main aim was to assess energy technologies in a large scope and integrated structure. Although it had implementations in 1980s, but MARKAL has evolved in time to a more dynamic scope that focused on energy supply–demand equilibrium. The model involves all phases of energy market; starting from extraction, conversion, or process and end use in each related sector by various technologies that produce or consume energy carriers. These energy technologies compose a reference energy system (RES) with every single energy input and output of the energy system as regional or global level. RES also represents the energy economics of the relevant technologies, which take place in the entire energy system.

MARKAL Model aims to minimize energy system cost in analysis time period, while satisfying identified demands and other constraints such as GHG or limitations on fuel use. This objective of the model can also be expressed as minimizing the total cost or maximizing the surplus of the energy system, while satisfying a number of constraints mentioned earlier.

Analysis periods are generally applied as 5 or 10 years, and electricity demand is divided into three seasons as winter, summer, and spring or fall with two daily divisions as night and day. These time divisions compose six time slices. These time slices are identified for electricity or low-temperature heat producing technologies, both of which may be easily stored and require a finer time-period separation than the other energy carriers.

As a result, electricity or low-temperature heat is divided into time slices for each time period. The energy system of a region includes energy supply sources such as mining, imports and exports, as well as processing, conversion of the various energy forms and their transport, distribution, and end-use consumption by all sectors within the economy. These activities are expressed in MARKAL with these energy producing or consuming technologies. The end-use technologies produce special kinds of commodities that are physical products or services for which demands are specified. The set of resources, technologies, demands, and commodities determine the structural characteristics of the model's RES.

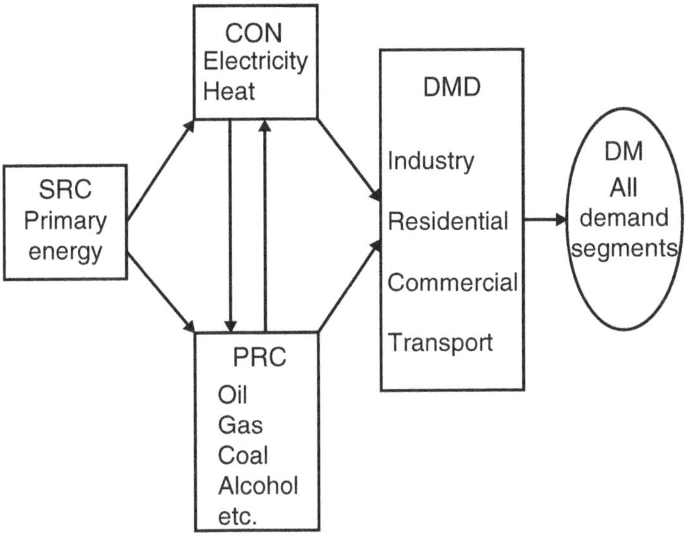

Fig. 1 Simplified representation of a typical MARKAL RES [11]

Figure 1 gives a simplified representation of a typical MARKAL RES showing the five main components usually recognized in each model structure: primary energy resources (SRC), energy conversion into electricity or low-temperature heat (CON), other energy processing (PRC) and energy end uses (DMD), and the demands (DM) for energy services and products.

Each "region" identified into model may be in a scale of district municipality, a city, a country, or a group of countries. The regions may also be interconnected by technologies that transport energy forms like transmission lines and pipelines. Time horizon includes at most nine periods, each having equal duration, usually 5 or 10 years, defined by the user. After the time horizon is determined and identified within the model, at least one of the initial periods' relevant data is fixed to their historical values, which the model has no freedom and for which all of the quantities of interest to represent energy balance of the realized part of energy system.

In order to determine the basis, this calibration to an initial period is one of the important tasks required for setting up a MARKAL model for an identified region. The main variables that must be fixed are the capacity and operating levels of all technologies as well as extraction, exports, and imports for all energy forms and materials. The calibration to an initial period also affects the model's reactions over several future periods because the "profile of residual capacities" is fully specified over the remaining lives of the technologies existing at the start of the model forecast horizon.

As defined its objective function, MARKAL minimizes the "total energy system cost," which includes the followings: investments, fixed and variable annual O&M costs, energy commodity import costs, minus export revenues and demand losses incurred from reduced product and service demands.

4.2 Data Parameters

MARKAL uses a number of data parameters to characterize each element of RES. The general categories of data required for a MARKAL model are as follows:

- System parameters,
- Energy demands,
- Energy carriers,
- Technology characterizations,
- Resource, process, conversion, and end-use (demand) technologies,
- Environmental variables,
- Emissions per resource, technology, or investment.

4.3 Simplified Mathematical Formulation of MARKAL Linear Program

The description of the objective function and constraints may be translated into a formal set of mathematical expressions. A streamlined formulation of the equations represents, which ignores exceptions and some complexities that are not essential to a basic understanding of the principles of the model.

An optimization problem formulation consists of three types of entities:

- Decision variables: the unknowns to be determined by the optimization,
- Objective function: expressing the criterion to minimize or maximize and
- Constraints: equations or inequations involving the decision variables that must be satisfied by the optimal solution [10, 11].

4.4 Objective Function

The objective function is the sum over all regions of the discounted present value of the stream of annual costs incurred in each year of the horizon, as given in Eq. (1).

$$\mathrm{NPV} = \sum_{r=1}^{R} \sum_{t=1}^{t=\mathrm{NPER}} (1+d)^{\mathrm{NYRS}\cdot(1-t)} \cdot \mathrm{ANNCOST}(r,t) \cdot \left(1+(1+d)^{-1}+1(1+d)^{-2}+\cdots+(1+d)^{1-\mathrm{NYRS}}\right) \tag{1}$$

- NPV: Net present cost of the system as defined in the region (MARKAL objective function),
- ANNCOST (*r*, *t*): *r* region, the annual cost for period *t*,
- NPER: Number of periods in the planning process (determined as 4),

- NYRS: Number of years in each period *t*, *r* is the number of regions. *T* value at each period of 5 years, *r*-value determined to be 1, Turkey defined as one region.

ANNCOST (*r*, *t*), as expressed in the total annual cost of all the **k** technologies, all **d** demand in the identified region, all **p** pollutants, all imported fuel **f** and annual investment of fixed and variable costs of technologies, fuel delivery costs, extracting the energy carriers including import costs, annual operating costs with the emission taxes applied. The cost of exported energy carriers is subtracted from this value.

Mathematically, ANNCOST (*r*, *t*) can be expressed as follows in Eq. (2):

$$\begin{aligned}
\text{ANNCOST}(r,t) = {} & £_k\{\text{Annualized_Invcost}(r,t,k) * \text{INV}(r,t,k) \\
& + \text{Fixom}(r,t,k) * \text{CAP}(r,t,k) \\
& + \text{Varom}(r,t,k) * £_{s,s}\text{ACT}(r,t,k,s) \\
& + £_c\left[\begin{matrix}\text{Delivcost}(r,t,k,c) * \text{Input}(r,t,k,c) \\ * £_s\text{ACT}(r,t,k,s)\end{matrix}\right]\} \\
& + £_{c,s}\{\text{Miningcost}(r,t,c,l) * \text{Mining}(r,t,c,t) \\
& + \text{Tradecost}(r,t,c) * \text{TRADE}(r,t,c,s,i/e) \\
& + \text{Importprice}(r,t,c,l) * \text{Import}(r,t,c,l)\} \\
& - \text{Exportprice}(r,t,c,l) * \text{Export}(r,t,c,l)\} \\
& + £_c\{\text{Tax}(r,t,p) * \text{ENV}(r,t,p)\} \\
& + £_d\{\text{DemandLoss}(r,t,d)\}
\end{aligned} \tag{2}$$

Meaning of the variables in the equation is given briefly as follows:

- Annualized_Invcost: Annual investments for technologies,
- Fixom: Fixed operation and maintenance costs,
- Varom: Variable operation and maintenance costs,
- Delivcost: Distribution costs,
- Miningcost: Mining extraction costs,
- Tradecost: Interregional trade costs,
- Importprice: Import costs,
- Exportprice: Export costs,
- Tax: Taxes,
- DemandLoss: Reduction in demand.

4.5 Methodology

In general, the modeling process typically consists of a number of interrelated steps. These steps can be grouped under seven main titles:

- Determining model objectives,
- Model formulation,

- Data collection,
- Data analysis,
- Model calibration,
- Model implementation,
- Sensitivity analysis.

While setting up and developing the country's model, a road map has been followed. The important steps of this study can be summarized in order as given as follows:

In the preparation phase, the main objective and background of the study is determined by

- Identifying the goals of study and scenario analysis,
- Studying on the background of the topic,
- Literature review,
- MARKAL+SAGE Manuals overview,
- Software Installation,
- Practicing on ANSWER user interface,
- Focusing on current energy system,
- Detailed research about the system components,
- Scanning and compiling the data resources,

Second phase of determining the main components and shaping the core of the model includes:

- Determining the supply items,
- Determining main demand sectors and each sectoral demand item,
- Reviewing the installed current conversion and process technologies,
- Determining consumption and demand technologies,
- Setting up the timeframe of the model,
- Setting a draft model by identifying each energy carrier, technology and demand as a cell, and relationship of each other in general terms,

The third phase of model structure study focused on providing and entering data and model calibration includes:

- Collecting useful data,
- Setting up the units according to model defaults,
- Identifying the user-defined variables according to naming conventions,
- Data entry and editing,
- Characterizing each of the technologies and resources used, including fixed and variable costs, technology availability, performance and pollutant emissions,
- Setting up the Reference Energy System,
- Setting up the energy equilibrium and creating an integrated model to calculate final energy consumption from primary energy supply,

After having the data-driven model sketch, implementation process includes:

- Running the model,
- Determining and eliminating the infeasibilities and errors,

- Obtaining the Base Scenario results,
- Finally, sensitivity analysis phase includes
- Charting and analyzing Base Scenario results,
- Practicing and training on model,
- Developing the base scenario,
- Verifying baseline data and testing consistency of the system,
- Comparing with the results of ENPEP modules,
- Improving the BASE Scenario by previously executed model studies,
- Developing alternative scenarios to analyze cost-effective technology selection options,
- Running each alternative scenario against the base scenario; analyzing model responses and scenario results to provide technical recommendations.

5 Reference Energy System of Turkey

The aim of this study is to establish Turkish MARKAL energy-system model for the period of 2005–2025 for analyzing alternative energy technology strategies in order to supply technical support for decision-makers.

Reference Energy System of Turkey represents the general structure of the energy flows from resource to demands throughout every step within the conversion, process, and end-user technologies. A generalized RES of Turkey is given in Fig. 2.

RES consists of six main columns. Two of these columns are energy carriers in primary and final forms, two of them are energy technologies of conversion/process and end-user technologies, the last column stands for all the demands met by these various energy types. Starting from the left-hand side, different resources of the country which are classified in sets as imports, domestic resources, and exports enter the system with primary energy carriers of lignite, hard coal, crude oil, natural gas, wind energy, geothermal electricity, geothermal heat, hydropower, renewable waste, biomass, animal waste, agricultural waste, solar energy, tree, wave energy, and uranium.

Some of the primary energy carriers converted to final energy carriers partially by conversion or process technologies. Final energy carriers are electricity, low temperature heat, industrial process heat, briquette, hydrocarbons for nonenergy feedstock, coke, petrocoke, conservation, oil products, and natural gas. End-use demands are identified in five main sectors. The sectors are agricultural, residential, commercial, industrial, and transportation. Demands are determined in time by analyzing these five sectors in detail. Amount of useful energy to meet these demands, sectoral fuel consumption statistics belonging to years 2000 and 2005, published by MENR, have been used. Additionally, primary energy conversions are all set as convenient outputs according to official energy balance outputs in calibration phase. Demands are satisfied by end-use technologies using various types of final energy carriers.

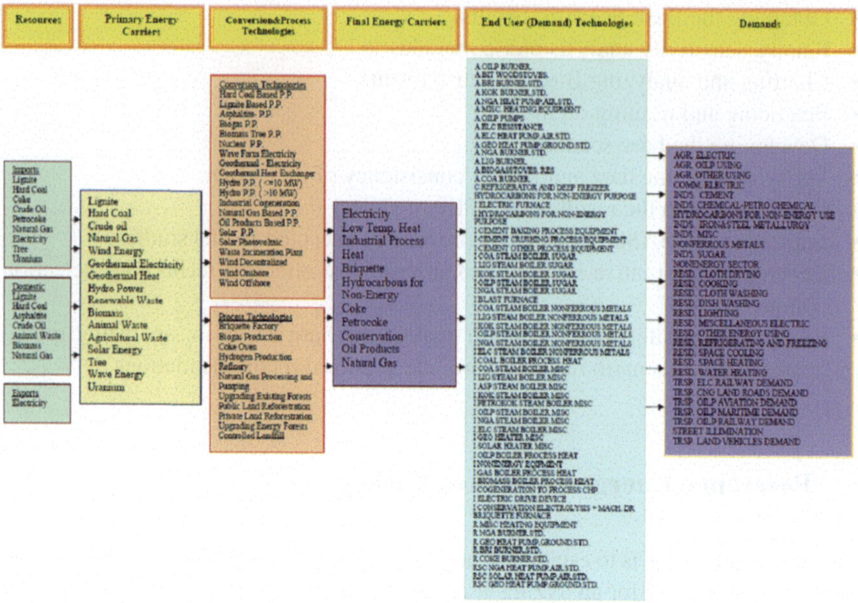

Fig. 2 Generalized RES of Turkish MARKAL model

6 Conclusion

RES has been evolved in time and continuously developed throughout the study period including the current and estimated resources, fuels, and technologies. Finally, 28 energy carriers, 29 resource technologies, 11 process and 20 conversion technologies, 139 end-user (demand) technologies, 30 demands in five sectors, 11 emission components, 11 tax/subsidies, and 12 global items (i.e., annual discount rate, GDP in first year, fraction of year for season, time of day) have been identified in appropriate positions according to MARKAL hierarchy and characterized with data by using ANSWER user interface. MARKAL is provided with menu-driven user interface of ANSWER that simplifies modification of input data and structuring of the model. Also demonstration cases with default data are provided. This interface also organizes file handling and execution of runs and includes result-processing menus for both tabular and graphical display.

However, RES components and structure may be developed for further analyses requirements in the future depending on the economical, technological, or environmental requirements.

References

1. Meadows DH, Meadows DL, Randers J, Behrens WW (1972) Limits to growth. Potomac Associates, New York
2. Daly HE (1991) Steady-state economics. Island Press, Washington, DC
3. Conzelmann G, Koritarov V (2002) Turkey energy and environmental review task 7: energy sector modeling. Final Report, Center for Energy
4. Altaş M, Sakaryalı Z, Sevgör S, Güler M, Bayrak Y, Conzelmann G (2006) PIMS: 3367, Turkey first national communication project reference case: energy sector modeling draft report
5. Turkish Electricity Transmission Company (TEİAŞ), Board of Research, Planning and Coordination (2006) Capacity coordination (2006–2015)
6. Turkish Electricity Transmission Company (TEİAŞ), Board of Research, Planning and Coordination (2007) Capacity coordination (2007–2016)
7. Ministry of Energy and Natural Resources (2005) Energy and natural resources research programme. Board of Research, Planning and Coordination
8. World Energy Council Turkish National Committee (2007) Turkey energy report (2005–2006). Ankara, ISSN: 1301-6318
9. MENR, General Directorate of Energy Affairs (2006) Workgroup report on greenhouse gas reduction in energy sector. Ankara. http://www.enerji.gov.tr/yayinlar_raporlar_EN/Enerji_Grubu_Raporu_EN.pdf
10. Kydes AS, Kanudia A, Loulou R (2004) National energy modeling systems. Encycl Energy 4:103–106
11. Loulou R, Goldstein G, Noble K (2010) Standard MARKAL user manual, vol I. http://www.etsap.org/documentation.asp

Reference Energy System Development for Turkish Residential Sector

Fatih Mutluel and Egemen Sulukan

1 Introduction

Turkey became an official party to the Kyoto Protocol in August 2009, and reducing the greenhouse gas emissions came into attention. In this context, Turkey developed the National Climate Change Strategy also in 2009 in order to contribute to the global efforts to reduce the impacts of climate change, and according to the National Climate Change Strategy, the main purpose is using clean and efficient energy resources in all sectors and implementing necessary alterations in the energy system in the short term and long term [1, 2]. Turkey as one of the developing countries has a huge desire to expand its economy. In the first three quarters of the 2013, the growing rate in GDP exceeded expectations and reached 4 % [3]. In this paper, the current situation of the residential sector of the Turkish energy market is modelled using an energy decision support tool.

Energy decision support tools are being used to develop the RES model of the country and carry out the economic analysis of different energy-related systems in different sectors. It represents the base scenario which is the current situation of the system and points out its potential evaluation as business as usual (BAU). Also we compare BAU with user-defined alternative scenarios over a long period of years and make comparisons in order to help the user to decide which scenario should be followed. The objective function of these tools is minimizing the total cost of the energy system.

F. Mutluel (✉) • E. Sulukan
Department of Mechanical Engineering, Marmara University, Istanbul, Turkey
e-mail: fatihmutluel@gmail.com; egemensulukan@gmail.com

T.S. Uyar (ed.), *Towards 100% Renewable Energy*, Springer Proceedings in Energy, DOI 10.1007/978-3-319-45659-1_17

2 Significance of the Residential Demand Technologies in the Energy Sector

In terms of energy supply, Turkey is highly dependent on the imports of the primary energy carriers. Dependency on imported sources increases the total cost and the risks of the system. As Turkey started to pay more attention on energy efficiency subjects, it is inevitable to focus on the residential sector because of its size of about 31 % of the total energy consumption in Turkey [4]. Other sectors somehow employ profit-oriented institutions and organizations; they utilize a variety of applications regarding energy efficiency and cost reduction policies. In this respect, residential sector is one of the sectors that need urgent attention most with its millions of independent users but no common interest.

Demands in the residential sector include energy for heating, cooking, cleaning, washing, drying, lighting, cooling and entertainment [5]. As a result, there is a clear incentive for government officials to implement programmes and measures that reduce the sector's demand for energy and increase the efficiency of the energy usage in residents. For instance, there are fundamental modifications like using double glass as an exterior insulation to buildings. Although that can be applied to nearly every household and reduces the energy consumption by 10 %, only 58 % of the dwellings are using this technology. Also, even though it can save up to 50 % energy, only 24 % of the residents are utilizing thermal insulation in their buildings [6].

3 RES for Turkish Residential Sector

Government institutions like the *Ministry of Energy and Natural Resources* and *World Energy Council Turkish National Committee* publish the total annual energy consumption and shares of the different sectors in detail, and these reports are called the energy balance table. The energy balance table of Turkey for the year 2012 is shown in Fig. 1. Data is very useful in terms of determining which energy carrier is consumed and what are the amount and the portion of it in the total consumption. However this data does not go into further detail in each sector such as how much energy is consumed for space heating and cooling, cloth washing and drying, etc. Swan et al. (2008) and Hamzaçebi (2006) investigated electricity consumption in their studies in terms of numeric analysis. The only study concerned modelling residential demand

Electricity energy consumption by sector (GWh)						Electricity energy consumption rates by sector			
Year	Industry	Residence	Agriculture	Transporation	Total	Industry	Residence	Agriculture	Transportation
1970	4690	2502	36	80	7308	64.18	34.24	0.49	1.09
1975	8745	4519	75	153	13,492	64.82	33.49	0.56	1.13
1980	13,008	7081	160	149	20,398	63.77	34.71	0.78	0.73
1985	19,608	9576	311	213	29,708	66.00	32.23	1.05	0.72
1990	29,212	16,688	575	345	46,820	62.39	35.64	1.23	0.74
1995	38,007	27,384	1513	490	67,394	56.40	40.63	2.25	0.73
2000	48,842	45,664	3070	720	98,296	49.69	46.46	3.12	0.73
2004	59,566	56,950	3895	731	12,1142	49.17	47.01	3.22	0.60

Fig. 1 Total electricity consumption on sectorial basis

Table 1 2012 Turkey energy balance table for residential sector (kTEP)

Hard coal	Lignite	Asphaltite	Wood	Bio. waste	Petroleum	Natural gas	Electric	Geo and other heat	Solar	Total
6662	3317	108	2195	211	804	8833	8084	1081	500	31,794

technologies conducted by Sulukan (2010). This study is intended to carry the modelling of the residential sector demand technologies phase further and update the data available. Table 1 states the energy balance table of Turkey for 2012 [7] (Table 2).

When the electricity energy consumption rates are assessed from the 1970s to 2000s in Fig. 1, it is clear that total share of the residential sector reached industrial sector level of about 60 k GWh, and predictions claim that it would be higher than the industrial sector in the 2020s [8–11]. The distribution of the electricity consumption inside the residential sector is done by applying the ratios given by the Ministry of Energy and Natural Resources [9].

While distributing the total natural gas consumption into end-use technologies, the following approach is carried out because of the lack of detailed data on this area; a study held by Aras et al. on the portion of space heating in total natural gas consumptions of the residential sector for Eskişehir, Turkey, is examined [10]. Since the data available is very limited, natural gas consumed in residents is assumed to be the same as Eskişehir in the rest of the Turkey, and for space heating its ratio is taken to be 90 %. The rest is divided equally into two: 5 % for cooking and 5 % for water heating demands.

Reference Energy System model for residential end-use demands and demand technologies is constructed with ANSWER-MARKAL. The end-use demands are determined to be space heating, space cooling, water heating, lighting, refrigeration, cooking, cloth washing, cloth drying, dishwashing and other miscellaneous electric energy uses. Also 23 demand technologies are determined and the RES model is constructed. The RES schematic is shown in the Appendix. The RES schematic represents the relations of the demand technologies with the primary energy carriers as input and residential end-use demands as output. Developing the RES schematic is the first step for the model construction in ANSWER software. The same mainframe structure is created in the software before any data entry. Afterwards, system parameters and values for the parameters should be entered for the model. Even though this study focuses on the residential demand technologies, in order to have meaningful results from the model, it is necessary to define inputs and outputs of the system. A portion of the model is represented in Fig. 2. To be able to run the constructed model, obligatory parameters of the software should be entered. Some of these parameters are capacity, lifetime of the technology, starting date, fix operating and maintenance costs and investment costs, energy carrier input and end-use demand output (Table 3).

3.1 Assumptions

When the required data is not sufficient to proceed, for the sake of easing the process, some assumptions are made:

Table 2 Electricity consumption in residents in 2012 (*source:* eie.gov.tr)

	Refrigeration	Washing machine	Dishwasher	Cloth dryer	Heater	TV	Lighting	Others	Total
Consumption (kTep)	242,520	72,756	32,336	24,252	72,756	56,588	97,008	210,184	808,400
Consumption (%)	0.30	0.09	0.04	0.03	0.09	0.07	0.12	0.26	1

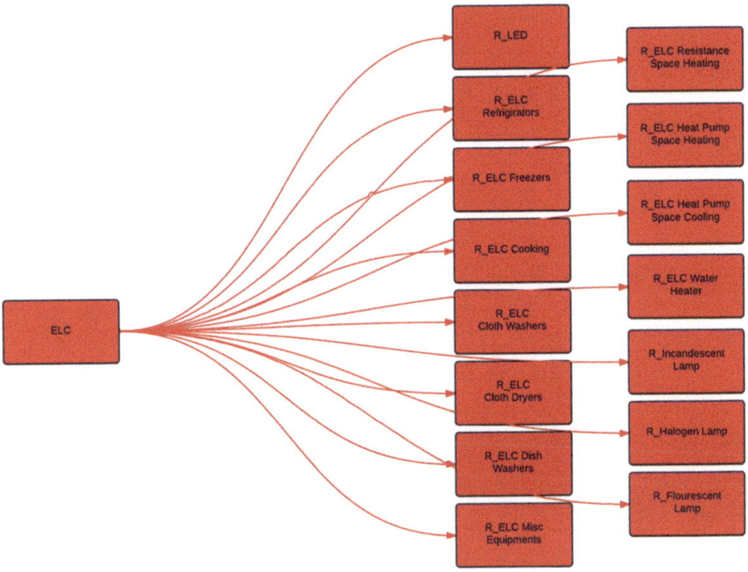

Fig. 2 Demand technologies use electricity as input

- The amount of natural gas used for water heating and cooking is the same and consists of 10 % of total consumption.
- The utilization of the heat pump technology is zero.
- The share of four lighting technologies is same.

4 Conclusion

In this paper, the modelling of the current situation of the residential sector demand technologies in Turkey is explored using ANSWER-MARKAL software. In the process of data entry and model construction, some assumptions are made. The model can get as complicated as the available data permits. Even so, the first step of building a base scenario for the demand technologies in the residential sector of Turkey is completed. The next step would be running the base scenario model and having BAU scenario results. After running the base scenario, the next task is constructing various alternative scenarios: one scenario with fully renewable energy supply, one with fossil fuel utilization setting to zero, etc. In addition to these, energy-saving and energy-efficiency-increasing activities should be able to import to the program as input.

The main purpose of this work is to have a projection of the next 50 years and help prospective alternative scenarios to be developed in time. The objective function is minimizing the total cost of the system; therefore, these results would be used to give technical support to decision-makers in this sector.

Table 3 Consumption and cost calculations for residential demand technologies [12–15}

Calculations per unit dwelling										
	NGA cooking		NGA water heating		NGA space heating		COA space heating		OSF space heating	
Consumption	0.0581	TEP/year	0.0581	TEP/year	0.3487	TEP/year	0.3506	TEP/year	0.3068	TEP/year
	2.4330	MJ/year	2.4330	MJ/year	14.5982	MJ/year	14.6802	MJ/year	12.8469	MJ/year
INV cost	328.8086	TL/MJ/year	657.6171	TL/MJ/year	109.6029	TL/MJ/year	34.0594	TL/MJ/year	58.3800	TL/MJ/year
FIXOM	123.3032	TL/MJ/year	205.5054	TL/MJ/year	34.2509	TL/MJ/year	6.8119	TL/MJ/year	7.7840	TL/MJ/year
	SOL water heating		ELC water heating		ELC resistance SH		ELC lightings		ELC refrigeration	
Consumption	0.0263	TEP/year	0.0128	TEP/year	0.0128	TEP/year	0.0511	TEP/year	0.1362	TEP/year
	1.1018	MJ/year	0.5344	MJ/year	0.5344	MJ/year	2.1376	MJ/year	2.1376	MJ/year
INV cost	1815.2288	TL/MJ/year	1122.7294	TL/MJ/year	2993.9449	TL/MJ/year	748.4862	TL/MJ/year	935.6078	TL/MJ/year
FIXOM	272.2843	TL/MJ/year	187.1216	TL/MJ/year	187.1216	TL/MJ/year	46.7804	TL/MJ/year	93.5608	TL/MJ/year
	ELC cloth washing		ELC cloth drying		ELC dishwashing		ELC TV and others		ELC cooking	
Consumption	0.0340	TEP/year	0.0128	TEP/year	0.0128	TEP/year	0.1404	TEP/year	0.0128	TEP/year
	2.1376	MJ/year	2.1376	MJ/year	2.1376	MJ/year	2.1376	MJ/year	0.5344	MJ/year
INV cost	467.8039	TL/MJ/year	7017.0584	TL/MJ/year	374.2431	TL/MJ/year	1169.5097	TL/MJ/year	1122.7294	TL/MJ/year
FIXOM	116.9510	TL/MJ/year	233.9019	TL/MJ/year	93.5608	TL/MJ/year	140.3412	TL/MJ/year	187.1216	TL/MJ/year

Appendix: RES for Turkish Residential Demand Technologies

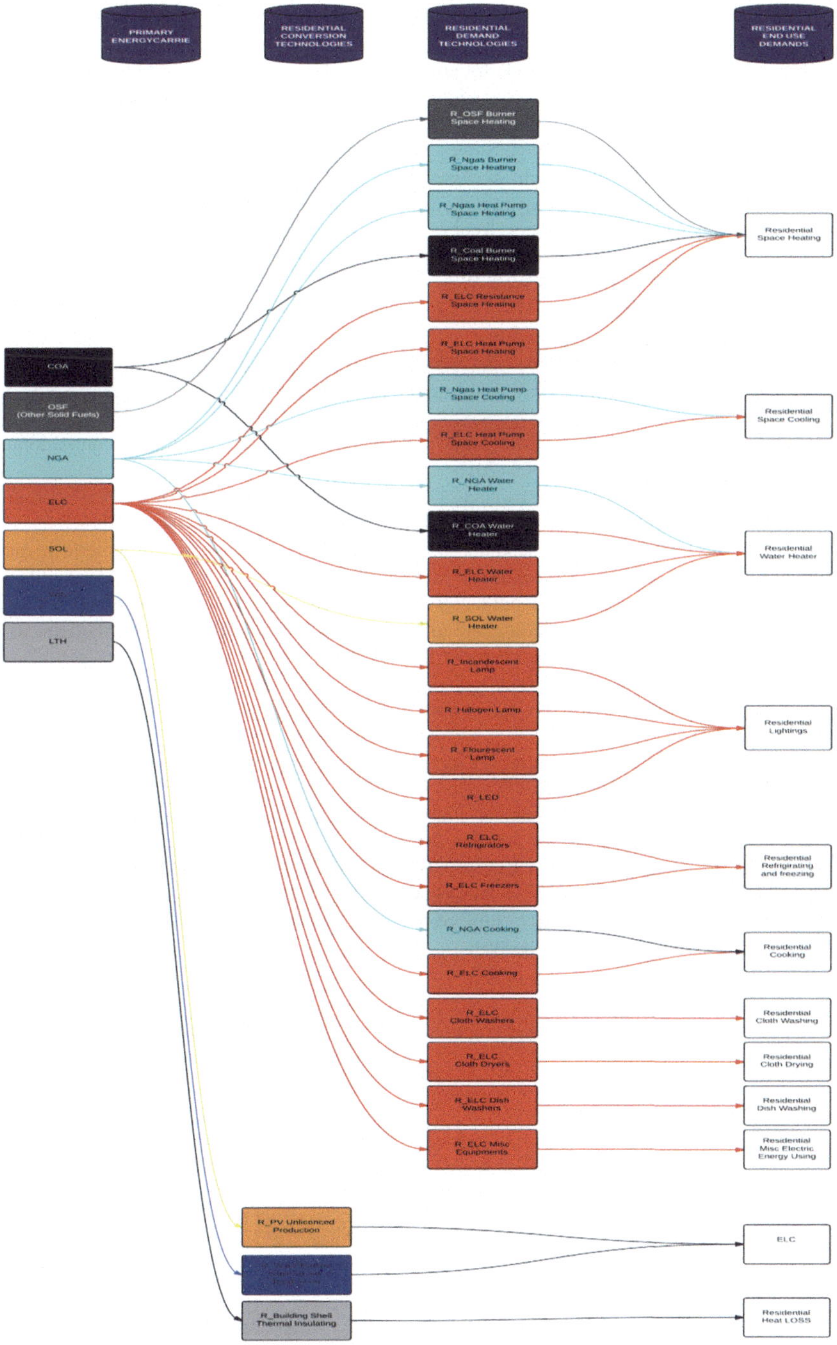

References

1. Global Environment Facility Evaluation Office (2010) GEF country portfolio evaluation: Turkey (1992–2009)
2. Sulukan E (2010) A preliminary study for post-Kyoto period for Turkey by MARKAL model. In: 5th international ege energy symposium and exhibition (IEESE-5), Denizli, 27–30 Jan 2010
3. Republic of Turkey Climate Change Strategy 2010–2020. http://iklim.cob.gov.tr/iklim/Files/Stratejiler/%C4%B0DES_ENG.pdf. Accessed 5 Jan 2014 from World Wide Web
4. Swan LG, Ugursal VI (2008) Modeling of end-use energy consumption in the residential sector: a review of modeling techniques. Renew Sustain Energy Rev 13(2009):1819–1835
5. Turkish Statistical Institute (2013) http://www.tuik.gov.tr/PreHaberBultenleri.do?id=15843. Accessed 5 Jan 2014 from World Wide Web
6. ISI YALITIMI PLANLAMA RAPORU (2010) http://www.izoder.org.tr/tr/dokumanlar/isi_yalitimi/izoder_isi_yalitim_raporu.pdf
7. 2012 Turkey Energy Balance Table (2013) Ministry of Energy and Natural Resources. http://www.enerji.gov.tr/EKLENTI_VIEW/index.php/raporlar/raporVeriGir/72222/2. Accessed 5 Jan 2014 from World Wide Web
8. Hamzaçebi C (2006) Forecasting of Turkey's net electricity energy consumption on sectoral bases. Energy Policy 35(2007):2009–2016
9. Ministry of Energy and Natural Resources—Renewable Energy Institute (n.d.) Yenilenebilir Enerji Genel Müdürlüğü. http://www.eie.gov.tr/verimlilik/b_en_ver_b_1.aspx
10. Aras H, Aras N (2014) Eskişehir'de konutsal doğal gaz talebine ekonomik göstergelerin ve diş ortam sicakliğinin etkileri. Mühendis ve Makina 46–540. http://www.mmo.org.tr/resimler/dosya_ekler/728efbda8169228_ek.pdf?dergi=70. Accessed 5 Apr 2014 from World Wide Web
11. Dilaver Z, Hunt LC (2010) Modelling and forecasting Turkish residential electricity demand. Surrey Energy Economics Discussion paper Series131
12. UCTEA the Chamber of Mechanical Engineers (2011) UCTEA the Chamber of Mechanical Engineers. http://www.mmo.org.tr/resimler/dosya_ekler/297644a5f701766_ek.pdf?dergi=1148
13. World Energy Council Turkish National Committee (2012) http://www.dektmk.org.tr/upresimler/TURKIYEENERJIVERILERI2012.pdf
14. UCTEA the Chamber of Mechanical Engineers (2014) http://www.mmo.org.tr/resimler/dosya_ekler/8a83eeb494a308f_ek.pdf?dergi=185
15. IDEP (2010) TÜRKİYE'NİN ULUSAL İKLİM DEĞİŞİKLİĞİ EYLEM PLANI'NIN GELİŞTİRİLMESİ PROJESİ BİNALAR SEKTÖRÜ MEVCUT DURUM DEĞERLENDİRMESİ RAPORU. http://iklim.cob.gov.tr/iklim/Files/IDEP/%C4%B0DEP_TR.pdf. Accessed from World Wide Web

Energy Management Performance in Country Scale: A Data Envelopment Analysis

Egemen Sulukan, Mumtaz Karatas, and Ilker Akgun

1 Introduction

The relationship between energy consumption structure, economic structure and energy intensity in a country is now arising as a current discussion. Most of the world has agreed with the common strategies to increase energy efficiency for a sustainable economy with more savings. Besides, the increasing share of energy efficiency to countries' economies at different levels has already changed this perspective to more efficient energy technology implementations with the economic and—lately—environmental concerns. Besides, the main challenge is the necessity of controlling the total demand levels and naturally total energy consumption. More efficient technologies won't help if the energy demand levels tend to rise with respect to increasing populations. Demand side management in the energy sector is a multidimensional issue with its primarily technical and social aspects.

Energy is used by every sector in a country's economy. The subsectors that form a national economy continuously replace the obsolete technologies with more efficient technological options under the effect of rising population and energy demand. In this historical loop, energy efficiency lowers the cost of production and naturally

E. Sulukan (✉)
Department of Naval Sciences, Turkish Naval Academy, Istanbul, Turkey
e-mail: egemensulukan@gmail.com

M. Karatas
Department of Industrial Engineering, Turkish Naval Academy, Istanbul, Turkey
e-mail: mkaratas@dho.edu.tr

I. Akgun
Faculty of Business Administration, Department of Production Management and Marketing, Marmara University, Istanbul, Turkey
e-mail: iakgun@marmara.edu.tr

T.S. Uyar (ed.), *Towards 100% Renewable Energy*, Springer Proceedings in Energy, DOI 10.1007/978-3-319-45659-1_18

increases total production with energy consumption as well. Simply, if you bought a new car with a better fuel efficiency than your current one, would you use it less or more [1]?

Energy efficiency improves when a given level of service is provided with reduced amounts of energy inputs or services are enhanced for a given amount of energy input, but energy intensity is measured by the quantity of energy required per unit output or activity, so that using less energy to produce a product reduces the intensity. For this reason, decreases in energy intensity brought about because of technological change rarely reduce total energy consumption [2].

In addition to the effects on the economy, energy production and consumption also contribute to climate change as a key player. Countries are taking the environmental aspects of energy usage into consideration as an anthropogenic activity for their long-term development plans, while some alternative clean development mechanisms allow a country with an emission-reduction or emission-limitation commitment under the Kyoto Protocol (Annex B Party) to implement an emission-reduction project in developing countries. These efforts are organized and still operational under the United Nations Framework Convention on Climate Change (UNFCCC) [3].

Turkey, as a member of the Organisation for Economic Co-operation and Development (OECD) and an old candidate for membership to the European Union, faces periodical challenges in its highly dynamic economy, partially influenced by the energy production and consumption in all sectors. Energy consumption, specifically in conventional combustion processes, is linked to environmental problems, as a result of emitted greenhouse gases, especially carbon dioxide (CO_2), with its increasing atmospheric concentration.

In this study, we applied a holistic methodology, data envelopment analysis (DEA), for evaluating the efficiency of ten selected countries in terms of energy consumption, carbon dioxide (CO_2) emissions and energy intensity. The organization of the paper is as follows. First we give an overview of DEA methodology in Sect. 2. DEA application and results are discussed in Sect. 3. Finally, Sect. 4 summarizes our conclusions.

2 Data Envelopment Analysis

DEA technique, regarded as a powerful management science tool, was introduced by Charnes et al. [4] in 1978. It has widely and successfully been employed to benchmark and evaluate the efficiency of similar entities, known as decision-making units (DMUs), in industries such as banking, health care, transportation and education [5]. The popularity of this non-parametric technique basically lies in its flexibility to integrate multiple inputs and outputs without any requirement about the functional form of a production function and a priori information on the importance of inputs and outputs [6, 7]. DEA considers processes as a black box and analyses the relationships between its inputs and outputs.

DEA offers a variety of models that use multiple inputs and outputs to compare the efficiency of two or more processes. The ratio model is a common method based on the definition of efficiency as the ratio of the weighted sum of outputs to the weighted sum of inputs. In particular, assume that there are data on A inputs and B outputs for N DMUs. For the ith DMU, these are represented by the vectors x_i and y_i, respectively. $(A \times N)$ input matrix X and the $(B \times N)$ output matrix Y represent the data for all DMUs. From that point, the efficiency of a certain DMU is measured by

$$\omega_i = \frac{u_i y_i}{v_i x_i} \tag{1}$$

where u_i and $v_i > 0$ are the weight vectors corresponding to the outputs and inputs of the ith DMU. DEA provides an evaluation of the relative efficiency of a DMU compared to the set of other DMUs. In other words, the efficient DMUs may not necessarily form a "production frontier" but rather lead to a "best-practice frontier" in a [0,1] scale which forms the basis for a benchmark analysis. As an empirical rule in [8], it is suggested that the number of DMUs be at least twice the number of inputs and outputs combined, whereas in [9] it is stated that the number of DMUs should be at least three times the number of inputs and outputs combined. Interested readers can find a detailed survey of DEA applications in [4].

3 DEA Application and Results

In this study, we selected ten countries, the USA, Canada, Brazil, Denmark, France, Germany, Italy, Turkey, the UK and China, as DMUs. As input parameter, we selected total primary energy consumption (quadrillion Btu), and as outputs we selected total carbon dioxide (CO_2) emissions from the consumption of energy (million metric tons) and energy intensity—total primary energy consumption per dollar of GDP (Btu per year 2005 US dollars (market exchange rates)). Primary energy is an energy form found in nature that has not been subjected to any transformation process. It can be non-renewable or renewable. CO_2 is the primary greenhouse gas emitted through human activities. While CO_2 emissions come from a variety of natural sources, human-related emissions are responsible for the increase that has occurred in the atmosphere since the Industrial Revolution [10]. Energy intensity is a measure of the energy efficiency of a nation's economy. It is calculated as units of energy per unit of GDP. High energy intensities indicate a high price or cost of converting energy into GDP, whereas low energy intensity indicates a lower price or cost of converting energy into GDP.

By using the open-source statistical data in the US Energy Information Administration website, http://www.eia.gov, we apply DEA technique to perform a benchmark analysis of the above-mentioned ten countries for time horizon 1995–2011. Assuming equal weights for all outputs for each year, we show the best-practice frontier by 100 % efficiency and the efficiency of other countries with respect to that efficient frontier. Figure 1 gives the main structure of our DEA model in terms of inputs, outputs and DMUs.

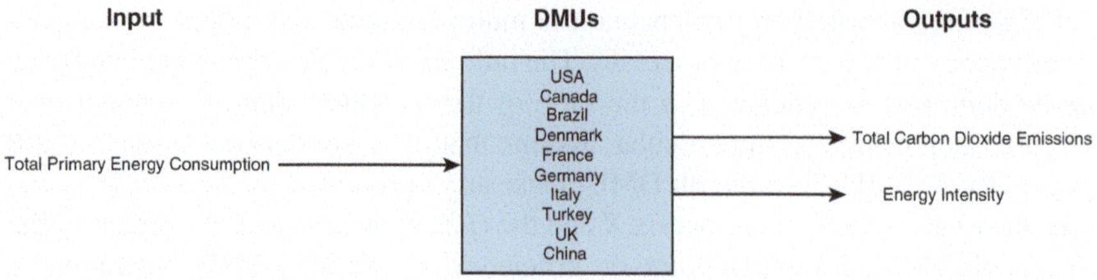

Fig. 1 DEA model structure

Table 1 provides DEA efficiency scores of countries for years between 1995 and 2011. The USA and France should be highlighted by their full relative DEA efficiencies, which can be evaluated as these two countries are based on a perfect equilibrium with respect to their energy input and CO_2 emissions and energy intensities. Turkey stands near the bottom of this ranking list, with its stable profile including the analysis parameters, namely, huge energy efficiency potential, which the country has a way ahead to cut its energy consumption, but with its relatively low levels of CO_2 emissions and total primary energy consumption per dollar of its GDP when compared with the other countries in the list.

Figure 2 illustrates the DEA efficiency results of countries for the time domain. This stable profile indicates the DEA efficiencies; with an average of 59 %, Turkey experienced midstream energy policies in this analysis period in terms of energy consumption, CO_2 emissions and energy intensity. This trend also may arise from the imbalance of economic development, facing economic crisis in the past periodically, and relatively low energy and environmental efficiencies.

Figure 3 illustrates the efficient frontiers and efficiency plots for the time domain. The relative leap for Denmark, China and Turkey indicates a significant negative change in terms of energy efficiencies, resulting in these countries to place as the last three ranked ones in the analysis scope. While Denmark experienced a decreasing performance in terms of both emission and energy intensity issues, Turkey and China just recorded a significant negative jump in energy intensities.

In general terms, every economic activity requires various amounts of energy, material and labour input. If the consumer chooses to conserve aforementioned amount of energy, then he/she shouldn't do something that requires energy. But this approach is not so realistic for developing economies with sustainability targets.

In these terms, energy intensity may sound as a more reasonable measure to indicate the energy efficiency on an entire economy, as it takes to produce one monetary unit of economic output by using one standard unit of energy. Therefore, progress in energy intensity can be achieved by technological and structural shifts in economies, by using more efficient production technologies or outsourcing the labour force with cheaper options. But this mechanism may also influence the total production in negative manner, resulting to less GDP levels in the end. Then the sector targeting sustainability with increasing productions may be headed to decrease the total energy consumption at the first stage.

Table 1 DEA efficiency scores of countries for years 1995–2011

	DEA efficiency																	
Years	1995 (%)	1996 (%)	1997 (%)	1998 (%)	1999 (%)	2000 (%)	2001 (%)	2002 (%)	2003 (%)	2004 (%)	2005 (%)	2006 (%)	2007 (%)	2008 (%)	2009 (%)	2010 (%)	2011 (%)	Average (%)
USA	100	100	100	100	100	100	100	100	100	100	100	100	100	100	100	100	100	100
France	100	100	100	100	100	100	100	100	100	100	100	100	100	100	100	100	100	100
Brazil	90	90	89	97	93	91	89	90	92	91	92	92	95	92	92	88	85	91
Canada	89	90	86	87	86	85	84	84	83	83	83	85	89	89	86	84	85	86
Germany	77	77	77	80	80	79	79	79	79	79	79	80	80	82	79	79	78	79
UK	72	73	74	76	76	75	75	76	76	75	74	74	75	74	73	72	72	74
Italy	69	70	71	72	71	70	71	70	69	70	70	69	71	71	70	70	69	70
Turkey	60	60	60	63	60	57	57	59	59	60	59	59	60	60	57	59	58	59
China	48	50	51	53	54	54	55	56	57	58	59	61	62	64	63	64	63	57
Denmark	47	50	47	59	58	59	58	59	54	55	59	56	58	59	58	64	58	56

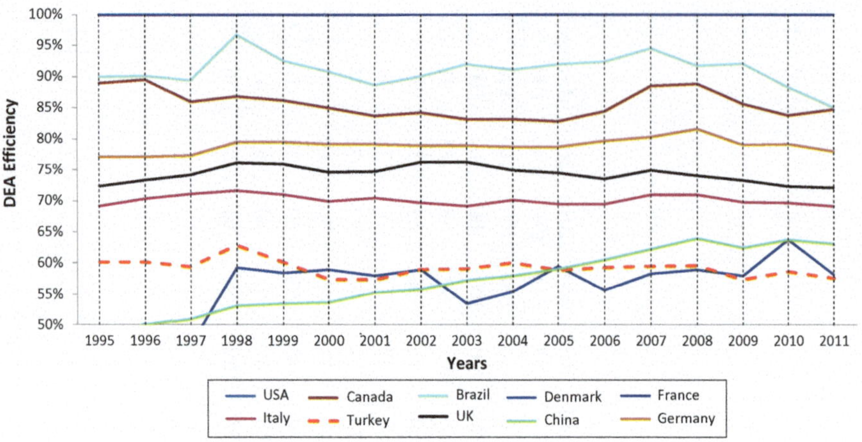

Fig. 2 DEA efficiency results of countries for years 1995–2011

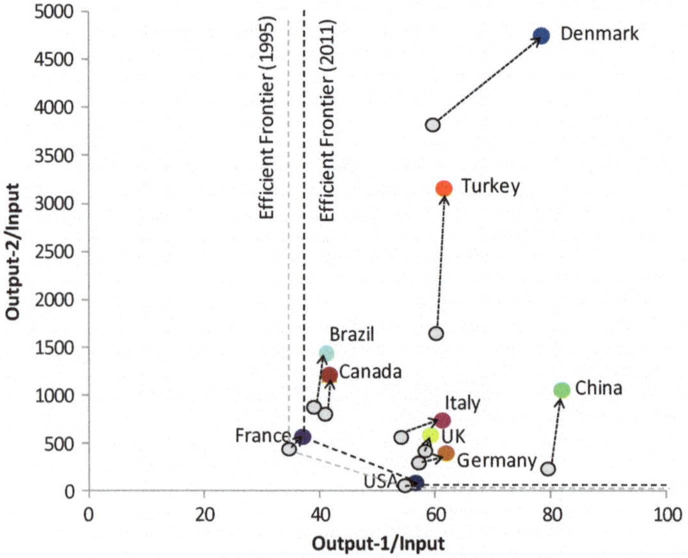

Fig. 3 Efficient frontiers and efficiency plots for year 1995 and 2011. (*Grey* and *coloured circles* represent the relative positions of DMUs in years 1995 and 2011, respectively)

Turkey, as a founding member of the International Energy Agency (IEA), a member of OECD and an "experienced" candidate for membership to European Union, faces a broad range of energy challenges. Ensuring sufficient energy supply to a growing economy remains the government's main energy policy concern. Turkey has also evaluated as it progressed significantly in all other areas of energy policy over the past few years, but that "improving energy efficiency is essential for responding to Turkey's energy policy challenges, and considerable potential remains in all sectors", was highlighted in the in-depth country review published in 2009 [11].

4 Conclusion

Economic development is a fundamental process of structural transformation including technological and environmental aspects. Current profile may be shifted to better positions by applying more efficient technology measures both improving technical efficiencies and CO_2 emission reductions and naturally increasing the economic standards, by higher GDP levels. However, effective energy and environmental protection policies issued and implemented by the Turkish government may have contributed a lot in the improvement of Turkish energy and environmental efficiency with an increasing GDP in the past decade.

Turkey may improve this profile by:

- Developing alternative paths for renewable energy system investments
- Increasing the deployment of renewable energy for heating/cooling and transportation to match the national targets
- Evaluating the options for efficiency improvement of thermal power plant expansion plans or technologies used in all sectors, which affects the energy generation, consumption and also greenhouse gases emission levels
- Analysing the possible effects of increasing the alternative potentials of hydraulic, wind, solar and wave energy resources to the national energy system
- Utilizing the cogeneration in all sectors, especially power generation and industrial subsectors
- Analysing the candidate nuclear power plants' effects in the energy system
- Analysing the CO_2 reduction options to project the total emissions starting from the regional scale to country level

References

1. Greer T (2013) The recovery that never was: energy, debt, and the U.S. GDP. The Scholar's Stage, 10 March 2013
2. US Department of Energy Website (2014) Energy intensity indicators: efficiency vs. intensity. http://www1.eere.energy.gov/analysis/eii_efficiency_intensity.html. Accessed May 2014
3. United Nations Framework Convention on Climate Change (UNFCCC) Website (2014) Clean Development Mechanism (CDM). https://unfccc.int/kyoto_protocol/mechanisms/clean_development_ mechanism/items/2718.php. Accessed May 2014
4. Charnes A, Cooper WW, Rhoes E (1978) Measuring the efficiency of decision making units. Eur J Oper Res 2(6):429–444

5. Liu JS, Lu LYY, Lu W-M, Lin BJY (2013) A survey of DEA applications. Omega 41(5): 893–902
6. Charles V, Kumar M (2014) Satisficing data envelopment analysis: an application to SERVQUAL efficiency. Measurement 51:71–80
7. Lee H, Kim C (2014) Benchmarking of service quality with data envelopment analysis. Expert Syst Appl 41(8):3761–3768
8. Golany B, Roll Y (1989) An application procedure for DEA. Omega 17(3):237–250
9. Banker RD, Charnes A, Cooper WW, Swarts J, Thomas D (1989) An introduction to data envelopment analysis with some of its models and their uses. Res Gov Nonprofit Account 5:125–163
10. NRC (2010) Advancing the science of climate change. National Research Council, The National Academies Press, Washington, DC
11. International Energy Agency (2010) Energy policies of IEA countries, Turkey 2009 review

An Alternative Carbon Dioxide Emission Estimation for Turkey

Mustafa Sağlam, Egemen Sulukan, and Tanay Sıdkı Uyar

1 Introduction

Anthropogenic CO_2 is one of the greenhouse gas (GHG) emissions which is mostly realized by the combustion of the carbonaceous fuels. A small amount of it comes from the nonfuel use of energy inputs, electricity generation using municipal solid waste, geothermal energy, and cement production. CO_2 is the most dominant component of the GHGs causing climate change [1].

It is clear that Turkey will be involved in environmental discussions hereafter. Turkey needs an environmental policy initiative to achieve energy-related emission control target, and she needs to have her data to control her part of carbon dioxide in the climate change negotiations. A tool using the overall Turkish energy system with Turkish energy data can only provide the right control mechanism to the decision makers [2].

The same tool can provide to establish scenarios to see the results of necessary legislative arrangements for emission reduction by using some numerical values representing developed combustion techniques and high-efficiency operation ensuring less fuel consumption in the production of energy [2, 3].

Scenarios include Tier 1 national consumption-based CO_2 emission estimations directly from Inventories for primary energy carriers. The model estimated additionally CO_2 amounts from the fuels used by technologies in sectors and it produced carbon intensities (grams of carbon dioxide released per megajoule of energy produced) for many technologies that resulted in the reference energy system.

M. Sağlam (✉) • T.S. Uyar
Department of Mechanical Engineering, Marmara University, Istanbul, Turkey
e-mail: fulbars@yahoo.com; tanayuyar@marmara.edu.tr

E. Sulukan
Department of Naval Sciences, Turkish Naval Academy, Istanbul, Turkey
e-mail: egemensulukan@gmail.com

T.S. Uyar (ed.), *Towards 100% Renewable Energy*, Springer Proceedings in Energy, DOI 10.1007/978-3-319-45659-1_19

2 Methodology

The CO_2 emission rate methodology based on the calculation method based on fuel use described in the IPPC guidelines is used in the estimations [4].

The fuel analysis approach consists of calculations based on summation of results of multiplications of the amount of fuel consumed and the fuel characteristic data including heat content, carbon content oxidation factor, the molecular weight of carbon dioxide (44)/the molecular weight of carbon (12), and conversion factor from kilograms to tons (1/907.2) in residential, industrial, transportation, and agricultural sectors in the base scenarios.

The generalized approach that benefits from emission factors and activity data is used in the calculation method based on fuel use by multiplying the amount of fuel type combusted on mass/volume/energy basis (PJ or petajoule) with default fuel-specific emission factors (Kton/PJ or kilotons/petajoule) and sums the results (Kton/year) for all type of fuels. Here in the model, GHG emissions are estimated from the entire resources including exports and imports, assuming as if consumed being combusted as solid, gas, or liquid fuels, and forestry impact on the system is evaluated in the analysis in the nonbase scenarios [4].

Fuel analysis and generalized approaches are compared with an example of estimating CO_2 emission from sample a fuel consumption in a technology.

It is possible to use different methodologies in the MARKAL model providing the total amount of emission generated in each period from all sources (resources and technologies). But a comparison of the methodologies is important from the point of the accuracy of the CO_2 emission estimation.

3 Base Scenario CO_2 Emission Results and Projection

The Turkish energy and emission data set is used in base scenario (business as usual or *BAU*) including various parameters and some assumptions of the Turkish MARKAL energy system to develop policies on how to decrease CO_2 emission and how to explain the relationship among the economic growth of Turkey, energy consumption, and GHG emissions by comparing with other scenarios. Thus we set up and run various scenarios with the base scenario and make detailed analyses for economic composition change and change of carbon intensity and energy intensity by using new energy sources in the energy mix in sectoral basis. We estimate the projections in the periods from 2005 up to 2025 in the scenarios.

In the base scenario, we used the 2000 and 2005 data compiled by the Ministry of Energy and Natural Resources (MENR), The Turkish National Committee of the World Energy Council (WEC-TNC), and General Directorate of Energy Affairs (EIGM) in the MARKAL (market allocation) optimization model. *Carbon dioxide coefficients* had been obtained from the calibration of two pairs of tables in our studies (Table 1) [2].

Table 1 CO_2 emission coefficients/activity (ENV_ACT) in the base scenario (Kton/PJ)

Technology	2000	2005
Hard coal power plant	193.24	243.71
Geothermal electricity	45.32	45.64
Lignite power plant	220.82	278.42
Natural gas power plant	178.5	130.88
Oil product power plant	362.72	360.64
Waste incineration plant	552.58	698.33

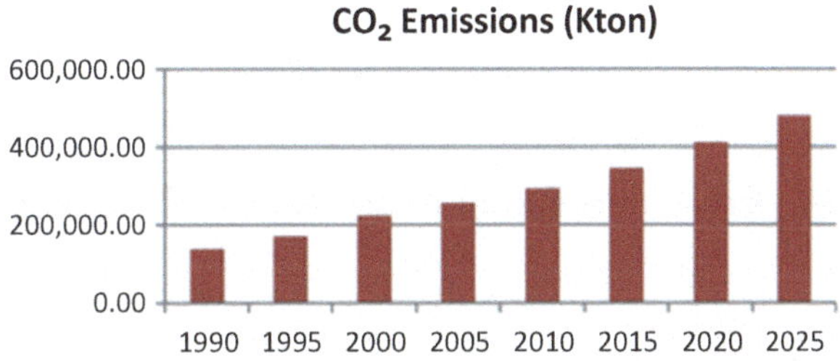

Fig. 1 CO_2 emission inventory projections between 1990 and 2025

The model gives CO_2 emission projection of 5 years period where the emission result of the base scenario is added to the TUIK data in Fig. 1. Herein, especially the projection of the demands, domestic reserves of fossil fuels and the bound capacities have been assumed proportional to the economic growth of Turkey in each sector of 3.3 % as forecasted by the World Bank [6].

Long-term CO_2 emission forecasts from energy-related usage are escalating continuously from 1990 up to 2025 with a rate of 183.49 %, from 139.59 Kton to 479.95 Kton [5, 7, 8].

The realized part of the projection of energy-related CO_2 emission gives hints about the economic growth rate of a country. We suggested 293,979.36 Kton of CO_2 emissions in 2010. Comparing the TUIK data 299.11 Kton CO_2 in 2009 with this study made in 2005, for example, one may get an impression on the economic growth of Turkey in energy demand between 2005 and 2010 which is not much more than 3.3 %.

Ratios of CO_2 to electricity production are found for Turkish power plants. The increase in emission is 153.21 % from 2000 to 2025. The CO_2 emission rate increase is less than the emission sourced from other activities in the country (Fig. 2).

What we also reached in the base scenario after calibration is a list of CO_2 emission coefficients for fuels used in each activity in Table 1. The difference between the values 2000 and 2005 comes from the data officially declared to the international institutions.

CO_2 emission rates from the consumption in sectors, such as agriculture, residential, transportation, industry, forestry, and electricity production, are estimated in the base scenario. Electricity production is the most important sector among them with the rate of 40 % in 2025 (Fig. 3).

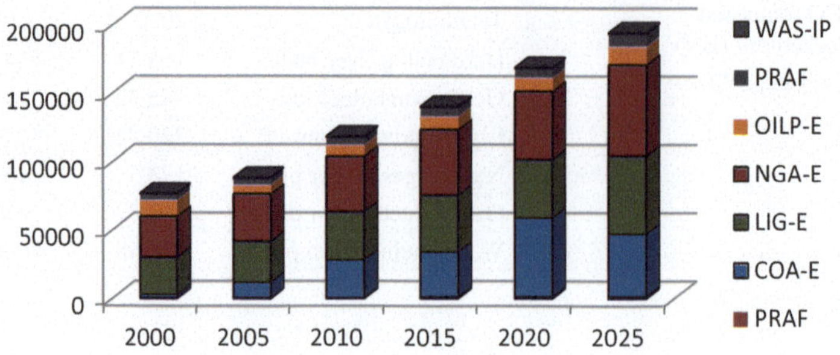

Fig. 2 CO_2 emitted from electricity production (Kton)

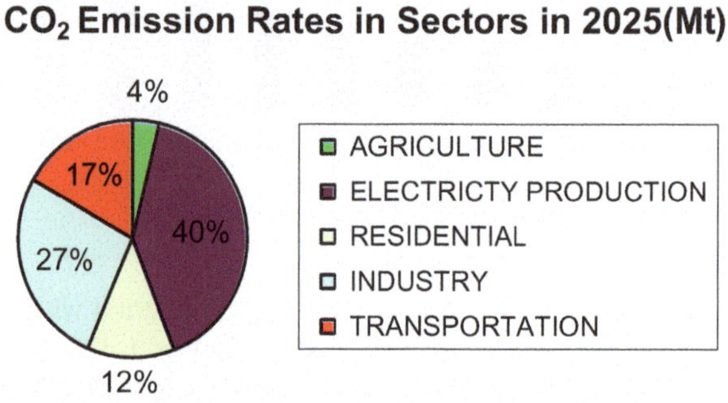

Fig. 3 CO_2 emission rate forecasts in 2025

In the base scenario, the emission coefficient/activity (ENV_ACT) parameter is applied to the output variable of the process or conversion technologies for activity-related emissions.

The values in Table 1 include any efficiency loss caused by the type of technology in which commodity is used.

The model has provided emission intensity for each technology. (Table 2) That is the volume of any emission coming from a given source to compare the environmental impact of different fuels.

4 Carbon Dioxide Emitted from Resource Scenario (GEMIS and TCR)

It is possible to estimate the amount of the CO_2 by using default carbon factors calculated from the combustion of the resources at the beginning in case there is no detailed data related to process or conversion or combustion in plants or other end-use technologies such as trucks, steam boilers, stoves, etc. (generalized approach). Here

Table 2 Emission intensity for each technology (gr/Mj)

Technology	2000	2005
Hard coal power plant	90.84	114.54
Lignite power plant	99.37	125.29
Natural gas power plant	80.32	58.90
Oil product power plant	145.08	144.25
Waste incineration plant	82.83	104.68

Table 3 CO_2 coefficients (ENV_SEP) (Kton/PJ) used in GEMIS and TCR scenarios

Technology	TCR	GEMIS
Animal biomass waste	83.1568	262.5253
Briquette	98.33	98.33
Coal	98.22	92.36
Coke	107.74	92.54
Lignite	91.4028	91.25
Natural gas	50.57	51
Crude oil	70.654	66.86
Oil products	67.39	69.92
Petro coke	96.8	74.6
Tree	89.55	102.94
Asphaltite	88.59	93.35
Waste	83.1568	103.017

in this study, emission coefficient/activity (ENV_ACT) has been set to zero in the Scenarios, while emission coefficient/resource activity (ENV_SEP) is being used as a carbon dioxide emission factor. This a more accurate result than what my previous studies has put forward [2].

4.1 *Carbon Emitted from Resource Scenario (GEMIS and TCR) Results*

Carbon coefficients in the BAU scenario had been calibrated in the model from the consumption given in the sheet of balances and the inventories of emissions from each activity in 2000 and in 2005 (Table 1). The values of CO_2 coefficients obtained for each technology for 2005 had been used for the following periods (2010–2025).

In GEMIS and TCR scenarios, CO_2 emission is estimated from the resource technologies, that is, before combustion in the conversion or end-use technologies. All GHG emission factors for all sectors and technologies come from the Intergovernmental Global Emission Model of Integrated Systems (GEMIS) and The Climate Registry (TCR) General Reporting Protocol (Table 3).

Table 4 Comparison of CO_2 emissions in the GEMIS, TCR, and BAU scenarios (Kton)

Case	2000	2005	2010	2015	2020	2025
Base	207,236.17	236,841.69	293,979.36	346,746.85	411,642.87	479,954.04
GEMIS	223,368.72	246,621.61	301,476.34	355,197.95	446,768.16	499,362.81
Change	7.8 %	4.1 %	2.6 %	2.4 %	8.5 %	4.0 %
Base	207,236.17	236,841.69	293,979.36	346,746.85	411,642.87	479,954.04
TCR	220,206.92	246,696.23	303,384.26	357,445.24	449,354.78	501,780.03
Change	6.3 %	4.2 %	3.2 %	3.1 %	9.2 %	4.5 %

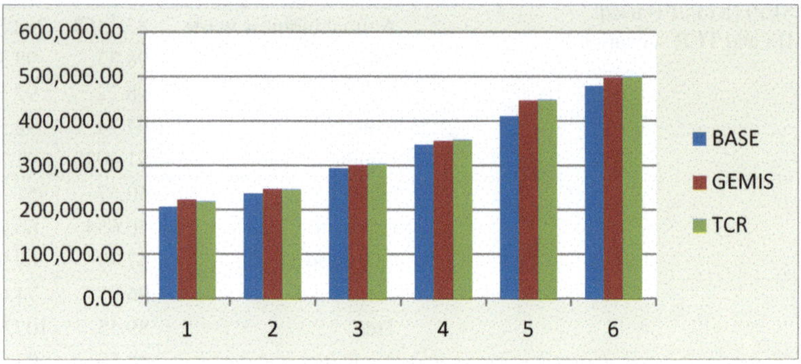

Fig. 4 CO_2 emissions in the GEMIS, TCR, and BAU scenarios (Kton)

Table 5 CO_2 emission estimation differences in BAU and fuel analysis approach scenarios (Kton) (2005)

	BAU	TCR (fuel analysis approach)	Change (%)
OIL	85,477.14	78,809.73	−7.8
LIGNITE	45,290.07	34,135.53	−24.6
HARD COAL	38129.66	51879.41	36.1

In GEMIS and TCR scenarios, as it is seen in Table 4, the results are meanly 4.9 % and 5.1 % more than the results of BAU, respectively, in the overall periods (Fig. 4).

When Fuel Analysis Approach is used for sample commodities extracted and imported in the RES system, different results come out (Table 5).

5 Conclusion

Activity data of the technologies is the most important part of reliable statistics some of which based on assumption or estimation.

Although it requires arduous work to develop emission coefficients for fuel combustion technologies employed in Turkey, they should be calculated for each of the fuels used in the Turkish sectors since they must be reliable both in tracking the actual carbon dioxide emission and in constituting the emission inventories based on the measurement and scientific methods.

The Turkish energy model should be detailed with the fuels and current or probable future technologies. The actual rational growing rates of the sectors should be used instead of the whole economic growth rate used in Turkish Energy System MARKAL Model. The successful implement of aforementioned will permit us to obtain healthier results in conjectural demand guess. Thus the results of the scenarios approach the higher level rightness.

References

1. Hofmann DJ, Butler JH, Dlugokencky EJ, Elkins JW, Masarie K, Montzka SA, Tans P (2006) The role of carbon dioxide in climate forcing from 1979 to 2004: introduction of the Annual Greenhouse Gas Index. Tellus Ser B 58:614–619
2. Sağlam M (2010) Establishing mitigation strategies for energy related emissions for Turkey using the MARKAL family of models. Ph.D. Thesis 2010, Marmara University, Istanbul
3. Sulukan E (2010) Establishing energy efficient utilization and cost-effective energy technologies selection strategies for Turkey using MARKAL family of models. Ph.D. Thesis 2010, Marmara University, Istanbul
4. Greenhouse Gas Emission Guidelines Stationary Combustion Sources April 2009 (The Climate Registry (TCR) General Reporting Protocol, Version 1.0, March 2008 http://www.theclimateregistry.org/downloads/GRP.pdf World Resources Institute (WRI) GHG Protocol – Stationary Combustion Guidance, Version 3.0, July 2005 http://www.ghgprotocol.org/calculation-tools/all-tools USEPA Climate Leaders GHG Inventory Protocol – Direct Emissions from Stationary Combustion Sources, November 2007 http://www.epa.gov/climateleaders/documents/resources/stationarycombustionguidance.pdf)
5. Energy Balance Reports (2009) http://www.dektmk.org.tr/incele.php?id=MTAw
6. Global Economic Prospects 2010, the International Bank for Reconstruction and Development / the World Bank, (2010)
7. Sectoral Report for Energy Turkey Inventory 2000 and 2005 Submission 2009 v1.1
8. http://www.tuik.gov.tr/Gosterge.do?id=3662&metod=IlgiliGosterge

Variability Analysis of Wind and Wind Power in Turkey

Zehra Yumurtacı, A. Yasin Demirhan, and Yüksel Malkoç

1 Introduction

Growth in the potential development of wind power facilities in Turkey has resulted in a need for understanding their impact on reliability of the Turkish interconnected electrical system. As the share of wind energy increases in electricity systems, negative effects of the fluctuations in wind energy on grid become important. These fluctuations can be observed at all timescales in seconds, minutes, hours, months, seasons, and years. When wind energy is added to a utility system, its natural variability and uncertainty are combined with the natural variability and uncertainty of loads. This increases the need for flexible resources such as hydro, gas-fired power plants, or dispatchable loads to maintain utility system balance and reliability across several different timescales. The demand for this flexibility increases with the amount of wind in the system [1]. The predictability of wind energy fluctuations is highly important for electrical systems integration and in terms of utilization of the highest degree of wind energy.

According to Turkey's Energy Markets and Supply Security Document, installed capacity of wind power plants will reach the level of 20.000 MW by 2023. With targeted measures for improving usage of renewable energy resources, the share of natural gas in electricity production is to be reduced below 30 %. This value of wind penetration corresponds to 25 % of peak power demand of the grid by year 2023.

Z. Yumurtacı (✉)
Makine Mühendisliği Bölümü, Yildiz Teknik Üniversitesi, Beşiktaş, 34349 İstanbul, Türkiye
e-mail: zyumur@yildiz.edu.tr

A.Y. Demirhan
Makine Mühendisliği Bölümü, Doğuş Üniversitesi, Acıbadem, 34722 İstanbul, Türkiye
e-mail: ademirhan@dogus.edu.tr

Y. Malkoç
Enerji Kaynakları Dairesi, Elektrik İşleri Etüd İdaresi, Çankaya, 06520 Ankara, Türkiye
e-mail: ymalkoc@eie.gov.tr

T.S. Uyar (ed.), *Towards 100% Renewable Energy*, Springer Proceedings in Energy, DOI 10.1007/978-3-319-45659-1_20

2 Turkey Wind Energy Potential

The gross wind energy potential of Turkey is 400 billion kWh/year, of which 120 billion kWh is a technically feasible potential, in case of utilizing regions in which average wind speeds in excess of 7.0 m/s, at least 48.000 MW of technical power potential, have been identified. 10.013 MW of this technical potential comes from offshore areas [2] (Table 1).

Wind power plants can be installed on land as well as at sea. All installed wind power plants in our country are located at terrestrial areas, but there are also some offshore license applications that exist. Turkey however has an important offshore wind energy potential in terms of power plants that can be installed on the sea. Turkey is surrounded by seas on its three sides. Excluding the Sea of Marmara, the length of coastline is 8.210 km. Aegean, Mediterranean, and Black Sea coasts are great wind energy potential. Especially, the Aegean coast, Black Sea coasts of Sinop Province, and Mediterranean coasts in the province of Hatay surround high energy density areas.

According to Electrical Power Resources Survey and Development Administration (EIEI), Balıkesir has the greatest wind energy capacity with average wind speed in excess of 6.8 m/s. Balıkesir allows for wind power installation of capacity 13.827 MW with 13.021 and 11.854 MW technical power potential; Çanakkale and Izmir follow Balıkesir, respectively.

Although the energy potential of the Marmara and Aegean region appears to be prominent, it is also possible to install wind power plants in other areas; one such a region is Samsun with capacity of 5.222 MW. Tokat follows Samsun with capacity of 2.002 MW; on the Mediterranean side, Içel and Hatay have, respectively, 3.531 MW and 3.441 MW technical power potential [3].

3 Current Status of Wind Power Plants

In Turkey, electricity production through wind energy was first realized at Cesme Altınyunus Facilities in 1986. Wind turbines with a nominal power of 55 kW were used [4].

Since the first wind-based power plant started operation in Izmir-Cesme region in 1986, the total installed capacity reached 1332 MW at the end of 2010 and 1406 MW by May 2011 with a total of 41 operating wind power plants. Nineteen

Table 1 Turkey wind energy potential (>7 m/s–50 m)

Average wind speed: 50 m (m/s)	Energy density: 50 m (W/m^2)	Power potential (MW)
7.0–7.5	400–500	29.259
7.5–8.0	500–600	12.994
8.0–9.0	600–800	5.400
>9.0	>800	196
	Total	47.849

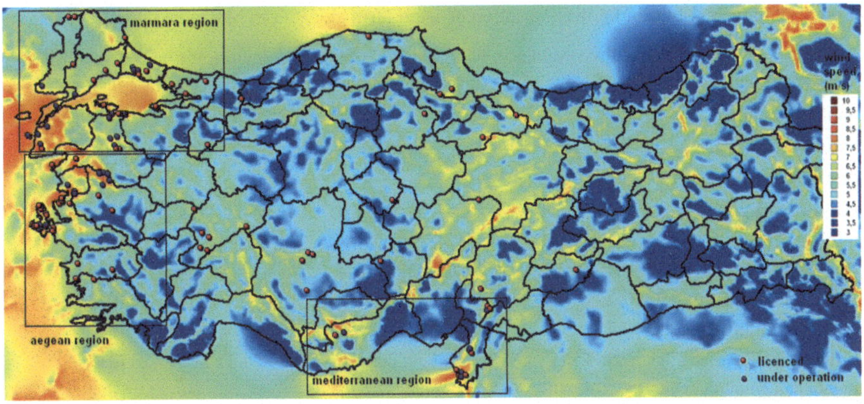

Fig. 1 Location of wind power farms on Turkey energy potential map

power plants with a total capacity of 750 MW are under construction [5]. The installed capacities of the wind power plants vary from 0.85 to 135 MW. Wind power plants represent 3 % of the total installed power capacity of Turkey. By the end of 2011 the number of EMRA licensed projects was 131 with a total capacity of 4991. Eighty-four of 131 licensed projects which were given a capacity of 2838 MW are located in the Marmara and Aegean regions. One thousand two hundred and eleven megawatts of this total capacity is in operation [5, 6].

4 Wind Energy Production Variability Analyses

In this study, wind power farms are considered as installed in three geographically different regions. The variability of wind power generation hourly bases is identified in each region. Using the wind power data for the 33 existing power farms that adds up to the total capacity of 1332 MW was analyzed for the first 10 days of January 2011. Location of the power farms that are examined is shown in Fig. 1. The raw data is in the forms of power time series in 1-h resolution. Installed capacity, number of turbines, capacity factors associated with examined period, and standard deviation of hourly energy production of each farms are given in Tables 2, 3, and 4.

4.1 Wind Power Fluctuations

The first step in the analysis was to determine the variability of the power output for the existing wind power farms. Fluctuations are defined as the difference in average power outputs between consecutive time intervals. As shown in Fig. 2, wind power behavior is presented in terms of ratio of the power output to the nameplate capacity of wind farms.

Table 2 Characteristics and power variation of wind farms located in Marmara region

Wind farm name	Installed capacity (MW)	Number of turbine	Standard deviation of power fluctuations (%)	Capacity factor
Şamlı wind farm	90	30	6	0.17
Sares wind farm	22.5	9	5	0.12
Ayyıldız wind farm	15	5	8	0.26
Bandırma-3 wind farm	25	10	6	0.20
Bandırma wind farm	30	20	9	0.34
Çamseki wind farm	20.8	20+1	8	0.25
Intepe wind farm	30.4	38	6	0.11
Keltepe wind farm	20.70	23	9	0.16
Boreas wind farm	15	6	7	0.34
Burgaz wind farm	14.90	13+5	7	0.20
Çatalca wind farm	60	20	9	0.24
Bandırma wind farm	60	20	6	0.24
Sarıkaya wind farm	28.8	14+1	11	0.26
Lodos wind farm	24	12	6	0.21
Overall	**457.1**	**247**	**3**	**0.22**

Table 3 Characteristics and power variation of wind farms located in Mediterranean region

Wind farm name	Installed capacity (MW)	Number of turbine	Standard deviation of power fluctuations (%)	Capacity factor
Serenoba wind farm	30	15	8	0.47
Belen wind farm	36	12	7	0.25
Mersin wind farm	33	11	8	0.41
Gökçedağ wind farm	135	54	5	0.20
Şenbük wind farm	15	5	9	0.11
Ziyaret wind farm	35	14	6	0.34
Overall	**284**	**111**	**3**	**27 %**

All over Turkey, during the time period under consideration, the lowest wind energy generation occurred on January 1 at 5:00 p.m. (36 MW) and the highest generation observed on January 4 at 5:00 p.m. (860 MW). On 4 January between 00:00 and 06:00 power ramp with a maximum value 400 MW. This increase corresponds to 30 % of the total installed wind power capacity.

With target at 20.000 MW wind generation capacity, 30 % of power fluctuation corresponds to 6000 MW power. Such a power fluctuation will be serious threat for stable grid operation.

Table 4 Characteristics and power variation of wind farms located in Aegean region

Wind farm name	Installed capacity (MW)	Number of turbine	Standard deviation of power fluctuations (%)	Capacity factor
Akbük wind farm	31.5	15	14	0.58
Aliağa wind farm	90	36	8	0.29
Düzova wind farm	30	12	11	0.30
Datça wind farm	29.60	36	12	0.40
Kocadağ-2 wind farm	15	6	12	0.21
Kuyucuk wind farm	25.6	14	9	0.13
Mazı-1 wind farm	39.20	49	9	0.23
Mazı-3 wind farm	30	12	9	0.27
Sayalar wind farm	34.2	38	6	0.15
Soma wind farm	90	36	13	0.15
Soma-I wind farm	88.2	98	6	0.18
Soma-II wind farm	29.2	32	10	0.23
Turguttepe wind farm	22	11	11	0.39
Yuntdağı wind farm	42.5	17	15	0.31
Overall	**587**	**412**	**3**	**0.25**

4.2 *Wind Power Fluctuation Frequency Distributions*

When the frequency distribution of all the installed farms' output power variation in the first 10 days of January is examined, total hourly output of 33 wind farms varied less than 1 % (change in average power as a fraction of installed capacity) with 20 % probability, and maximum hourly output power variation has been 6 % of the total installed capacity. Standard deviation of all installed wind power plants hourly average power fluctuation was realized 2 % (Fig. 3).

The same analysis was done for the province of Osmaniye where Turkey's largest WPF is installed. Installed capacity is 135 MW. The hourly average wind power change of this farm reached the level of 20 % of installed capacity. This wind farm hourly average output power changes reach 20 % of installed capacity. If the same study has been done for a small wind farm that consists of a couple of wind turbines, power variation could be reached up to 100 %.

5 Conclusion

An installation of wind power farms in different wind regimes will reduce power fluctuations significantly. In the case of distributing wind farms to areas with the different wind characteristics, wind farms will back up each other's unbalanced energy production. This will result in smoother power generation.

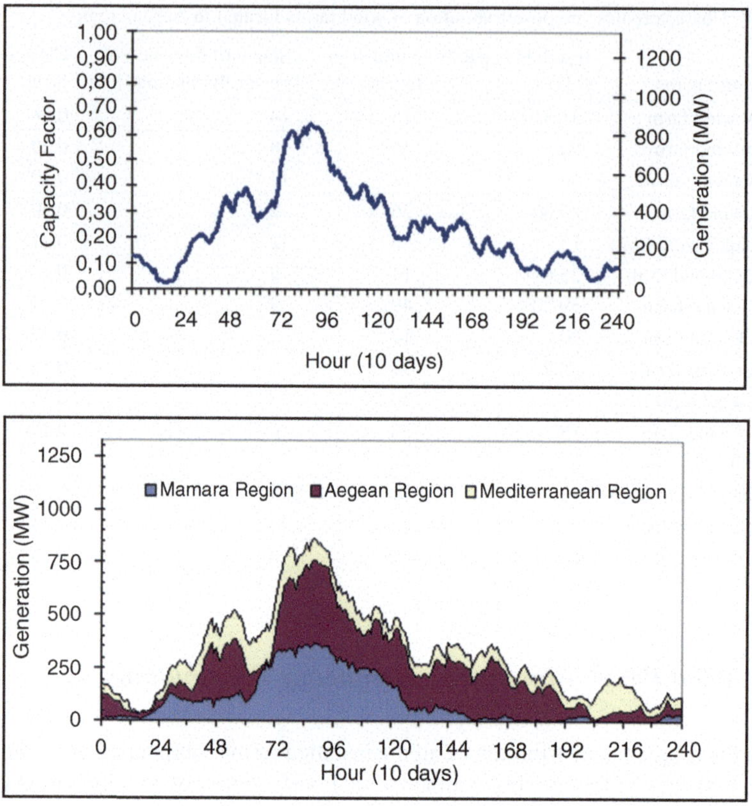

Fig. 2 Hourly variations of 33 wind power farms with 1332 MW capacity, for the first 10 days of January, and the distribution of power generation over the regions

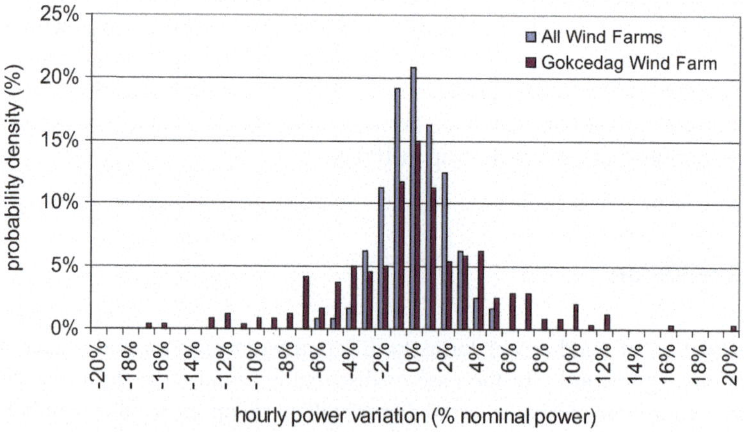

Fig. 3 Frequency distribution of hourly output power variation

This study was done with short period of real generation data for illustrating variability of wind as an energy resource. Important issue is set up and manages the energy system with variable energy source. For this reason, to improve percentage of wind energy on energy system, power plant locations should be based on wind regime analyses.

For reaching higher wind penetration rate, licensing should be done on a limited base. Licenses should be issued only for regions which are a priori identified. This type of legislation which is considering wind regimes and wind power variations has positive effects on sustainable electrical system operation conditions, and also with the decrease of reserve power demand, consequently wind integration cost shall be declined significantly.

References

1. Karier T, Stephan J (2007) Northwest wind integration action plan. http://www.nwcouncil.org/energy/wind/library/2007-1.htm. Accessed 10 May 2011 from the World Wide Web
2. Ayaz F, Çalışkan M, ve Malkoç Y (2006) Turkish wind energy potential, legislation and investment possibilities. In: Hamburg wind energy fair and conference, 21–25 Sept 2006
3. Wind Energy Departments of the Electrical Power Resources Survey and Development Administration (EIEI) (2011) http://www.eie.gov.tr/duyurular/YEK/YEKrepa/REPA-duyuru_01.html. Accessed 10 May 2011 from the World Wide Web
4. Hepbaslı A, Ozgener O (2003) A review on the development of wind energy in Turkey. Renew Sustain Energy Rev 2004(8):257–276
5. Turkish Wind Energy Association (TUREB) (2011) http://www.ruzgarenerjisibirligi.org.tr/index.php?option=com_docman&Itemid=86. Accessed 10 May 2011 from the World Wide Web
6. Energy Market Regulator Authority (EPDK) (2011) http://www2.epdk.org.tr/lisans/elektrik/lisansdatabase/verilentesistipi.asp. Accessed 10 May 2011 from the World Wide Web

Models of Solar Deployment: Decentralised Versus Centralised Generation

Andrea Bodenhagen, Engin Yaman, Julia Kusay, and Egemen Seymen

Nomenclature

kWh	kilowatt hours
kW	kilowatt peak
LCOE	levelised cost of electricity
MW	megawatt peak
PV	photovoltaic
TWh	terawatt hours

1 Introduction

The Turkish economy and with it the demand for and consumption of electricity are growing tremendously. While the economy grew 8.9 % in 2010 and 8.5 % in 2011, the energy consumption is expected to double over the next 10 years. The IEA predicts that Turkey will see the fastest medium- to long-term growth in energy demand

A. Bodenhagen (✉) • J. Kusay
EMEA Marketing/CSUN—China Sunergy Europe GmbH,
Schillerstr. 42-44, 60313 Frankfurt, Germany
e-mail: andrea.bodenhagen@gmx.de

E. Yaman
Central Europe & Turkey Sales/CSUN—China Sunergy Europe GmbH,
Schillerstr. 42-44, 60313 Frankfurt, Germany
e-mail: engin.yaman@chinasunergy.com

E. Seymen
Business Development/CSUN Eurasia Energy Systems Industry & Trade Inc.,
Istanbul Industrial & Free Trade Zone, Akif Kopuz St. No.: T-10 B Block, Tuzla, 34957
Istanbul, Turkey
e-mail: egemen@ceeg.cn

T.S. Uyar (ed.), *Towards 100% Renewable Energy*, Springer Proceedings
in Energy, DOI 10.1007/978-3-319-45659-1_21

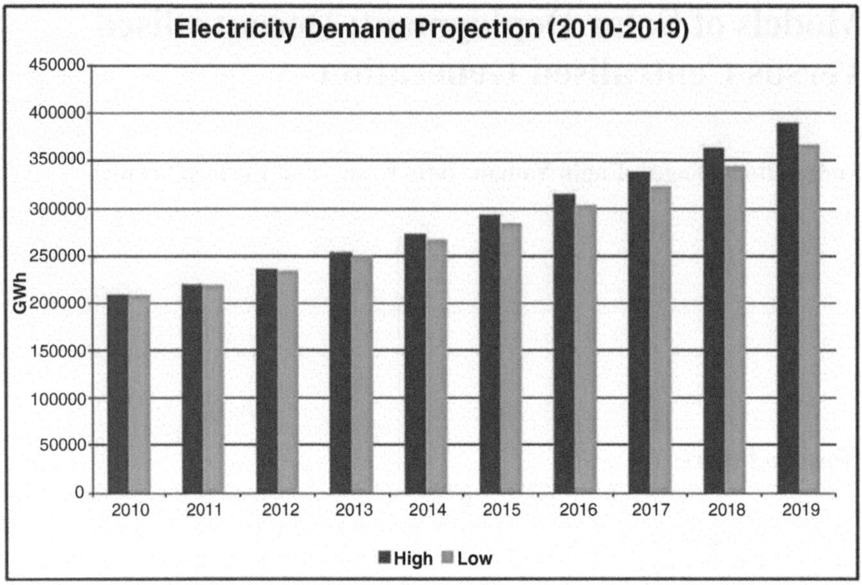

Fig. 1 Electricity demand projection, *Source*: TEIAS [2]

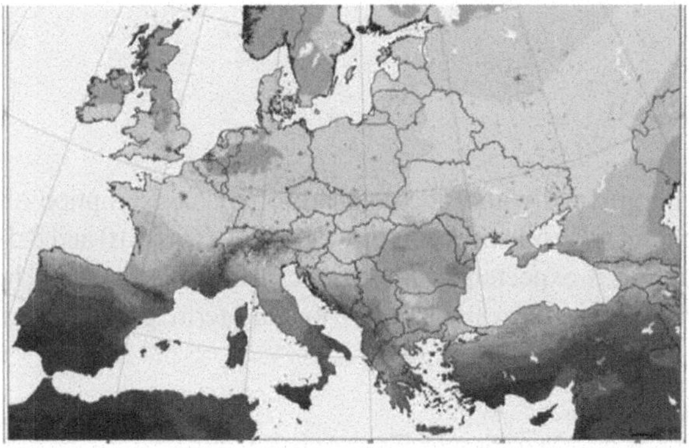

Fig. 2 Solar irradiation in Turkey (2010–2019), *Source*: PVGIS [3]

among its member countries [1]. As currently less than one-third of energy is generated from domestic resources, renewables can play an important factor in reducing imports and taking advantage of Turkey's natural resources, particularly high irradiation levels. Additionally, the European Union, which Turkey is aiming to join, requires that member states reach a renewable energy target of 20 % by 2020.

Turkey has a high solar energy potential comparable with that of Spain, due to its high irradiation, making the employment of solar electricity generation particularly suitable and interesting as can be seen from graphic 2 below (Figs. 1 and 2).

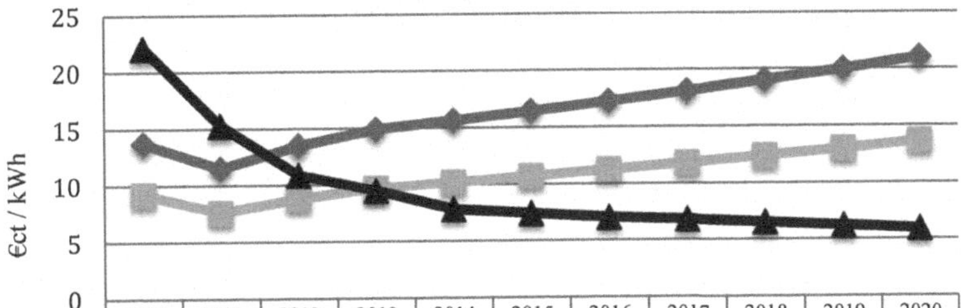

	2010	2011	2012	2013	2014	2015	2016	2017	2018	2019	2020
Household	13,7	11,5	13,5	14,9	15,6	16,4	17,2	18,1	19,0	20,0	21,0
Industry	9,2	7,6	8,8	9,7	10,2	10,7	11,2	11,8	12,4	13,0	13,6
PV LCOE	22,1	15,3	10,9	9,5	7,8	7,4	7	6,8	6,5	6,2	5,9
PV System cost in €	2,83	2	1,4	1,2	0,97	0,92	0,87	0,83	0,79	0,75	0,71

Fig. 3 Electricity prices compared to PV LCOEs, *Source*: Eurostat [7], own calculations

Despite this promising background, solar technology has not taken off in Turkey. Reasons are mainly the comparably low and short duration of the feed-in tariff, which is 10 years compared to 15–20 years in European countries, and protection of the newly privatised energy sector. In 2015, the cumulative installed PV power was about 55 MW, albeit its energy yield potential of 1300–16,000 kWh/kWp [4].

2 Economic Attractiveness of Solar Within the Turkish Electricity Market

Turkey has a solar potential of 300 TWh per year, which is yet to be exploited [1]. Taking into account the latest electricity price increases in October 2012 of 9.81 % for households and 4.03 % for industry [5] and the announced increase in July 2013 by 21 % [6] as well as an average increase of 5 % annually, it becomes clear that solar electricity generation is on its way to competiveness in Turkey.

Assuming a PV system of 1 MW, 20 years of employment duration, 1400 kWh/kW solar electricity generation, 80 % digression rate, one inverter replacement during the systems' lifetime, 0.5 % operating cost and a discount rate of 4 %, PV is already today an economically viable source of electricity for large-scale industrial users (Fig. 3).

Accordingly, while the focus has been on the feed-in tariff scheme provided by the Turkish government, it has been neglected that solar electricity is becoming more and more attractive in its own right, supported by the current increase in general electricity and gas prices.

As in major European markets such as Italy and Germany, a business model based on grid parity and self-consumption needs to be developed as an additional method to speed up solar employment. If standards such as priority grid access and net metering are integrated and promoted, the growth of the Turkish market will be further supported, allowing the reduction of costly fossil fuel imports.

3 Distributed Versus Centralised Solar Electricity Generation

Principally, there are two types of PV installations: distributed and centralised and utility-scale. Distributed installations mostly encompass residential and commercial rooftop applications. Utility-scale installations are often far larger than 1 MW and can supply thousands of homes with green power.

The advantages of utility-scale installations are numerous and are particularly attractive in rural areas, where space for solar parks is available. Utilities need peak power generators to provide additional power during peak times, e.g. in summer afternoons due to air conditioning. PV is ideal to satisfy this demand as it can achieve maximum outputs during these times. Moreover, large power plants have a cost advantage. First of all, distributed systems are a lot more expensive per watt due to fixed costs, and secondly the cost of PV panels in general has continuously decreased in the past years, while performance and efficiency have steadily increased. Utility-scale PV plants also contribute to reducing grid cost. Compared to other forms of electricity generation such as coal and nuclear, even utility-scale PV plants are distributed. Hence, the power generated can be supplied to the end users through the local utilities, and there is less need to expand expensive high-voltage grids.

On the other hand, solar's charm lies in its clear advantage of distributed energy generation, allowing electricity generation and consumption at one location. Distributed electricity requires even less grid expansion than centralised PV power generation and allows the installation of solar on different areas, roofs, etc., turning unused space into assets.

Accordingly, PV can produce electricity wherever it is needed without necessarily requiring additional space. This is a particular advantage in urbanised locations, in which available rooftop space can be turned in solar power generators, thereby reducing the daytime need of diesel generators and providing clean energy either into the grid or for direct in-house consumption. There are manifold options from PV electricity generation for direct own consumption to the establishment of a mini-grid to provide access electricity to neighbours as well as provision into the grid to take advantage of feed-in tariffs.

Another option based on solar's ability to realising distributed power generators is the combination of several small installations into a larger system to increase the overall buying and negotiation power of the end user. An interesting form of such an approach was developed in the Netherlands. Here a citizen-based initiative promoted solar PV for consumers, by bundling consumer demand to allow buying in bulk at rates usually applicable for large-scale installations. Reducing the cost of a solar kit by at least one-third of normal residential system costs allowed affordable solar electricity generation without feed-in tariffs based on net metering. The first phase of this project led to around 2500 families signing up for the programme—resulting in 5 MW of installations [8].

Of course, PV power plants are also highly beneficial from an environmental point of view. Utilities and industrial players are encountering increasing pressure to reduce their carbon emissions, and PV is a zero emission form of power generation, supporting green manufacturing efforts. Furthermore, local power generation reduces the dependence on power from fossil resources and foreign suppliers, an important factor in Turkey, which needs to import more than 60 % of the material for conventional energy generation like coal, oil and gas.

4 The Role of Different PV Systems in Turkey

4.1 Utility-Scale Solar Parks

The role of utility-scale solar parks in Turkey is basically twofold: It supports the production of inexpensive clean and local energy generation, helping to meet the growing energy demand and reducing the dependency on fossil fuel imports. As such it is an important pillar in the Turkish electricity mix.

Secondly, these installations help to jump-start the Turkish solar market and promote the development of a local solar industry by providing a sufficient demand for investment by domestic or foreign entities and allowing achieving certain scale effects. The gained experience will help a new Turkish industry to gain international competitiveness and can allow it to enter additional markets.

Particularly local utilities can benefit from including solar parks into their portfolio as they can either supply the power produced directly to entities with demand or combine solar power generation with other complementary forms, e.g. wind energy or technology forms, that allow inexpensive storage of PV power, e.g. hydro or gas.

Due to the high administrative hurdles for licensed projects, only 8 MW received a licence by 2015 [9]. In January 16, a 5 MW project was awarded by a Turkish government enterprise [10]. To speed the process up, large-scale projects were achieved by covering a few unlicensed plants, accumulating to 40 MW in total by 2014 [4]. With this legal loophole closed on March 23, 2016, large ground-mounted installations can only be realised based on a licence, which comes with a fee. Additionally stricter self-consumption requirements were introduced, further affecting the market negatively [11].

4.2 Industrial and Commercial Rooftop Systems

These systems support the further development of Turkey's solar market and, depending on the availability of electricity and the price range for conventional grid-related electricity, develop a growing demand for solar electricity.

Particularly, in order to foster economic growth, the access to and availability of electricity are mandatory. With PV LCOEs moving into economic competiveness, generating own electricity supports core business success.

For consumer brands, an additional advantage lies in the environmental friendliness, a potential competitive advantage particularly on international markets.

While one model is of course the ownership by the user itself, another option is a provision or rent out of the rooftop space to investors and a repurchase of the solar electricity produced through power purchase agreements (PPAs). This is interesting for the end user to avoid upfront investment costs, if production and demand are not in-line (remaining PV power sold by the investor to a third party) or if demand exceeds the generation capacity on a particular industrial site (particularly if the investor is able to provide additional PV power from a different but geographically close site).

4.3 Residential Systems

While residential systems are small in size, they support market growth and development tremendously as the German example shows, which was based on this market segment.

Additionally, it supports the population's education on electricity generation, consumption and the setting of preferences for ecologically viable energy production.

With costs of storage systems coming down, residential application based on self-consumption will see increased interests, as this will allow to better match production and demand cycles.

5 Conclusion

Solar is an attractive and necessary addition to the Turkish power portfolio with an untapped potential of up to 300 TWh per year. Making use of this resource will help reduce dependence on costly fossil fuel imports and allow developing a new local industry.

As solar system prices come down, while conventional electricity rates go up, photovoltaics are becoming a competitive source of electricity based on grid parity terms, particularly interesting to foster further economic growth.

While conventional electricity generation is generally centralised, solar offers further deployment opportunities by allowing distributed energy generation. As such it reduces the requirements and costs for high-voltage grid extension as it mainly uses medium- to low-voltage grids.

Distributed power generation opens up different market segments with particular own interests in solar power generation. A well-balanced combination and development of all segments, i.e. utility-scale solar parks, large industrial and commercial rooftops as well as residential systems, will allow Turkey to generate utmost benefit from this form of energy generation.

However, based on the slow administrative processes and further market hurdles introduced, a fast take-off of the Turkish PV market cannot be expected.

References

1. Energy Market Regulatory Authority (2012) Turkish energy market: an ınvestor's guide
2. Turkish Electricity Transmission Corporation, Research Planning and Coordination Department (2010) Turkish electrical energy 10-year generation capacity projection (2010–2019)
3. Šúri M, Huld TA, Dunlop ED, Ossenbrink HA (2007) Potential of solar electricity generation in the European Union member states and candidate countries. Solar Energy 81:1295–1305. http://re.jrc.ec.europa.eu/pvgis/
4. Müller-Russo G (2015) Solar market Turkey, ıntersolar global 2015. http://www.intersolar-global.com/de/summits/turkey/market.html
5. Coskun O, Butler D, Fahmy M, Fenton S (2012) Update 1-Turkey raises gas, electricity prices by 9.8pct. Thomson Reuters
6. Hurriyet Daily News (2013) Turkey hikes electricity prices 21 percent
7. Eurostat (2013)
8. "Wij Willen Zon" – We want Sun – initiative of the Urgenda foundation
9. Cankorel T (2016) Turkish solar update. http://www.renewableenergyworld.com/articles/2016/01/turkish-solar-update.html
10. Tranov G (2016) 2016 Turkish solar market report. http://www.pvbuzz.com/2016-turkish-solar-market-report/
11. Tsagas I (2016) Turkey amends its unlicensed market regulation. http://www.pv-magazine.com/news/details/beitrag/turkey-amends-its-unlicensed-market-regulation_100023979/#axzz4BXfUWi43

Testing, Product Certification, and Inspection of Photovoltaic Modules for Local Production

Yusuf Biçer, Cevat Özarpa, and Y. Erhan Böke

1 Introduction

Turkey recently focuses on renewable energy-based power production including PV power plants. According to the current regulations, there are two ways for the installation of renewable energy-based power plant, namely, licensed or unlicensed. The licensed power plants require a license from EMRA (Energy Market Regulatory Authority), whereas unlicensed renewable energy power plants are allowed to be built without a license but limited up to 1 MW for each user. The unlicensed option brings important advantages for stand-alone power production especially for residential applications. In order to have a reliable photovoltaic energy system in residential buildings, each unit of the system such as inverters, modules, and transmission units needs to work in harmony. However, the quality and reliability of the PV modules are the crucial features which lead to longer lifetimes and higher power generation. Turkey keeps progressing for the licensed and unlicensed solar power plants where in June 2013 approximately 9000 MW applications were made by 496

Y. Biçer (✉)
Faculty of Engineering and Applied Science, University of Ontario Institute of Technology, 2000 Simcoe Street North, Oshawa, ON, Canada L1H 7K4

Faculty of Mechanical Engineering, Istanbul Technical University, İnönü Caddesi, No. 65, Gümüşsuyu, 34437 Beyoğlu/İstanbul, Turkey
e-mail: yusuf.bicer@uoit.ca

C. Özarpa
İGDAŞ (Istanbul Gas Distribution Corporation), Cumhuriyet Mah. N. Fazıl Cad. No: 3 Sultangazi, Istanbul, Turkey
e-mail: cozarpa@igdas.com.tr

Y.E. Böke
Faculty of Mechanical Engineering, Istanbul Technical University, İnönü Caddesi, No. 65, Gümüşsuyu, 34437 Beyoğlu/İstanbul, Turkey
e-mail: boke@itu.edu.tr

T.S. Uyar (ed.), *Towards 100% Renewable Energy*, Springer Proceedings in Energy, DOI 10.1007/978-3-319-45659-1_22

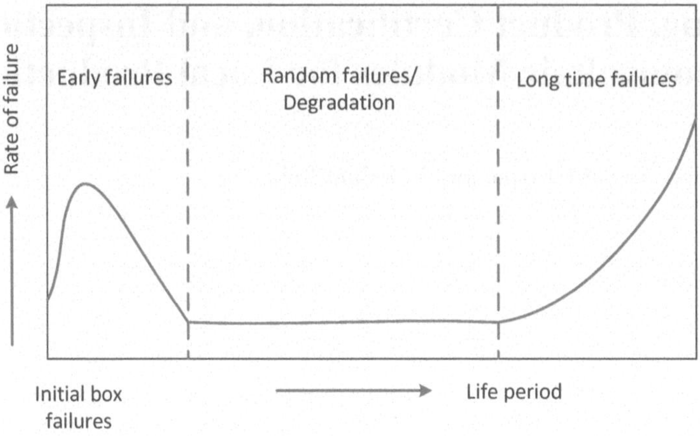

Fig. 1 Life cycle failure analysis of PV modules

different applicants for the permissible limit of 600 MW licensed solar photovoltaic power plants.

Most of the PV module manufacturers guarantee the minimum performance of their modules for 20–25 years, whereas recently 30-year warranties are also introduced. The warranty typically guarantees the module performance of at least 90 % capacity in the first 10 years and to at least 80 % in the following 10–15 years [1]. If the PV modules are removed or replaced from the field before the warranty period expires due to any type of failure, including power drop beyond warranty limit, these failures are classified as hard failures or reliability failures. If the performance of PV modules degrades but still meets the warranty requirements, then those losses are classified as soft losses or degradative losses [2].

The long-term failures which can be named as wear-out failures occur after several years mainly because of climatic conditions. Wear-out failures occur at the end of the working lifetime of PV modules. They determine the maximum working life of a PV module. In contrast, early failures as shown in Fig. 1 generally occur due to design and production defects. Minimizing the early failures can be satisfied using product-specific tests and certification, whereas long-term failures can be compensated using inspection services.

2 Certification and Accreditation

The certification process includes the conditions to be met by the manufacturer such as tests during production line, installation requirements, and training of the personnel. The Turkish Accreditation Agency (TÜRKAK) conducts the audits for certification bodies in terms of international standard TS EN ISO/IEC 17065 (Conformity assessment—Requirements for bodies certifying products, processes and services) [3]. In addition, according to the Turkish regulations, in order to certificate the domestic production and benefit from additional financial incentives, components that are used in renewable energy power plants are required to carry a conformity certificate of international or national standards given by organizations which carry the TS EN ISO/IEC 17065 Product Certification System [3]. One of the important stages in the certification process is test laboratories. The test laboratories are eligible to be utilized when they satisfy the standard of TS EN ISO/IEC 17025 (General requirements for the competence of testing and calibration laboratories) [4], where this standard regulates the reliability of the test results. Product certification satisfies PV-specific standards and other standard and norm documents to provide assurance for compliance. Product certification systems include initial experimental evaluation of the supplier's quality system, subsequent periodic inspection of factories, and testing of sample products of the suppliers taken from free market. A certified PV module by accredited institutions assures that the PV module is manufactured according to the international standards of IEC 61215 (Crystalline silicon terrestrial photovoltaic (PV) modules—Design qualification and type approval), IEC 61646 (Thin-film terrestrial photovoltaic (PV) modules—Design qualification and type approval), and IEC 61730 (Photovoltaic (PV) module safety qualification) [5–7]. These standards are also available as EU norms. There is a new and updated standard announced in March 2016 which includes both thin-film and crystalline silicon PV modules as IEC 61215:2016 (Terrestrial photovoltaic (PV) modules—Design qualification and type approval).

3 PV Modules in Turkey

PV module life cycle inspection services can be satisfied using product-specific tests and certification, whereas long-term failures can be compensated with continuously decreasing module costs due to technological and manufacturing advancements. Naturally, as the costs of PV module production continue to drop, solar power grows even faster, which further brings down the cost of solar investments. Production of any technological equipment brings the research and development activities together with some innovations. Therefore, manufacturing of solar photovoltaic modules in Turkey is significant for the Turkish solar energy market in the near future. There are many PV modules in the world market which are not compatible to international standards although they bear specific quality labels. In Turkey, there are currently about 10–15 PV module manufacturers ranging from small to large production capacities [8]. These producers need to test their products before marketing and certificate them according to international standards in order to be competitive in the PV market. Since the testing and certification capabilities in Turkey are new and limited, the producers tend to utilize external institutions such as accredited European PV certification bodies to get the required certificates. PV power plant investors in Turkey are not so keen on using locally produced PV modules because of their price and recentness. In general, manufacturers purchase the raw materials in Far Eastern or EU countries. In addition, the usage of external institutions for testing and certification located in Europe or Asia brings additional costs and responsibilities in the transportation. The introduction of various domestic

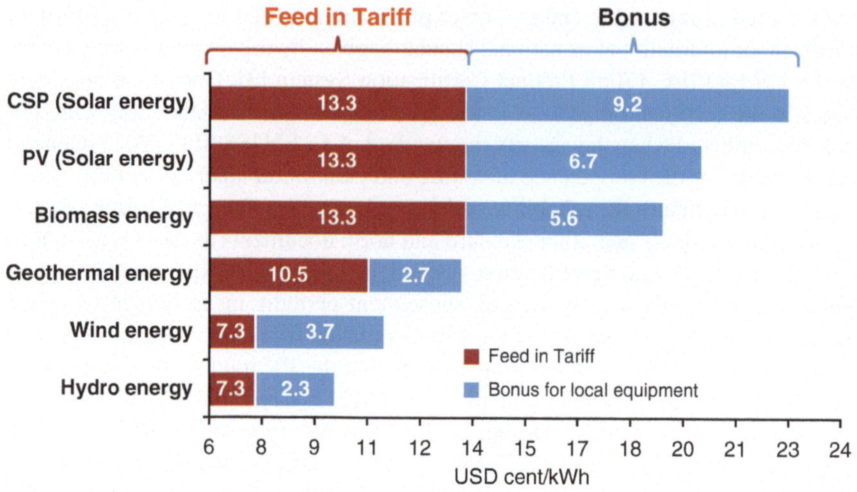

Fig. 2 Feed-in tariff for renewable energy-based power plants in Turkey and additional financial incentives for local equipment utilization

PV raw material manufacturers can decrease the unit costs of locally produced PV modules. Furthermore, the availability of local testing and certification institutions for PV modules leads to an advanced technology utilization and further improvements.

According to published regulation [9], renewable energy power plants which use domestically manufactured components are provided up to 6.7 USD cents/kWh additional incentives for solar photovoltaic plants as illustrated in Fig. 2. This incentive was for the plants to be commissioned by 2015 and valid for 5 years. By the recent legislation, it has been extended for the plants which will be commissioned by 2020. Considering the cost reductions in the renewable energy market, any additional incentives provided by the government facilitate the investments. In this context, investors try to benefit from additional incentives by using locally produced and certified products. Developed countries all over the world in their respective industries have production-oriented work which leads to being pioneers in the market. Therefore, it has great importance for Turkey in the coming years to encourage domestic production and its use, constituting itself as having high solar energy potential in order to be a pioneer in renewable energy sector. It is one of the key points in the legislation to emphasize domestic production by requiring product certificate taken from accredited institutions having TS EN ISO/IEC 17065 (Conformity assessment—Requirements for bodies certifying products, processes and services) [3]. There are three documents to be prepared and presented to the ministry for having additional incentives: domestic manufacturing status certificate issued by the Chamber of Commerce and Industry, type certificate, and product certificate issued by accredited institutions having TS EN ISO/IEC 17065 (Conformity assessment—Requirements for bodies certifying products, processes and services) [3]. Thus, the relevant domestic component manufacturers do not only focus on their own technologies; they have the chance to follow the most current national and international standards at the same time and fulfill the requirements of this certification which brings higher quality and reliable products. Power plant inspection bodies are encouraged to bear accreditation according to international standard TS EN ISO/IEC 17020 (Conformity assessment—Requirements for the operation of various types of bodies performing inspection) [10].

4 Results and Recommendations

In the PV certification applications, the most common problem encountered during the certification process is seeking for a reliable testing laboratory and certification institution. The distance of the laboratory plays an important role for the producers. In case there are local laboratories around the production area, the utilization of the laboratory is more common allowing new advances based on the test results. The manufacturers are able to improve the PV modules by using local laboratory and certification body which ends up with decreased costs. Transportation of the test samples is also significant during testing process. In case there are some failures during transportation from factory to testing laboratory, it affects the results of the tests and causes negative consequences. Nearby laboratories shorten the testing and certification period which brings customer satisfaction. Especially for situations of material changes, some additional tests are required in the international standards to keep current PV module certificate- which creates a new certification cycle for the producers. In this case, having a change in the raw material becomes inflexible. The manufacturers tend to keep using the current materials although they have low-cost and high-quality alternatives from various resources. Figure 3 shows the scattering of the failure types at the start of the service period given by a German distributor. Transportation damages constitute 5 % of all failure cases. The most important failures in the field are junction box failures, glass breakages, defective cell interconnections, loose frame, and delamination. The rates are presented relative to the total number of failures, and the PV modules are supplied by a German distributor in the years 2006–2010. The statistic is based on a total volume of approximately two million delivered PV modules.

Climate conditions of Turkey are dissimilar to European regions which affect the module performances. In Turkey, the average yearly temperatures are higher compared to Europe. In addition, there are dusty and desert areas in the south Anatolian regions. In such cases, the PV modules may need to have additional tests and certification procedures. In this respect, salt mist corrosion testing of photovoltaic modules and environmental testing—test of dust and sand—are two of the significant tests which can indicate the reliability of the PV modules used in Turkey.

Figure 4 shows the failure distribution of PV modules that have been in the field for 8 years. Two percent of the PV modules are predicted to not meet the manufacturer's warranty after 11–12 years of operation. The rate is given relative to the total number of failures. Approximately 2 % of the entire fleet are predicted to fail after 11–12 years.

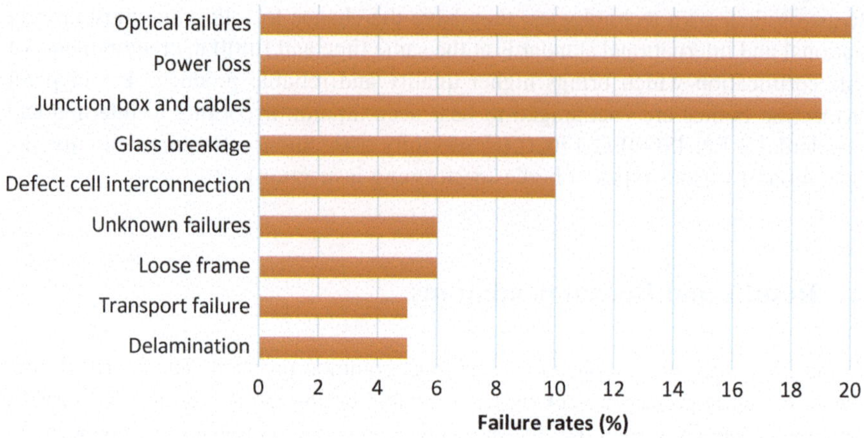

Fig. 3 Failure rates of PV modules in the first 2 years after delivery based on customer complaints (adapted from [11])

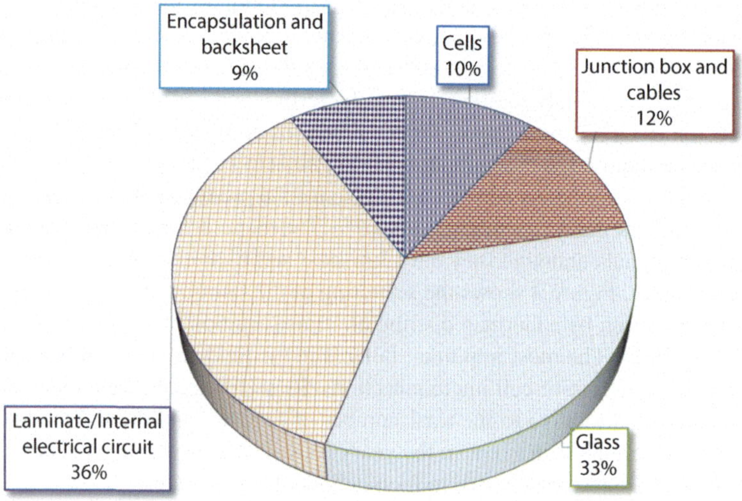

Fig. 4 PV module failures for various PV modules from 21 manufactures installed in the field (adapted from [12])

Factory inspection is one of the crucial stages in the certification process. If the certification body is nearby the manufacturer, the inspection process becomes more frequent and accurate. In the necessity of repeating a specific type of tests, the process time shortens and the cost decreases. A high-quality production starts with raw material input check. Especially PV cells need to be controlled cautiously. While combining the cells as strings, automated robots are advised to be used. Each product type should have quality control forms for records. In case there are materials which are out of calibration limits, they are marked and not used in the production line. Any measuring equipment used during or after production need to be calibrated and renewed periodically. Calibration and maintenance plans are updated and presented to certification body regularly. Local certification bodies suggest improvements of module production lines by considering experiences in the market. Each manufacturer has a solar simulator to label the performance values of PV modules. Each solar simulator test report is suggested to be backed up and given to certification body in order to have clearance under any complaint circumstances. Solar simulator test results can be delivered to end users to satisfy the transparency. Labeling of PV module should include minimum required information stated in the standards. The solar simulator unit is critical for the PV production line because the manufacturer can find out the performance of modules and defines the market price accordingly since the prices are defined based on cost per watt. Therefore, measurement certainty of this device plays an important role. The solar simulator includes a reference module which is calibrated by certified bodies. There might be primary and secondary references in the factory. Calibration period of these modules is dependent on production rate of manufacturer and needs to be defined carefully.

Some producers in the world have their own testing mechanism in their production facilities for some safety tests present in the IEC 61730 standard. Among those, reverse current overload test and ground continuity test are favorable in the production line. Besides, these tests are suggested to be implemented during yearly factory inspection processes. For each PV module produced in the factory, the flash tests are carried out and stored. The presentation of these test results is suggested during planned factory inspections. In the PV production line, an electroluminescence (EL) tester or laser tester is a type of checkpoint in which defective modules are determined and classified. Each PV module is recommended to be tested in these devices after or before lamination. The criteria for the evaluation of EL test should be determined by the producer and approved by the certification body.

5 Conclusions

In this study, the importance of local PV testing, certification, and inspection bodies is emphasized. The PV applications in Turkey continue to develop by increasing investments. The certification bodies need to define specific type of procedures for PV module certification processes and factory inspections. There are many critical points in a PV module production line which affect the performance and efficiency of PV modules. It is suggested for PV module manufacturers to fulfill the international standards and to apply for accredited testing and certification services in order to sustain high-quality and more efficient PV modules. Inspection, installation, and manufacturing processes require trained employees. Therefore, personnel certification through vocational qualification authorities is very critical in the near future. Local certification and testing institutions are a background for high-quality PV modules. The additional incentives for local product utilization are certainly an important factor for the investors. By realizing the local certification bodies and laboratories, it is assured that PV power plants operate in a reliable manner. As a result, having local testing, certification, inspection, and maintenance institutions brings significant technological and economic advantages for the developing solar energy market in Turkey.

Acknowledgment The authors acknowledge the support provided by UGETAM and Istanbul Technical University.

References

1. TamizhMani G, Kuitche J (2013) Accelerated lifetime testing of photovoltaic modules. Arizona State University
2. IEA (International Energy Agency) Photovoltaic Power Systems Programme (2012) PV module reliability issues including testing and certification. Jahn, U., Tüv Rheinland, Frankfurt, 24 Sept 2012
3. TS EN ISO/IEC 17065, Conformity assessment—requirements for bodies certifying products, processes and services
4. TS EN ISO/IEC 17025 General requirements for the competence of testing and calibration laboratories
5. IEC 61215 Crystalline silicon terrestrial photovoltaic (PV) modules-Design qualification and type approval
6. IEC 61646 Thin-film terrestrial photovoltaic (PV) modules-Design qualification and type approval
7. IEC 61730 Photovoltaic (PV) module safety qualification—Part 1: requirements for construction Part 2: requirements for testing
8. Bicer Y, Ozarpa C, Boke YE (2015) Certification issue of solar photovoltaic modules for local production, problems, and solutions in Turkey. In: Dincer I, Colpan CO, Kizilkan O, Ezan MA (eds) Progress in clean energy, vol 2. Springer International, New York, pp 929–935
9. Legislation, Official Gazette Nr. 27969 (June 19, 2011) Local manufacturing of the equipment used in power plants producing electrical energy from renewable energy sources, Ankara, Turkey
10. TS EN ISO/IEC 17020 Conformity assessment-Requirements for the operation of various types of bodies performing inspection
11. Richter A (2011) Schadensbilder nach Wareneingang und im Reklamationsfall, 8. Workshop Photovoltaik-Modultechnik, TÜV Rheinland, Köln, 24–25 Nov 2011
12. DeGraaff D, Lacerda R, Campeau Z (2011) Degradation mechanisms in Si-module technologies observed in the field, their analysis and statistics. In: Presentation at PV module reliability workshop, NREL, Denver, Golden. http://www1.eere.energy.gov/solar/pdfs/pvmrw2011_01_plen_degraaff.pdf

Aerodynamic and Performance Analysis of Drag-Driven Vertical-Axis Wind Turbines

Emre Alpman, Zafer Canal, and İbrahim Baysal

1 Introduction

Environmental pollution generated by fossil fuels and the fact that sources of such fuels will diminish considerably in the near future has forced scientists to search for clean and renewable energy sources. One such energy source is wind from which energy can be obtained using devices called wind turbines. Depending on the axis of rotation of the blades, wind turbines can be divided into two categories: horizontal axis wind turbine (HAWT) and vertical axis wind turbine (VAWT). Although the horizontal axis type is more widely used and commercialized, vertical axis wind turbines may be more advantageous in urban areas where the wind direction may change considerably due to surrounding buildings [1]. Vertical axis wind turbines can be divided into two categories: lift-driven and drag-driven. Despite being less efficient than lift-driven ones, drag-driven turbines may be easier to design and manufacture since they do not require sophisticated lifting surfaces to generate power. Main disadvantage of drag-driven VAWTs that they can generate power less efficiently and have much larger blade surface area compared to lift-driven ones [2]. Drag-driven VAWTs rotate due to the nonsymmetric drag force distribution on the blades which can be almost always obtained thanks to the nonsymmetric geometry of the turbines. As a result they may have better self-starting capabilities, and they may be used as a starter device for lift-driven turbines [2].

In this study aerodynamic analysis of drag-driven VAWTs was performed using the open-source computational fluid dynamics software OpenFOAM® [3]. The turbines studied consisted of circular-arc-shaped blades attached to a central shaft. A four-bladed turbine can be seen in Fig. 1. In order to see the effect of number of blades two-, three-, and four-bladed turbines were analyzed and compared. Results

E. Alpman (✉) • Z. Canal • İ. Baysal
Mechanical Engineering Department, Marmara University, 34722 Kadikoy, Istanbul, Turkey
e-mail: emre.alpman@marmara.edu.tr

T.S. Uyar (ed.), *Towards 100% Renewable Energy*, Springer Proceedings in Energy, DOI 10.1007/978-3-319-45659-1_23

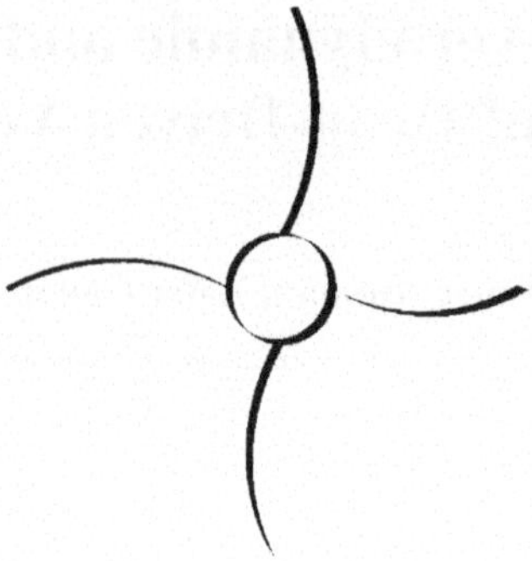

Fig. 1 A four-bladed turbine

include pressure distributions in the flow field, force distribution on the blades, and the power coefficients of the turbines analyzed. Numerical solutions were performed for two-space dimensions assuming that blade shapes and orientation do not change in the vertical direction.

2 Methodolody

In this study blade shapes and orientation were assumed to be fixed in the vertical direction. Thus, numerical solutions were performed for two-space dimensions. The solution domain consisted of a rotating zone which contained the turbine and a non-rotating surrounding zone. These zones can be seen in Fig. 2. For numerical solutions MRFSimpleFoam [3] solver of OpenFOAM® was used. This solver was designed for incompressible turbulent flows which contain multiple rotating reference frame regions. Solutions were performed using block-structured meshes generated using the blockMesh utility of OpenFOAM® [3].

Turbulent flow solutions were performed using the standard k-ε model [4]. Although this model is known for its weaknesses for unsteady complex flows, it was assumed to be sufficient for comparison purposes of this study. For all the numerical solutions freestream turbulence intensity was assumed to be 5 % and ratio of eddy viscosity to molecular viscosity was taken to be 10.

3 Results and Discussion

Numerical solutions were performed for two-, three-, and four-bladed versions of the turbine displayed in Fig. 1. Simulations were performed for 1000 s of real time, and Fig. 3 shows the variation of power coefficients (Cp) of the turbines with time. Here the turbines rotate with a tip speed ratio (λ) of 0.1. The power generated by the turbines clearly showed an unsteady behavior. Negative power shown in the figure meant that the torque developed on the turbine is negative and it was trying to stop

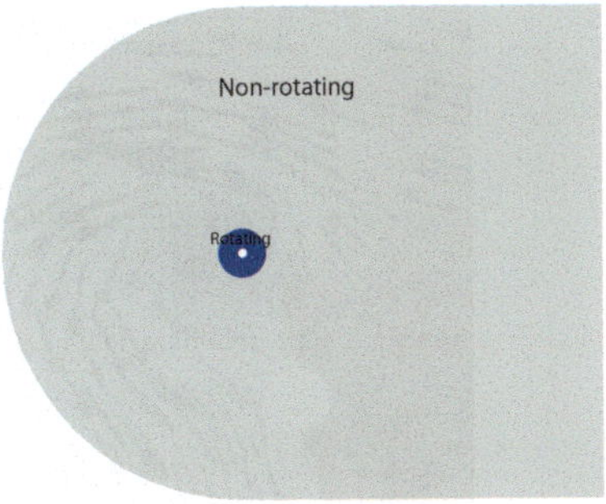

Fig. 2 Rotating and non-rotating (blue disk) zones

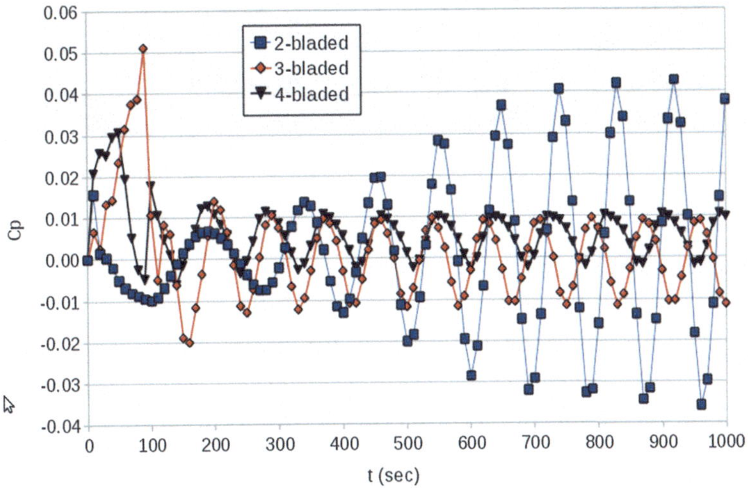

Fig. 3 Variation of power coefficient with time

the turbine rather than rotating it. The unsteady behavior observed in the figure was due to the continuous vortex shedding from the blades which lead to a highly unsteady pressure field downstream of the turbines. Pressure distribution and velocity vectors of flow around two-bladed turbine at 1000th second is displayed in Fig. 4 where Cp was used for pressure coefficient. In this figure shedded vortices and low pressure spots they generate are evident.

Table 1 displays the time averaged power coefficient values obtained for these turbines for tip speed ratios of 0.1 and 0.2. This table clearly showed that the turbines performed very poorly and two- and three-bladed turbines were useless at tip speed ratio of 0.2. The four-bladed one performed relatively better compared to the others.

Blades of the turbines analyzed above had blades with high radius of curvature. In order to increase the asymmetry of the turbines, which is desired for drag-driven turbines, the curvature of the blades were increased by decreasing the radius of curvature while keeping hub and tip locations fixed. A resultant four-bladed rotor is displayed in Fig. 5. Variation of power coefficients with time for two- and four-bladed

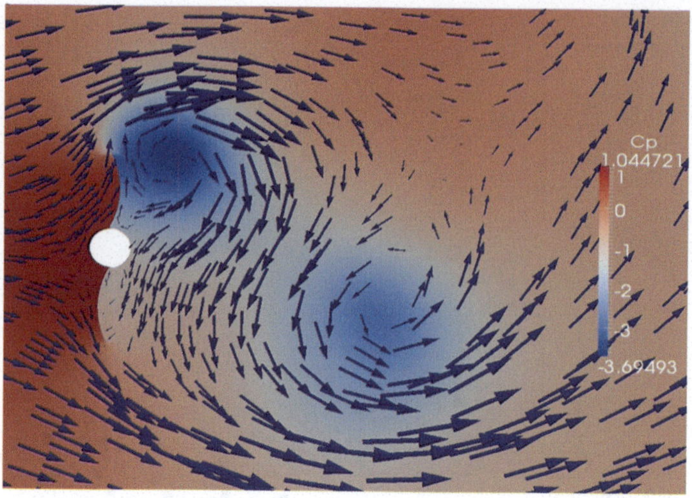

Fig. 4 Pressure distribution and velocity vectors of flow around two-bladed turbine at 1000th second

Table 1 Power coefficient vs tip speed ratio

	Cp		
λ	Two-bladed	Three-bladed	Four-bladed
0.1	0.0016	0.0016	0.0061
0.2	−0.109	−0.0055	0.0051

Fig. 5 A four-bladed turbine

rotors rotating at tip-speed ratio of 0.1 was displayed in Fig. 6. The average Cp for both cases increased considerably however, they were still very low. Four-bladed turbine yielded small amplitude oscillations after some initial transient phase while the amplitude of oscillations were large and increased with time for two-bladed case. In order to better understand the low power generation normal stresses on the

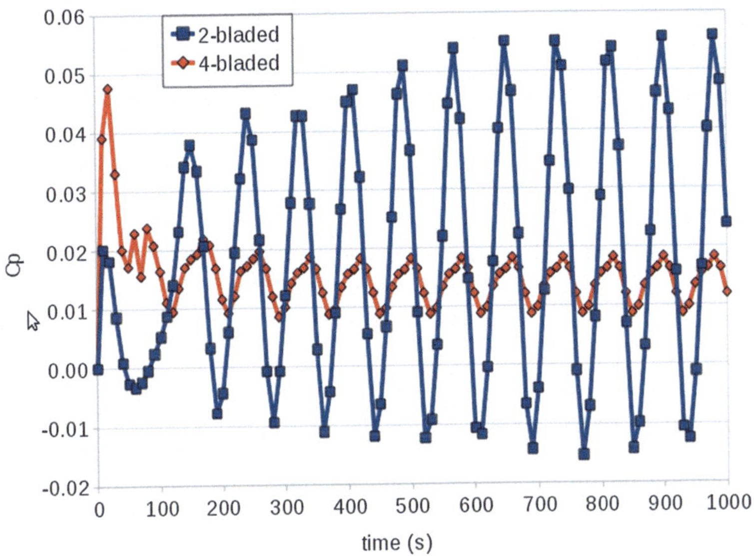

Fig. 6 Variation of power coefficient with time

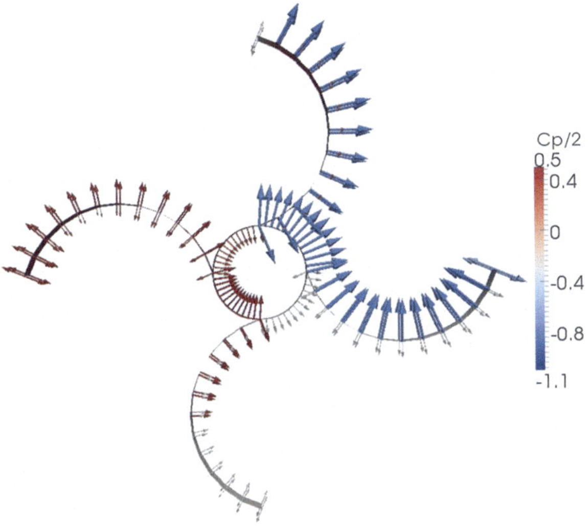

Fig. 7 Normal stress distribution on the four-bladed turbine

four-bladed turbine at 1000th second are plotted in Fig. 7. In this figure Cp was used for pressure coefficient. As expected top blade generated a positive torque while the bottom blade did the opposite. Blade on the right also generated a negative torque due to the low pressure region downstream of the top blade. This clearly decreased the net torque developed on the turbine and the net power extracted.

4 Conclusions

In this study, aerodynamic analysis of drag-driven wind turbines consisting of circular-arc-shaped blades attached to a central shaft was performed. Numerical solutions were obtained using the open-source computational fluid dynamics software OpenFOAM®. In order to see the effect of number of blades on the performance, two-, three-, and four-bladed turbines were analyzed and the resultant power coefficients were compared. Computed power coefficients showed an oscillatory behavior for all cases and amplitude of the oscillations was especially high for the two-bladed case. Increasing the curvature of the blades also increased the average power extracted.

References

1. Eriksson S, Bernhoff H, Leijon M (2008) Evaluation of different concepts for wind power. Renew Sustainable Energy Rev 12:1419–1434
2. Kirke BK (1998) Evaluation of self-starting vertical axis wind turbines for stand-alone applications. Ph.D. thesis, Griffith University, Gold Coast Campus, Australia
3. http://www.openfoam.com
4. Pope SB (2001) Turbulent flows. Cambridge University Press, Cambridge

Emre Alpman was born on April 10th, 1977 in Ankara, Turkey. He received his B.S. and M.S. degrees from Aeronautical Engineering Department of Middle East Technical University in 1999 and 2001, respectively. He then attended the Pennsylvania State University and received a Ph.D. degree from Aerospace Engineering in 2006. His research interests mainly include aerodynamics, computational fluid dynamics, parallel processing, and turbulence modeling.

Dr. Emre Alpman is currently a faculty member in the Mechanical Engineering Department of Marmara University.

Zafer Canal was born in Tokat, in 1987. After graduation from high school, he attended the Marmara University Mechanical Engineering Department from where he received his B.S. degree in 2011. The topic of this chapter was also his graduation project.

İbrahim Baysal was born in Adana, in 1988. After graduation from high school, he attended the Marmara University Mechanical Engineering Department from where he received his B.S. degree in 2011. The topic of this chapter was also his graduation project.

Part III
Applications, Case Studies

Optimal Control of Solar Heating System

Bin-Juine Huang, Wei-Zhe Ton, Chen-Chun Wu, Hua-Wei Ko, Hsien-Shun Chang, and Rue-Her Yen

1 Introduction

Forced-circulation solar heating system has been widely used in process and domestic heating applications. Additional pumping power during the solar energy absorption is required to circulate the working fluid through the collector banks to absorb the solar heat. The solar heat absorbed increases with increasing flow rate as well as pumping power. A lot of solar heat systems were usually designed to operate at a flow rate given by the assigned flowrate of collector test standards, 0.02 kg m^{-2} s^{-1}, which may be overestimated and causes large pumping power consumption.

Many researchers intended to develop an optimal control technique to reduce the pumping power at optimal solar heat collection [1–4] such as using exergy concept [3] or system optimization method [1]. The control algorithms however were very complicated and not easy to be implemented in field. In addition, no feedback control scheme was ever been applied to assure the optimal performance under variable solar radiation. In the present study, we intend to develop the technology maximum-power-point tracking control (MPPT) similar to MPPT of solar PV system and implement it in a solar heating system for field test.

2 Design of Solar Heating System

A solar heating system using flow-through vacuum-tube solar collector was designed and installed for experiment. The solar heating system consists of 24 sets of vacuum-tube solar collectors with 25.92 m^2 total absorber area, as shown in Fig. 1.

B.-J. Huang (✉) • W.-Z. Ton • C.-C. Wu • H.-W. Ko • H.-S. Chang • R.-H. Yen
New Energy Center, Department of Mechanical Engineering, National Taiwan University, Taipei, Taiwan
e-mail: bjhuang@seed.net.tw

T.S. Uyar (ed.), *Towards 100% Renewable Energy*, Springer Proceedings in Energy, DOI 10.1007/978-3-319-45659-1_24

The piping is designed with 8 collectors in series and with three parallel connections, for total 24 collectors. The reverse-return piping design is adopted to maintain a uniform flow through the collector banks. An inverter for frequency control of the circulation pump was installed and a PC-based control system was developed to control the circulation pump. The system configuration is shown in Fig. 2.

Fig. 1 Solar heating system used in the present study

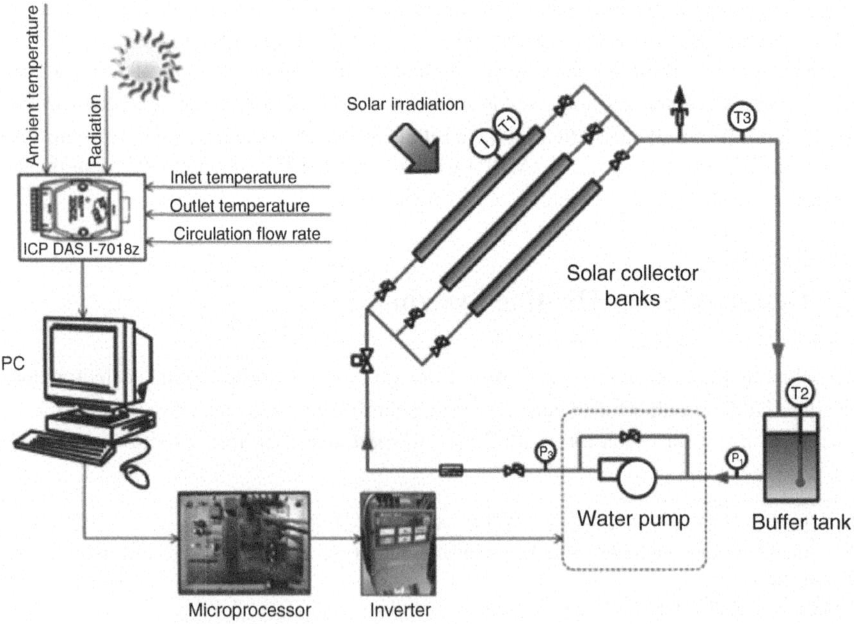

Fig. 2 Optimal control system

Most of solar heating systems design the flow rate according to the standard flow rate (0.02 kg m^{-2} s^{-1}) used in the standard test of solar collector. That is, the total flow rate for the present solar system will be rated at 0.02 L m^{-2} s^{-1} according to the test standard of collector, i.e., 31 LPM with 423 W pumping power. This may be too high and can be reduced without affecting the energy collection.

3 Cost Function for Optimal Pumping Power

In order to develop an optimal control for water pump, the cost function must be found first. The solar heat absorbed by the solar collector increases with increasing flow rate. The total solar energy collection Q_s may increase with flowrate. However, the associated pumping power may become unreasonably large. After certain range of high flowrate, the increase of flowrate no longer increases the total heat collection Q_s . The net energy gain Q_{net} can be defined as the total heat energy collected Q_s minus the pumping energy W_p which is converted into the primary heat energy:

$$Q_{net} = Q_s - \frac{W_p}{\eta_e} \tag{1}$$

where η_e is the primary energy efficiency of the electrical grid (0.365 in Taiwan). There seems exists an optimal flow rate at which the net heat collection is optimal and the pumping power is minimized. The definition of net energy gain Q_{net} is similar to the cost function of Kovarik and Lesse [1] or the exergy concept [4] and may have an optimal value as shown in Fig. 3. Since Q_{net} varies with solar radiation and ambient conditions (wind and temperature), there may not be an optimal value of Q_{net} all the time. The feasibility of using Q_{net} as the cost function in the MPPT of solar heating system needs to be verified experimentally.

A steady or quasi-steady state data were taken from field test at different mass flowrates to determine the instantaneous Q_s and Q_{net}. Figure 4 show that there exists an optimal flow rate at which the net heat gain Q_{net} is optimal. There exists an optimal Q_{net} (14.9 kW) at flow rate 22 LPM. The pumping power is 113 W with reduction of 73 % pumping power and the total heat energy collected Q_s is 15.2 kW. A very high electrical COP of the solar heating system ($Q_s/W_p = 134.5$) is obtained.

4 Feedback Structure of MPPT

The MPPT in feedback structure for the optimal control of cost function (Q_{net}) using step-up-step-down algorithm can be designed as shown in Fig. 5. The cost function at time instant i, denoted F_i, is feed back to predict the step-ahead cost function F_{i+1}. A predictive filter is used to predict the one-step-ahead cost function F_{i+1}, from

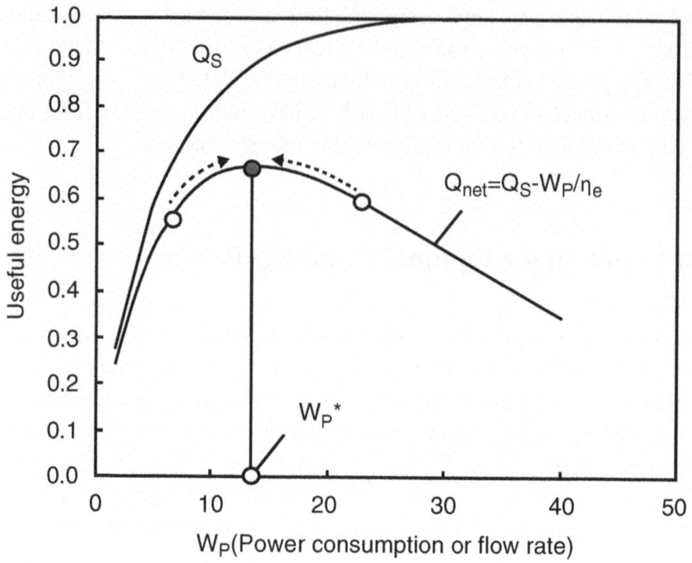

Fig. 3 Q_s and Q_{net} vs pumping power

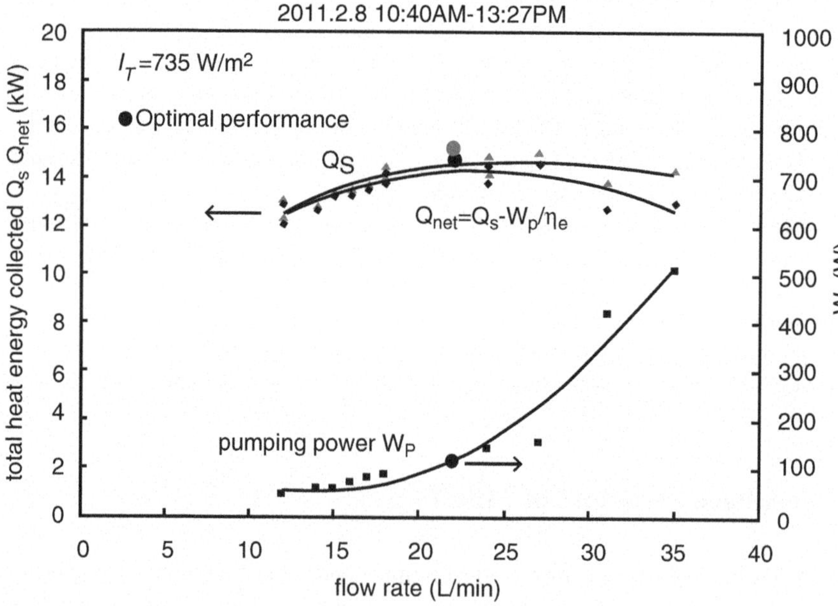

Fig. 4 Steady performance of a solar heating system

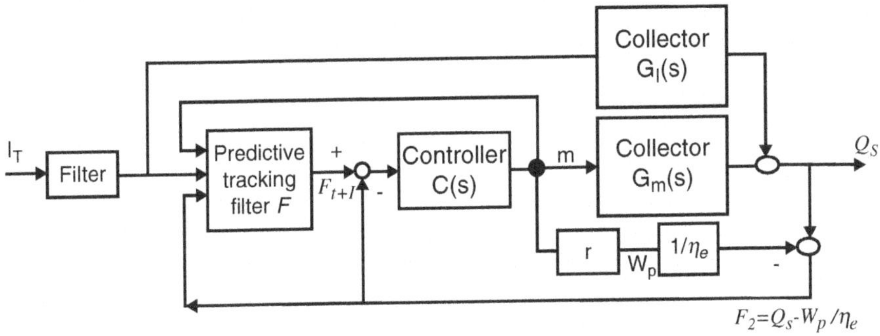

Fig. 5 Feedback structure of MPPT

which the increase or decrease of mass flowrate can be decided according to the difference $F_{i+1}-F_i$ to track the optimal cost function. The predictive filter includes the system dynamic model of the solar heating system:

$$\tau\frac{dq_s}{dt}+q_s=F_R(\tau\alpha)\cdot I_T-F_RU_L\cdot(T_i-T\alpha) \quad (2)$$

where $q_s=Q_s/Ac$, $Q_s=mCp(T_e-T_i)$, and ε is the time constant of the solar heating system which was determined from a step response test of the solar system (Fig. 6) by shading the solar radiation at a steady state. The test result shows that $\varepsilon=237$ s.

The two parameters $F_R(\tau\alpha)$ and F_RU_L are determined from the steady-state test data collected from field operation as shown in Fig. 7. The following correlations were derived:

$$\tau\frac{dq_s}{dt}+q_s=F_R(\tau\alpha)\cdot I_T-F_RU_L\cdot(T_i-T\alpha) \quad (3)$$

$$F_RU_L=2.97m^2+3.9m-0.14 \quad (4)$$

5 Test of Solar Heating System with MPPT

The solar heating system was tested with MPPT control. The test results of Fig. 8 show that the MPPT control works well. The pumping power is between 100 W and 200 W with total energy collection 14–18 kW during clear weather. The maximum instantaneous solar energy collection Q_s is 18.6 kW at pumping power 174 W and $I_T=875$ W/m^2 with COP=106.8. The maximum instantaneous COP

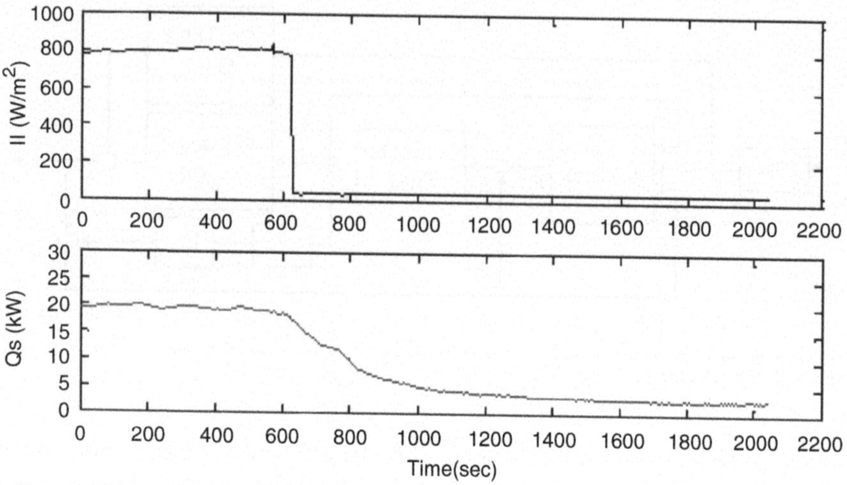

Fig. 6 Step response test of solar system

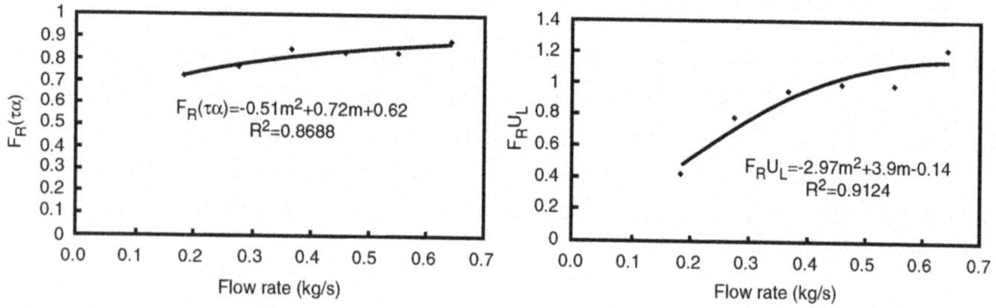

Fig. 7 Variation of $F_R(\tau\alpha)$ and F_RU_L with flowrate

is 188.1 at $Q_s = 8.8$ kW, pumping power 46 W and $I_T = 292$ W/m^2. For $Q_s = 14.2$ kW, $W_p = 116$ W and COP = 122.7. The overall COP is 111.0 between test period 10:02 am to 11:22 am with average $Q_s = 13.75$ kW and average pumping power 124 W.

The test was also carried out for the smaller collector area (17.28 m^2) by closing part of the solar system (Fig. 9). The maximum instantaneous solar energy collection Q_s is 13.7 kW at pumping power 89 W and $I_T = 892$ W/m^2 with COP = 153.8. The maximum instantaneous COP is 186.1 at $Q_s = 8.7$ kW, pumping power 47 W and $I_T = 450$ W/m^2. The overall COP is 114.6, for MPPT test period between 10:03 am to 13:15 pm, with average $Q_s = 9.27$ kW and average pumping power 81 W.

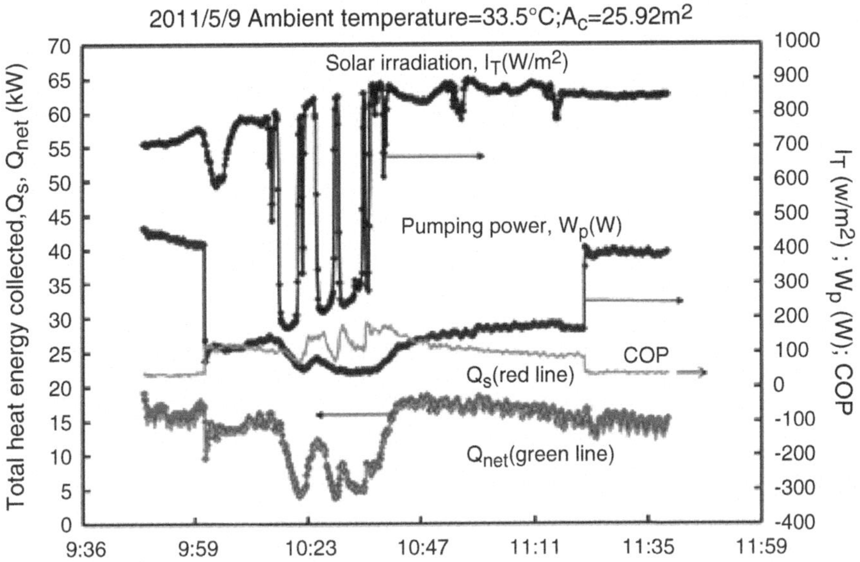

Fig. 8 Test result of MPPT (Ac = 25.92 m^2)

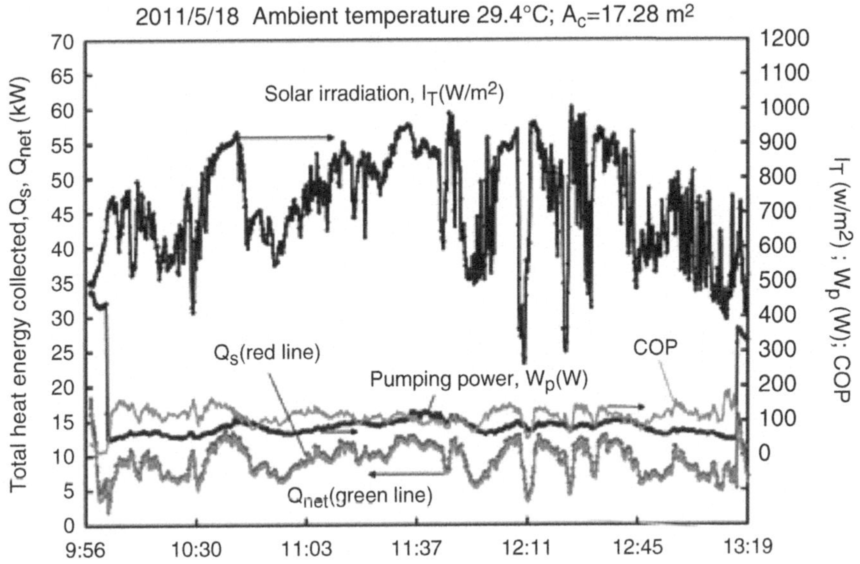

Fig. 9 Test result of MPPT (Ac = 17.28 m^2)

6 Conclusion

The present study developed a maximum-power point tracking control (MPPT) to obtain the minimum pumping power consumption at an optimal heat collection. The net heat energy gain Q_{net} ($=Q_s - W_p/\eta_e$) was found to be the cost function for MPPT. The step-up-step-down controller was used in the feedback design of MPPT on Q_{net}. The field test results show that the pumping power is 89 W at $Q_s = 13.7$ kW and $I_T = 892$ W/m^2. A very high electrical COP of the solar heating system ($Q_s/W_p = 153.8$) is obtained.

Acknowledgements This publication is based on work supported by Award No. KUK-C1-014-12, made by King Abdullah University of Science and Technology (KAUST), Saudi Arabia.

References

1. Kovarik M, Lesse PF (1976) Optimal control of flow in low temperature solar heat collectors. Solar Energy 18(5):431–435
2. De Winter F (1992) Comments on optimum flow rates in solar water heating systems with a counterflow exchanger. Solar Energy 49(6):557–558
3. Badescu V (2007) Optimal control of flow in solar collectors for maximum exergy extraction. Int J Heat Mass Transfer 50(21–22):4311–4322
4. Bejan A, Schultz W (1982) Optimum flow-rate history for cooldown and energy storage processes. Int J Heat Mass Transfer 25(8):1087–1092

Enhanced Geothermal Systems: The Soultz-sous-Forêts Project

Thomas Koelbel and Albert Genter

1 Introduction

The utilization of geothermal power from natural hot water or steam deposits is not rare, nor new. Today the installed capacity add up to more than 11 GW (electricity) and the first geothermal power production started in Italy at the beginning of the twentieth century. Unfortunately, geothermal sites with adequate geological conditions are limited. Contrariwise there is a high potential linked to low permeable, but deep and therefore hot rocks, which can be utilized by the so-called enhanced geothermal systems (EGS).

Since more than 20 years, scientists from universities and industry spent their effort at Soultz-sous-Forêts to build the first EGS power plant, which came into operation in 2009. During a first phase, a certain number of wells have been drilled down to more than 5000 m depth and hydraulic and chemical stimulation techniques have been used to reopen existing joints and tectonic faults with the aim to create an artificial heat exchanger in the crystalline basement rocks [1–3]. In a second phase, a binary power plant has been installed on site. The recent third phase of R&D activities is related to the operation of power plant and reservoir and their interaction. Major topics are for example corrosion, scaling and the testing of the whole equipment.

T. Koelbel (✉)
Research & Innovation/EnBW Energie Baden Wuerttemberg AG,
Durlacher Allee 93, Karlsruhe 76131, Germany
e-mail: t.koelbel@enbw.com

A. Genter
G.E.I.E. EMC Soultz-sous-Forêts, Route de Soultz BP 40038, Kutzenhausen 67250, France
e-mail: genter@soultz.net

T.S. Uyar (ed.), *Towards 100% Renewable Energy*, Springer Proceedings in Energy, DOI 10.1007/978-3-319-45659-1_25

2 Geology and Natural Reservoir

The Soultz site has a long lasting history of gas and oil exploration. In fact, the region is one of the oldest oilfields worldwide. Because of several thousand boreholes drilled for hydrocarbons, the geology and especially the temperature distribution in the depth was known quite well. A comparison to typical Central European conditions, anomalously high geothermal gradient—up to 11 K per 100 m—was the reason, why Soultz-sous-Forêts was chosen for the utilization of geothermal power.

The local caprock is formed by a suite of tertiary sedimentary rocks (light yellow in Fig. 1). Below a sequence of Mesozoic layers (Jurassic to Buntsandstein, blue and purple in Fig. 1) covers the crystalline basement (red in Fig. 1).

The crystalline basement at Soultz-sous-Forêts consists of Paleozoic granites. Microfissures at grain scale and cataclastic shear zones present various mineral alterations and are occasionally filled with highly mineralized brine with salt content up to 100 g/l.

Samples of the natural formation fluids in the crystalline basement from various wells and from different depths present a similar geochemical fluid composition (NaCl-dominated brine, pH 5). Isotope studies indicate that the whole brine volume has the same sedimentary origin and history. From gas analyses, carbon dioxide is the predominantly free-gas fraction. Geothermometry information derived from geochemical fluid data gave hints that the brine was once heated up to 220–240 °C temporarily [4–6].

Tectonic faults and joints are preferred flow paths. According to hydraulic tests the transmissibility of shear zones inside the granitic basement rocks can be determined

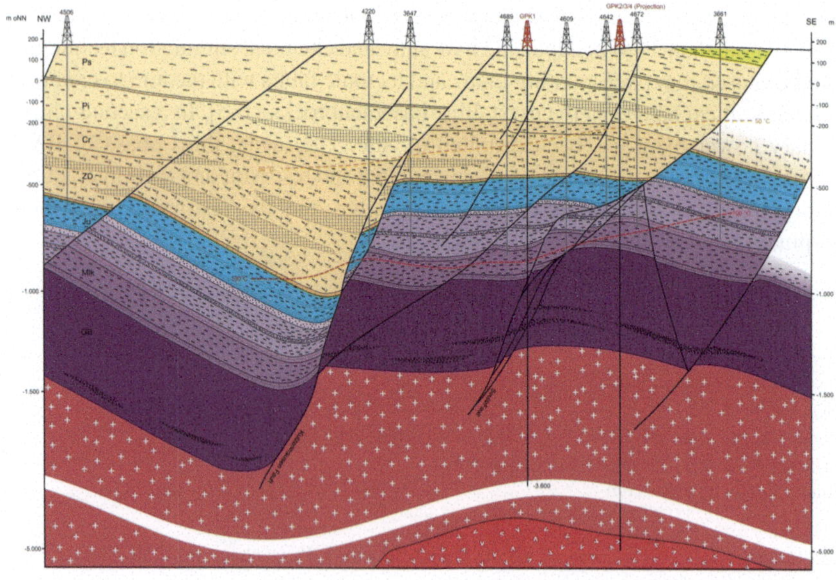

Fig. 1 Geological profile section (according to Cautru, 1989)

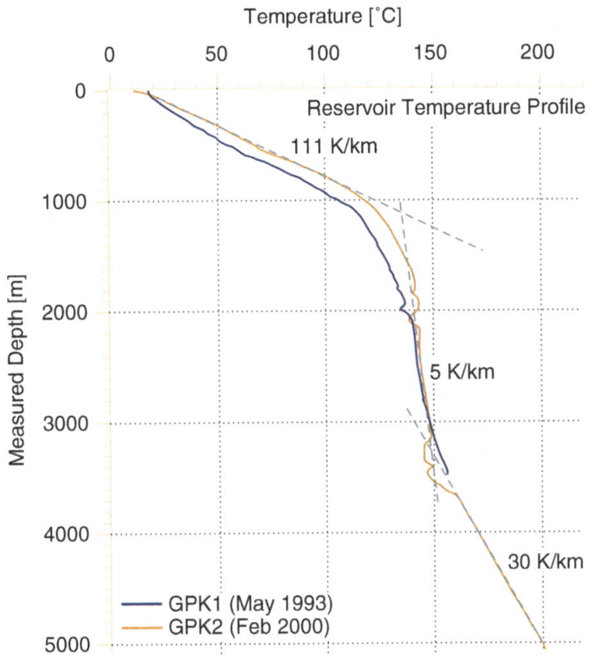

Fig. 2 Geothermal gradient in the Soultz wells (data from LIAG, Hannover)

to 10^{-11} m^2, while the non-sheared or alterated rock suites present a transmissibility of 10^{-16}–10^{-17} m^2 [7].

The average temperature gradient in Central Europe increases with 3 K per 100 m depth. As mentioned above, the temperature increase at Soultz-sous-Forêts amounts up to 11 K per 100 m in the sedimentary cover (Fig. 2). This leads to a geothermal heat flow which is in a range from 100 to 120 W/m^2. This is twice times higher than the average heat flow measured in Central Europe.

As presented in Fig. 2, the geothermal gradient drops significantly in a depth between 1000 and 3500 m (0.5 K per 100 m). The reason is a convection cell of natural brine which occurs at the interface of the sedimentary caprock and the underlying crystalline basement. Below this convection cell and therefore deeper than 3500 m the geothermal gradient at Soultz-sous-Forêts is equivalent to the average geothermal gradient in Central Europe.

3 Stimulation and Microseismicity

The main goal at Soultz-sous-Forêts was to enhance the natural permeability of the crystalline basement rocks. Therefore, several hydraulic and chemical stimulation tests have been performed. In Fig. 3 the principal of both stimulation techniques is

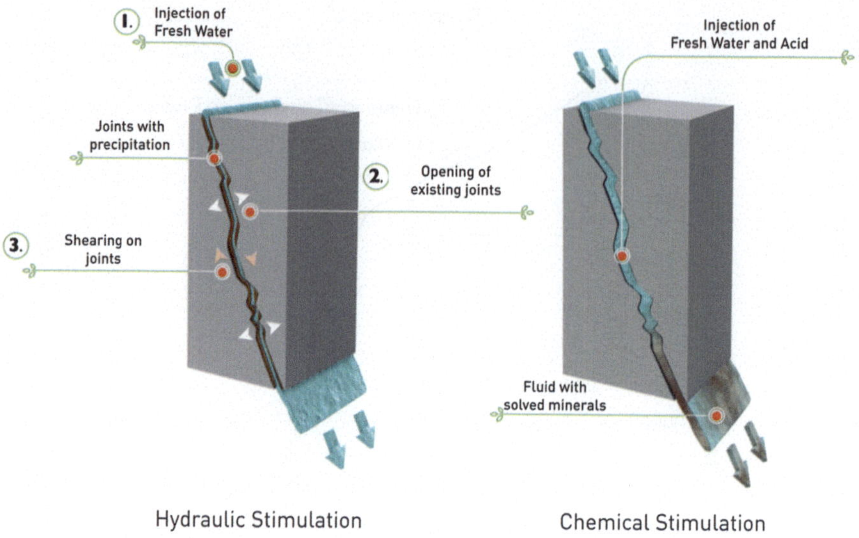

Fig. 3 Principal of hydraulic and chemical stimulation

presented. To open existing joints and tectonic faults, fresh water is injected under high pressure in deep wells (maximum 180 bar). As a consequence the joints is reopened and then sheared. After pressure reduction at the end of the hydraulic injection phase, the reopened joints and faults might tend to close again, but their uneven surface counteracts. For chemical stimulation, different additives can be used (alkaline or acidic). The target is to clean by mineral dissolution, joints, faults, and pore volume in the vicinity from precipitations or matrix cement.

After stimulation, the productivity/injectivity of the Soultz wells was improved 20 times and an artificial heat exchanger with a volume of 3 km^3 has been developed. While the chemical stimulation was nearly aseismic, the hydraulic fracturing caused ten thousand of microseismic events. They have been measured by an extending seismic monitoring network and located 3D in real time (Fig. 4). The highest magnitudes recorded while hydraulic stimulation were up to 2.9, but by far the largest proportion was below a certain level, which can be felt by the public [8, 9].

4 Recent R&D Activities

Since 2009, the geothermal power plant is in operation (Fig. 5). Today it is supported by the industrial consortium G.E.I.E. with its members EnBW and ES. After years of intensive R&D work the power plant unit together with the major part of the surface installation was retrofitted in springtime 2016 and will be used for commercial power production in future.

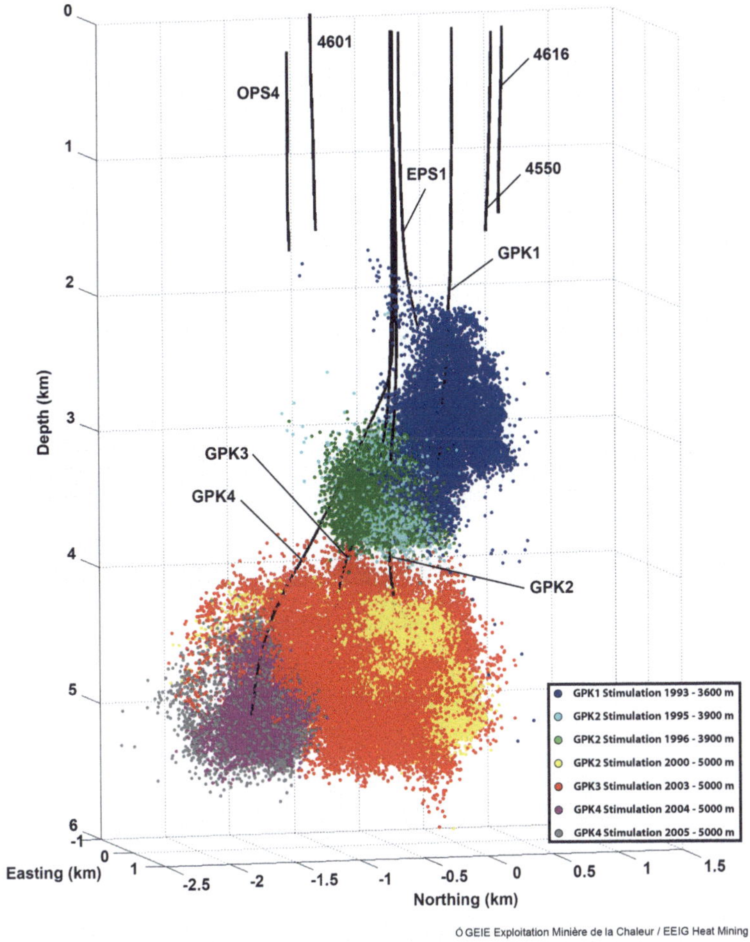

Fig. 4 Microseismic events at Soultz during hydraulic stimulation (©GEIE, 2011)

5 Conclusion

The Soultz-sous-Forêts project indicates clearly that geothermal power production is possible from an area without any active volcano within a low natural permeable reservoir (naturally fractured granite). This opens the opportunity to install geothermal power plants independent from sites with preferable geological conditions and—similarly important, because heat can only be transported for high cost—the utilization of geothermal power close to the client. Today the levelized cost of energy from an EGS plant is not low. But it can be compared with other renewable energy options like photovoltaic and it has a high cost reduction potential.

Fig. 5 Geothermal power plant at Soultz-sous-Forêts (©GEIE, 2009)

References

1. Gerard A, Genter A, Kohl T, Lutz P, Rose P, Rummel F (2006) The deep EGS (enhanced geothermal system) project at Soultz-sous-Forêts (Alsace, France). Geothermics 35(5–6):473–483
2. Baria R, Baumgaertner J, Gerard A, Jung R, Garnish J (1999) European HDR research program at Soultz-sous-Forêts (France) 1987-1996. Geothermics 28(4–5):655–669
3. Baumgaertner J, Gerard A, Baria R, Jung R, Tran-Viet T, Gandy T, Aquilina L, Garnish J (1998) Circulating the HDR reservoir at Soultz: maintaining production and injection flow in complete balance: initial results of 1997 experiment. In: Twenty-third workshop on geothermal reservoir engineering, Stanford University, Stanford, CA, pp. 11–20
4. Genter A, Fritsch D, Cuenot N, Baumgaertner J, Graff JJ (2009) Overview of the current activities of the European EGS Soultz project: from exploration to electricity production. In: 34th Stanford reservoir engineering geothermal workshop 2009, San Francisco, CA, pp. 163–169
5. Sanjuan B, Pinault J-L, Rose P, Gérard A, Brach M, Braibant G, Crouzet C, Foucher J-C, Gautier A, Touzelet S (2006) Tracer testing of the geothermal heat exchanger at Soultz-sous-Forêts (France) between 2000 and 2005. Geothermics 35(5–6):622–653
6. Dezayes C, Villemin T, Genter A, Traineau H, Angelier J (1995) Analysis of fractures in boreholes of Hot Dry Rock project at Soultz-sous-Forêts (Rhine graben, France). Sci Drill 5:31–41
7. Evans KF, Hopkirk R, Jung R, Nami P, Schindler M, Teza D, Tischner T (2008) Milestone events and key observations in thermics, stress and hydraulics at Soultz (1987–2002). In: EHDRA scientific conference, Soultz-sous-Forêts, France, 24–25 Sept 2008, p. 6
8. Nami P, Schellschmidt R, Schindler M, Tischner T (2008) Chemical stimulation operations for reservoir development of the deep crystalline HDR/EGS system at Soultz-sous-Forêts (France). In: Proceedings of the 33rd workshop on geothermal reservoir engineering, 28–30 Jan 2008, Stanford, CA
9. Cuenot N, Dorbath C, Dorbath L (2008) Analysis of the microseismicity induced by fluid injections at the hot dry rock site of Soultz-sous-Forêts (Alsace, France): implications for the characterization of the geothermal reservoir properties. Pure Appl Geophys 165:797–828

Field Study for the Determination of the Ratio of Convective to Total Energy Transport in Geothermal Systems

Ulvi Arslan and Heiko Huber

1 Introduction

For the design of smaller geothermal systems the geothermal, geological, and hydrogeological parameters can be estimated according to common literature. For the design of complex geothermal systems the subsoil has to be modelled with numerical approaches using values evaluated in laboratory or in situ. In saturated soils, especially in areas of groundwater flow, the energy transport of the fluid phase (convection) and the solid phase (conduction) has to be considered separately for a proper numerical modelling of geothermal systems. Even in the case of low groundwater flow velocity of about 10^{-7} m s^{-1}, the role of convective heat transport cannot be neglected [1–3].

Therefore, EGRTs were performed under natural and two different forced groundwater flows by the Technical University Darmstadt in combination with the Groundwater Flow Visualization (GFV) measurement technique with resulting groundwater flow velocities over borehole depth $v_f(z)$. $\lambda_{eff}(z)$ was correlated with the determined groundwater flow velocities $v_f(z)$. The ratio of conductive and convective energy transport to the whole transported heat energy was evaluated.

U. Arslan (✉)
Institute of Materials and Mechanics in Civil Engineering, Technical University Darmstadt, Petersenstraße 12, 64287 Darmstadt, Germany
e-mail: arslan@iwmb.tu-darmstadt.de

H. Huber
Institute of Materials and Mechanics in Civil Engineering, Technical University Darmstadt, Petersenstraße 12, 64287 Darmstadt, Germany

CDM Smith Consult GmbH, 64665 Alsbach, Germany

T.S. Uyar (ed.), *Towards 100% Renewable Energy*, Springer Proceedings in Energy, DOI 10.1007/978-3-319-45659-1_26

2 Site Description

The project area is located in Strausberg 40 km east of Berlin. On site, the energy demand of an office building is supported by four borehole heat exchangers (BHE). The depth of the BHE is 50 m each while the shape differs between single-U, double-U, and coaxial pipes.

In preliminary investigations, the groundwater flow direction on site was determined with the aid of hydrogeological maps as well as with GFV measurement technique at an existing groundwater standpipe B 0 (31 m depth; 25 m south-west to BHE 6). The groundwater direction is pointing south-west.

According to the groundwater flow direction four groundwater standpipes (B 1–B 4) were drilled to varying depths of 31 m (B 0) to 37 m (B 4) and in distances 1.7 m (B 1) up to 7.9 m (B 0) to BHE 6, see Fig. 1. The filter pipes are located in the last 3 m (B 1–B 3) respectively the last 6 m (B 4) of the borehole length. The geological conditions on site are shown in Fig. 1.

Confined groundwater was encountered in all groundwater standpipes with an energy level of about 10.8 m below ground. The measured natural hydraulic gradient i between B 1 and B 2 is 0.007.

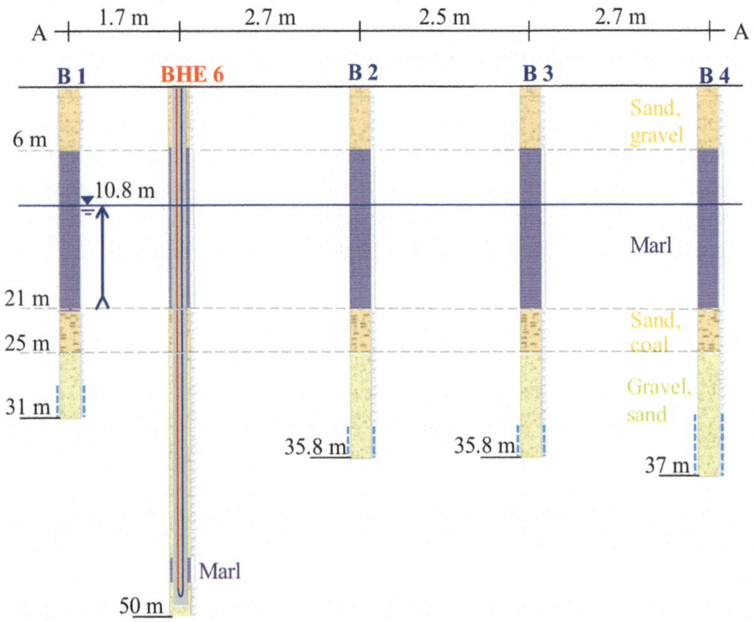

Fig. 1 Geological section

3 Performed Field Tests

3.1 *Pumping Tests*

Two pumping tests were performed on site. The duration of the tests was 5 days, each. Water was continuously extracted from B 4 with 3.4 $m^3\ h^{-1}$ (Pumping Test 1) and 7.5 $m^3\ h^{-1}$ (Pumping Test 2), while the groundwater level of the groundwater standpipes B 1–B 4 was measured. The extracted groundwater was transported to an infiltration area about 70 m north-west of B 4.

The Darcy permeability of the aquifer was determined to $3 \div 5 \ldots 10^{-4}$ m s^{-1}. Due to the confined groundwater and the thin layer of marl in a depth of about 46.20–47.50 m below ground, an almost horizontal groundwater flow can be assumed for BHE 6 in the depth of 21.0–46.2 m (≈50 % of the whole length of the BHE) due to the pumping rates, see Fig. 2.

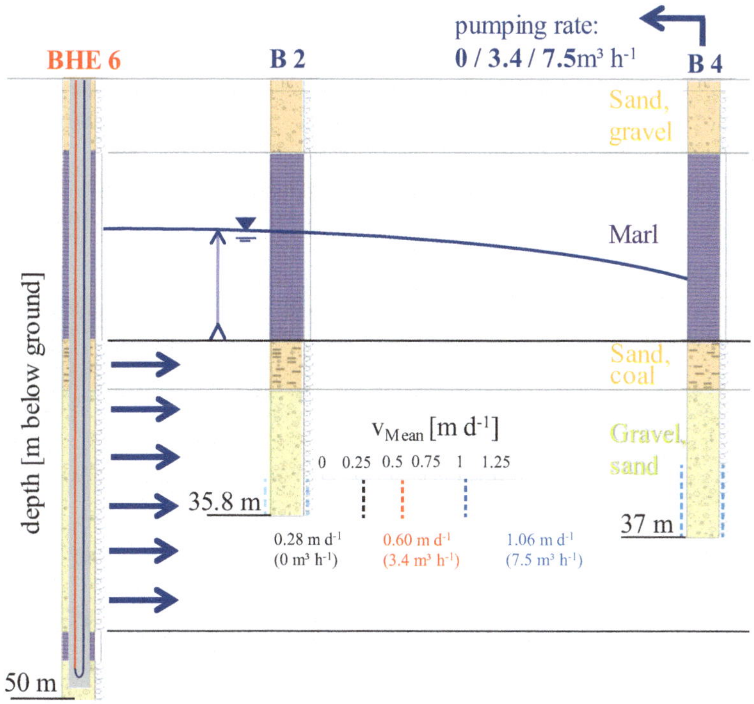

Fig. 2 Horizontal groundwater flow while pumping test

3.2 GFV Measurements

Three Groundwater Flow Visualization (GFV) measurements were performed at B 2. During a GFV measurement, a high resolution camera is located between two packers inside the filter pipe of a given groundwater standpipe. The camera recognizes suspended sediments transported by the groundwater and therefore, determines the groundwater flow direction and velocity. Extensive information on the theoretical background of the GFV measurement is given in [4].

While GFV 1 evaluated the natural groundwater flow velocity and direction, GFV 2 and GFV 3 were performed at the steady state of Pumping Test 1 (GFV 2) and Pumping Test 2 (GFV 3).

The results of GFV 1–GFV 3 performed in the filter pipe of B 2 (34.0–35.8 m) can be summarized as follows: The natural groundwater flow direction determined in GFV 1 (175°) is in good agreement with the groundwater flow direction according to hydrogeological maps. The natural groundwater velocity varies between 0.09 and 0.51 m day^{-1} with a weighted mean according to the quality of every single measurement of 0.28 m day^{-1}.

The groundwater flow direction determined in GFV 2 and GFV 3 differs from the natural groundwater flow direction for in almost 180°, see Fig. 3. The weighted mean of the groundwater flow directions is about to the north (349°), in direction of pumping well B 2.

According to the pumping rates the groundwater flow velocities rises in GFV 2 (3.4 m^3 h^{-1}) to 0.15–1.75 m day^{-1} (weighted mean = 0.6 m day^{-1}) and in GFV 3 (7.5 m^3 h^{-1}) to 0.24–3.09 m day^{-1} (weighted mean = 1.06 m day^{-1}).

More information about the theoretical background of the Groundwater Flow Visualization measurements as well as the detailed results of the performed tests on site is given in Ref. [5].

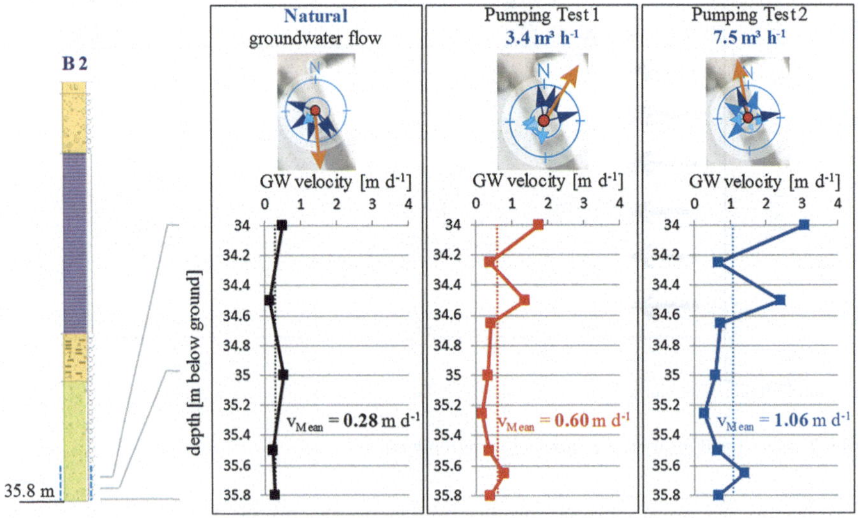

Fig. 3 GFV measurement

3.3 Enhanced Geothermal Response Test

Three Enhanced Geothermal Response Tests (EGRT) were performed at BHE 6 with a constant thermal load of 2617 W/52.3 W m^{-1} (EGRT 1), 2302 W/46.0 W m^{-1} (EGRT 2), and 2075 W/41.5 W m^{-1} (EGRT 3). While EGRT 1 was performed without any groundwater extraction, EGRT 2 was carried out at the steady state of Pumping Test 1 with a pumping rate of 3.4 m^3 h^{-1} and EGRT 3 was carried out at the steady state of Pumping Test 2 with a pumping rate of 7.5 m^3 h^{-1} (EGRT 3).

The increase of temperature k is inversely proportional to the effective thermal conductivity $\lambda_{eff}(z)$. Neglecting borehole resistivity and other effects the effective thermal conductivity of water saturated systems is primarily the sum of conduction and convection. A low increase of the temperature is caused by a high thermal conductivity of the ground or a high groundwater flow.

According to the slope k of the temperature development, the effective thermal conductivity over depth $\lambda_{eff}(z)$ can be determined by the Source Theory. The determined mean effective thermal conductivity $\lambda_{eff}(z=0\text{–}50\text{ m})$ increases according to the applied groundwater flow velocities from 2.11 W m^{-1} K^{-1} (EGRT 1) to 2.37 W m^{-1} K^{-1} (EGRT 2) and 2.49 W m^{-1} K^{-1} (EGRT 3). That means an increase of 12.3 % respectively 18.0 %, see Fig. 4.

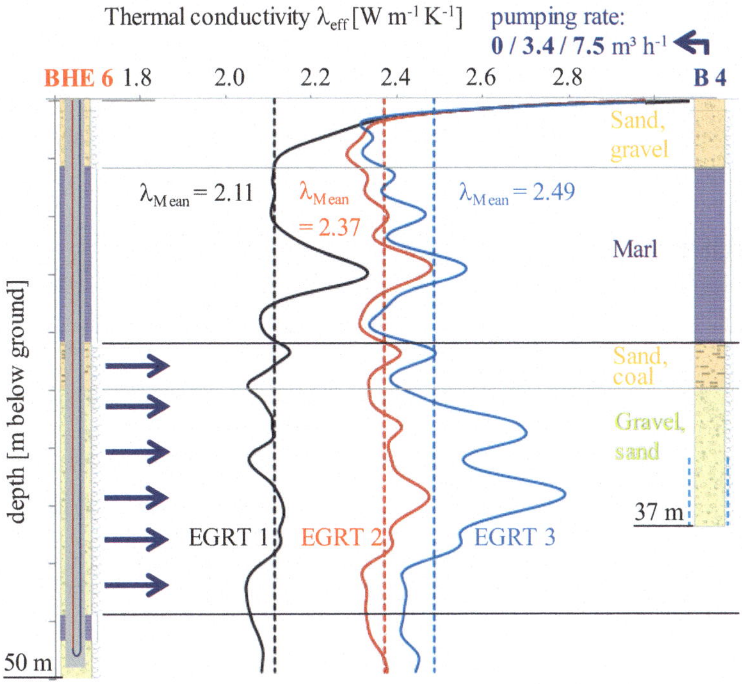

Fig. 4 Effective thermal conductivity over depth (EGRT 1, EGRT 2, EGRT 3)

Table 1 Determined effective thermal conductivity and its increase for different sections

	Effective thermal conductivity [W m^{-1} K^{-1}]		
Section	EGRT 1	EGRT 2	EGRT 3
0–50 m	2.11 (ref.)	2.37 (+12.3 %)	2.49 (+18.0 %)
21–46.2 m	2.1 (ref.)	2.38 (+13.3 %)	2.56 (+21.9 %)
34 m	2.11 (ref.)	2.46 (+16.6 %)	2.79 (+32.2 %)

Especially in the gravely, sandy aquifer layer (21–46.2 m), where groundwater flow occurs, a high increase of the effective thermal conductivity $\lambda_{eff}(z=21–46.2\text{ m})$ of 13.3 % (EGRT 2) and 21.9 % (EGRT 3) can be observed. According to singular flow paths in the aquifer an increase of the effective thermal conductivity $\lambda_{eff}(z=34\text{ m})$ in chosen incremental depths of even 16.6 % (EGRT 2) and 32.2 % (EGRT 3) can be determined, see Table 1.

4 Conclusion

After all, three EGRTs under natural and artificial increased groundwater flows were performed from July to November 2011. By the means of the Groundwater Flow Visualization measurement technique the effective groundwater flow velocities over depth were determined and compared to the effective thermal conductivities. An increase of the effective thermal conductivity up to 32.2 % according to the increased groundwater flow velocity was evaluated.

Currently all the results are compared to the performed laboratory tests and reanalyzed with numerical methods and compared to performed laboratory tests.

References

1. Katsura T, Nagano K, Hori S, Okawada T (2009) Investigation of groundwater flow effect on the thermal response test result. Proceedings of 11th energy conservation thermal energy storage conference, Stockholm, 15–17 Jun 2009
2. Spitler JD, Simon JR, Chiasson AD (1999) Incorporation of groundwater flow into numerical models and design models. Oklahoma State University, Stillwater, OK
3. Kleiner S (2003) Untersuchungen zur Optimierung von Erdwärmesondenanlagen – petrophysikalische Messungen und numerische Simulationen
4. Schöttler M (1997) Meßbarkeit der Grundwasserbewegung durch Visualisierung der Strömung in Bohrbrunnen, Aachen
5. Huber H, Arslan U (2012) Geothermal field tests with groundwater flow. Proceedings thirty-seventh workshop on geothermal reservoir engineering Stanford University, Stanford, CA, 30 Jan to 1 Feb 2012

A Simple Feedback Control Approach for Economic Measures to Deploy New Energy Technologies

Takanobu Kosugi

Nomenclature

C_t	PV system cost, assumed to be the same between residential and nonresidential systems (in thousand JPY/kW)
$F_{R,t}$, $F_{N,t}$	Ratio of residential/nonresidential consumers who have accepted the installation of PV systems to the systems' market potential
K_R, K_N	Ultimate market potential of residential/nonresidential PV system (GW)
$M_{R,t}$, $M_{N,t}$	Residential/nonresidential installed stock of PV systems considering their depreciation (GW)
N_t	Cumulative installed capacity of PV systems, total of residential/nonresidential systems (MW)
t, τ	Past and future time periods
$Y_{R,t}$, $Y_{N,t}$	Simple payback period of residential/nonresidential PV systems (years)
α_R, α_N	Sensitivity of residential/nonresidential consumer's acceptance of installing PV systems to simple payback period; that with a hat denotes the statistically estimated value
γ_R, γ_N	Growth rate coefficient of logistic growth curve of PV system stock capacity; that with an asterisk superscript denotes the estimated optimal value
λ	Progress rate of experience curve of PV systems cost; that with a hat denotes the statistically estimated value

T. Kosugi (✉)
College of Policy Science, Ritsumeikan University,
2-150 Iwakura-cho, Ibaraki, Osaka 567-8570, Japan
e-mail: kosugi@sps.ritsumei.ac.jp

T.S. Uyar (ed.), *Towards 100% Renewable Energy*, Springer Proceedings in Energy, DOI 10.1007/978-3-319-45659-1_27

1 Introduction

Economic measures such as subsidiary supports or feed-in tariffs (FIT) are effective for deploying new energy technologies like solar photovoltaic (PV) power generation systems. However, there is no straightforward way to determine the level of economic measures needed, such as, for example, controlling the purchase price of PV electricity. While generous economic support may result in the excess deployment of a new energy technology, an insufficient level of the needed measures will slow the pace of deployment compared to an optimal level of technological diffusion. The optimal future diffusion level is subject to revision as time passes, because, economically, the level is determined depending on the expected cost decline with an increase in production of the technology [1], and the estimated extent of the mass-production effect has to be revised in response to the newest statistical data observed during each passing time period.

Taking these issues into account, in this study I propose a decision support process for recursively determining an optimal amount of support needed, by using a simple feedback control approach, with the information about the newest cost and installed capacity of the technology added every year. This study also attempts a hypothetical application of the process to a case of policy formulation concerning PV deployment in Japan.

2 A Decision Support Process for Determining an Optimal Amount of Support to Deploy New Energy Technologies

The decision-making process for determining an optimal amount of support to deploy a new energy technology proposed in this chapter is a recursive dynamic process using a simple feedback control approach. The process is illustrated in Fig. 1. Applying the process is expected to enable us to provide policymakers with more rational information for determining the level of economic support to deploy a new energy technology every year. The process requires taking the following seven steps:

Step 1: Obtain actual historical data of the installed capacity, N, and cost, C, of the new energy technology together with the level of financial support, S, for the installation of the technology.

Step 2: Estimate/revise the parameter λ, which represents the effect of the experience curve of the technology, based on a correlation analysis using the historical data of N and C.

Step 3: Estimate/revise the parameter α, which determines the payback acceptance curve that indicates the percentage of potential consumers accepting a payback period of the technology, based on a correlation analysis using the historical data of N, C and S.

Step 4: Estimate/revise the optimal future trajectory of N so as to minimize the discounted sum of future net technology cost, given the (re-)estimated value of λ and the historical data of N and C.

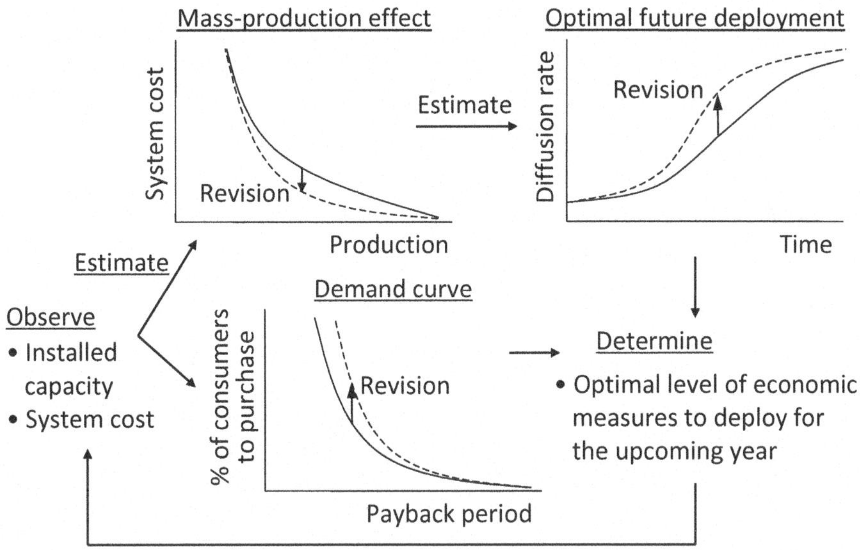

Fig. 1 Proposed decision support framework for determining the level of economic measures needed to deploy a new energy technology every year

Step 5: Project the value of C expected in the subsequent time period when the estimated optimal value of N is actualized, using the (re-)estimated value of λ and the latest data of N and C.

Step 6: Derive the value of S recommended to be set in the subsequent period from the (re-)estimated value of α, the estimated optimal value of N and the corresponding estimated value of C.

Step 7: Adopt the recommended value of S in the subsequent period, observe the up-to-date values of N and C by the end of the period. Repeat Steps 1–7 annually.

3 Hypothetical Application to Japanese PV Deployment Policy

3.1 Data Analysis Using the Latest (FY2010) and Earlier Data

With the intention of applying the process shown in the previous section to determine the future levels of supporting PV system deployment in Japan, the author has gathered actual statistical data from fiscal years (FY) 1994 to 2010 necessary for the application from several data sources [2]. This corresponds to Step 1 of the process. All monetary data are expressed in constant 2010 Japanese yen (JPY) hereinafter. The estimated currency exchange rate in 2010 was 1000 JPY = 17.1 Turkish lira (TL) = 8.59 EUR. Some critical assumptions for the application are shown in

Table 1 Critical assumptions for the decision support process application

Item	Value
Capacity factor of PV systems (%)	12
Depreciation rate of PV system stock (%/year)	3
Discount rate (%/year)	3
Ultimate potential of PV system installation (GW)	Residential: 112, nonresidential: 149
Surplus electric power rate of PV systems (%)	Residential: 55, nonresidential: 15
Marginal cost of utility electricity in FY2010 (JPY/kWh)	8.5
Growth rate of marginal utility power cost (%/year)	2
Consumer price of utility electricity in FY2011 (JPY/kWh)	Residential: 21.7, nonresidential: 12.9
Capacity of PV systems installed in FY2011 (MW/year)	Residential: 874 or 1748[a], nonresidential: 173 or 324[b]
Cost of PV systems in FY2011 (thousand JPY/kWh)	549 or 520[c]
Subsidy for PV system installers in FY2012 (JPY/kWh)	Residential: 88[d], nonresidential: 0

[a]100 or 200 % relative to the FY2010 installation of PV systems in residential sectors.
[b]80 or 150 % relative to the FY2010 installation of PV systems in nonresidential sectors.
[c]5 or 10 % drop from the cost of residential PV systems installed in FY2010; the cost of a nonresidential system is assumed to be the same as that of a residential system.
[d]Total of subsidies from national and local governments, which are the same as those in FY2011.

Table 1. PV systems are categorized into those used in residential and those used in nonresidential sectors; however, the system cost is assumed to be the same for the two types of system due to insufficient statistical data.

The experience curve and payback acceptance curve concerning the PV system installation are estimated based on statistical data analyses. They are shown in Fig. 2a, b, respectively, which also indicate the correlation equations of the curves and point estimates of the parameters. They represent the results of Steps 2 and 3.

Next, as Step 4, the optimal future deployment, i.e., trajectory of PV system capacity, is estimated from an economic viewpoint. We assume that the future deployment trajectory follows a logistic growth curve [3] and estimate the optimal value of the parameter expressing the growth rate that specifies the shape of the curve. The results are shown in Fig. 2c.

3.2 *Deriving FIT Price Recommended for the Subsequent Year (FY2012)*

Regardless of the above analysis, the financial support policy regarding the PV system installation, i.e., the levels of investment subsidy and FIT price, has already been established for FY2011. We thus attempt to derive the FIT price recommended to be set in

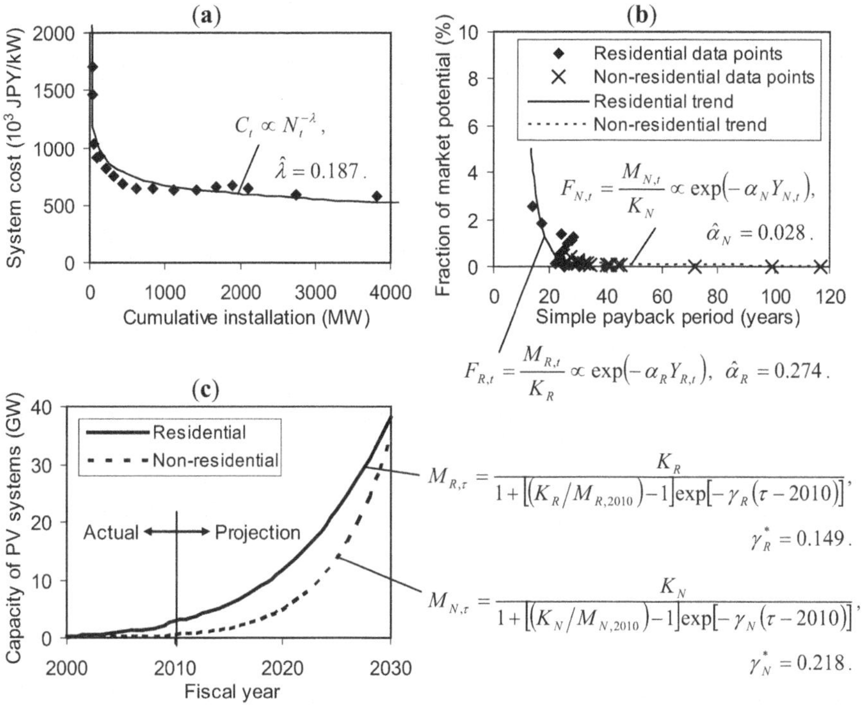

Fig. 2 Statistical analysis of PV system installations and costs in Japan. (**a**) System cost (experience curve). (**b**) Payback acceptance curve. (**c**) Optimal future deployment. *Note*: For the meanings of the variables and parameters in this figure, see the nomenclature list

FY2012 through the application of Steps 1–6 of the decision support process for four hypothetical cases concerning the FY2011 installed capacity, *N*, and cost, *C*, of the PV systems to be observed at the end of FY2011 as Step 7. We presume that surplus or all the electricity generated by the PV systems is purchased at a fixed FIT price in the first 10 years after installation and that the surplus electricity is purchased at the same price as the consumer price of utility electricity thereafter. The level of subsidy per unit of installed PV system capacity in FY2012 is assumed to be the same as that in FY2011.

The results are summarized in Table 2. The FIT price recommended to be set for residential PV systems installed in FY2012, when the same FIT for surplus electricity is adopted as before, falls within the range from 41.7 to 41.9 JPY/kWh, with the difference of only 0.2 JPY/kWh at most among the four cases. If the FIT policy is shifted to apply to all the electricity produced by the PV systems, the recommended FIT price becomes lower by 22 %, and the difference in the FIT price among the cases is curtailed accordingly.

As for the FIT price for nonresidential PV systems, owing to their small ratio of surplus electricity to all the electricity generated, as shown in Table 1, it is recommended to be set at more than double what it was calculated for residential systems when the

Table 2 Results of decision support process application

Cases for FY2011		Parameter estimates (post-revision)					Variable projected for FY2012		Recommended FIT price for FY2012[a]	
Installed capacity, N (MW/year)[b]	System cost, C (10^3 JPY/kW)	λ	α_R	α_N	γ_R	γ_N	Optimal N (MW/year)[b]	C (10^3 JPY/kW)	Residential (JPY/kWh)	Nonresidential (JPY/kWh)
1047	549	0.183	0.275	0.029	0.144	0.214	852	533	41.9 [32.8]	95.9 [25.4]
1047	520	0.185	0.268	0.029	0.168	0.249	979	503	41.7 [32.7]	103.5 [26.5]
2072	549	0.180	0.278	0.029	0.127	0.191	930	534	41.9 [32.8]	89.9 [24.5]
2072	520	0.183	0.272	0.029	0.145	0.219	1053	505	41.8 [32.8]	96.1 [25.4]
Pre-revision estimates[c]		0.187	0.274	0.028	0.149	0.218				

Note: For the meanings of λ, α_R, α_N, γ_R, and γ_N, see Fig. 2 and the nomenclature list.

[a]Figures outside brackets correspond to the case of FIT applied to the surplus electricity of PV systems, while figures inside brackets are for FIT applied to all the PV electricity produced.

[b]Total of residential and nonresidential installations.

[c]Estimates based on FY1994-2010 data. See Fig. 2 for more detail.

FIT for surplus electricity is adopted. On the other hand, when the FIT is applied for all the PV electricity produced, the recommended FIT price can be nearly quartered and even lower than that recommended for residential systems. While the biggest difference in the FIT price among the four cases is as large as 13.6 JPY/kWh for the FIT for surplus electricity, it can be greatly reduced to 2.0 JPY/kWh for the FIT for all the electricity generated.

4 Concluding Remarks

This study proposed and demonstrated a decision support process to determine the economically optimal level of financial support to deploy a new energy technology such as PV systems via a recursive dynamic use of a simple feedback control approach. Though the system needs further refinement, the author hopes the considerations in this study contribute to the establishment of a decision-making process to set an appropriate level of economic support for diffusing new energy systems.

This study was performed as part of a research project of environmental economic policy study entitled "Evaluation of the short, medium and long term effects of various economic policy measures and the effects of combinations of these measures" (Term: FY2009–2011) supported by the Ministry of the Environment, Japan. All opinions and any errors in this chapter are my own.

References

1. Messner S (1997) Endogenized technological learning in an energy systems model. J Evol Econ 7(3):291–313
2. Kosugi T (2011) Research, development, and deployment strategies of photovoltaic and stationary fuel cell energy systems considering the effects of system mass production. Environ Res Q 161:150–155 (in Japanese)
3. Mansfield E (1961) Technical change and the rate of imitation. Econometrica 29(4):741–766

Economic Impacts of Renewable Energy Increase in Germany

Ulrike Lehr and Philip Ulrich

1 Introduction

In 2010, the German government decided to transform the country's energy system and signalled a long-term commitment to an increasing role for renewable energy to play in the country's energy mix [1]. Around the world, new players have entered the market—on both the supply and demand side. 2011 in the aftermath of the Fukushima incident, created new challenges from the nuclear phase-out and the fact that eight nuclear power plants were switched off immediately. For renewables, it was again a year of records. After a final spurt at the end of the year, photovoltaics once more achieved a record level of new installations in Germany, while cell and module manufacturers worldwide had to struggle with drastic overcapacity. This led to an unexpectedly sharp drop in prices, which meant that many manufacturers in Germany, and also internationally, suffered falls in profits or losses. Meanwhile, PV production is mainly based in East Asia, despite EU tariffs and other measures. The wind industry is still successful on the markets, but the markets become more and more regionally distributed. If the expected market size in a region allows it, generator production is established abroad, saving transport costs and fulfilling local content requirements.

This study focuses on employment attributable to the renewable energy sector in Germany. Gross employment refers to the total number of people employed directly in the manufacturing, operation or maintenance of renewable energy facilities or the supply of fuel for them, as well as to people indirectly employed as a result of the demand from these activities for supplies of goods and services. This contribution describes the latest trends in the industry and gross employment for 2014. It is based upon an annual short-term analysis for the German Ministry of the Environment,

U. Lehr P. Ulrich (✉)
Gesellschaft für Wirtschaftliche Strukturforschung mbH, Osnabrück, Germany
e-mail: lehr@gws-os.com

T.S. Uyar (ed.), *Towards 100% Renewable Energy*, Springer Proceedings in Energy, DOI 10.1007/978-3-319-45659-1_28

Nature Conservation and Nuclear Safety. The analysis is part of a comprehensive study of employment generated by the expansion and operation of renewable energy installations up to 2030 and beyond. Whereas the annual short-term analysis focuses on the significance of the manufacture and operation of renewable energy facilities in terms of numbers of people employed there, the more comprehensive analysis also looks in detail at the future net effects for the economy as a whole, taking all additional costs and production shifts into account.

While the positive impacts of an increasing share of renewable energy (RE) on the mitigation of climate change as well as on reduced energy import dependency are indisputable, such are currently still the additional costs of heat and electricity generation from most renewable energy sources (RES). Therefore, the overall balance of positive and negative effects under different possible future development pathways of fossil fuel prices, global climate policies and global trade is of interest.

This contribution is organized as follows. Chapter 2 briefly explains the methodology, Section 3 presents results on gross employment and Section 4 gives an outlook on net employment and concludes. An Annex gives more details of the technical aspects and the model.

2 Methodology

For the analysis of gross effects we apply input–output analysis with I/O tables which have been specifically extended with the production sectors of renewable energy technologies. These RE sectors are represented by technology-specific vectors derived from a survey using the base years 2004, 2007 and 2012. Relevant benchmark data, such as the productivity of individual sectors, are regularly adjusted. A similar procedure is used to estimate the employment provided by operating and maintaining plant and equipment installed in Germany. Employment arising from the supply of biogenic fuels is also determined using input–output analysis. This methodology has also been confirmed by recent reviews from IEA-RETD [2].

For the net analysis, we include negative impacts on the economy, which stem from two different sources: firstly, investment in renewable energy technologies crowds out investment in fossil fuel technologies such as coal fired power plants, oil fired heating systems and maybe at some future point gasoline driven cars. This substitution effect leads to profit losses in the respective economic sectors.

The second negative effect is larger than the substitution effect and comes from the additional costs of RE systems. Germany supports RE electricity with a feed-in tariff, which leads to electricity price increases for households and firms. This so-called budget effect reduces the available budget for other expenditures resulting in job losses in the respective sectors. The direction of the budget effect can sometimes change, as high PV electricity production during midday already avoids price peaks.

With the further reduction of production costs and the better integration of RES into the electricity system, the average of future budget effects will tend to become less negative or even getting positive in the long run.

Positive and negative impacts can induce additional indirect impacts throughout the economy (so-called second round effects): additional employment results in additional expenditure on consumption and additional jobs in the respective sectors as well as additional taxes and therefore increases in the governmental budget. To account for all effects in a consistent framework, an econometric simulation model is employed. Economic impact of RES expansion is measured via the comparison of economic indicators such as GDP and employment from different simulation runs. Overall net positive effects can be seen for instance as higher employment in one simulation run compared with the other. The model consistently links energy balance data to economic development on sector level. It is enlarged by detailed data on 10 RES technologies based on comprehensive survey data. Additionally, the sector disaggregation of our model leads to a wide array of interesting results in terms of winners and losers of policies to support renewable energy.

For this analysis we employ the economy-environment-energy model PANTA RHEI. PANTA RHEI is an ecologically extended version of the 58 sector econometric simulation and forecasting model INFORGE. The extension comprises of a deeply disaggregated energy and pollution model, including 30 fuels which are used in households and the production sectors. From the modeling aspects, PANTA RHEI belongs to the class of econometric input–output models. Its advantages are the ability to model bounded rationality decisions and the wide empirical database. PANTA RHEI is built fully integrated and bottom up, leading to each sector of the economy being modeled in great detail. The macroeconomic aggregates are calculated by explicit aggregation. The model consists of more than 40,000 equations describing the inter-industry flows between the 58 sectors, their deliveries to personal consumption, government, equipment, investment, construction, changes in stocks, exports, as well as prices, wages, output, imports, employment, labor compensation, profits, taxes, etc. and describes income redistribution in full detail. One further strength of the model is its high level of interdependence, for instance between prices and wages or between prices and volumes.

Final demand is determined from the disposable income of private households, the interest rates and profits, the world trade variables and the relative prices for all components and product groups of final demand. For all intermediary inputs, imports and domestic origins are distinguished. Givens final and intermediary demand, final production and imports are derived. Employment is determined from the production volume and the real wage rate in each sector, which in return depends on labor productivities and prices.

The new detailed structure of the renewable energy sector has been integrated in the model and the future energy scenarios provide information on investment in the sector and the financial burden on the economy by way of additional costs of renewable energy (budget effects). The effects of a certain policy measure are calculated by comparing different runs of the model, one run using a reference development without the measure and one—or several—that includes a policy measure. The comparison of the effects on the macroeconomic indicators then shows the net economic effects, e.g., on the labor market and on GDP.

3 Results: Gross Employment from Manufacturing RE Technologies, Installation Operation and Maintenance and Provision of Biofuels

In 2014, investment in renewable energy installations in Germany totalled € 18.9 billion, 20 % higher than the year before. This estimate [3] is based on provisional results for 2014 published by the interministerial Working Group on Renewable Energies—Statistics (AGEE-Stat). On closer examination, it becomes clear that the rise in investments is driven by very successful wind energy. Onshore installations amounted to almost € 7 billion.

For the calculation of jobs from production in Germany, turnover of German companies is the decisive quantity. With € 21.8 billion, the turnover of German-based manufacturers of RE facilities and components remained roughly at the same level as the previous year. A positive trend was observed in wind energy, with 11 % increase in turnover. Wind energy was successful also in exporting generators (Table 1).

Table 1 Employment from renewable energy in Germany in 2014

	Investment (incl. export)	Maintenance and operation	Fuel supply activities	Total no. of jobs in 2014	Total no. of jobs in 2011
Onshore wind	109,700	20,800		130,500	119,000
Offshore wind	15,400	3300		18,700	18,800
Photovoltaics	27,200	11,100		38,300	56,000
Solar thermal	9000	1300		10,300	11,400
CSP	700			700	1100
Hydropower	7000	4800		11,800	13,100
Deep geothermal	800	300		1100	1500
Heat pump	13,400	2700		16,100	15,800
Biogas	15,700	11,700	20,900	48,300	49,200
Small-scale biomass	9100	4000	12,300	25,400	28,600
Biomass-fired heating/power stations	5600	8900	8600	23,100	23,000
Biofuels for transport			23,100	23,100	25,600
Total	213,600	68,900	64,900	347,400	363,100
Research/administration				8000	8300
Total				355,400	371,400

The relevant drivers of employment from operations and maintenance (O&M) of existing installations are the costs of operation (without fuel costs), which are calculated as a percentage of the investment costs. More detailed data has changed the breakdown of employment in the different areas in which biomass is used. This statistical effect caused a leap in employment in operation and maintenance of biogas facilities in 2011 by comparison with the previous year that was greater than would have been expected on the basis of the increase in numbers of installations. The correction caused a disproportionate decrease in employment in biomass-fired heat/power plants by comparison with the previous year. With growing installation numbers, employment in O&M overall is becoming increasingly relevant. Meanwhile, the development is more steady.

Overall employment from renewable energy in 2014 went down to 355,400, but still more than twice as high as in 2004.

Since the gross employment was first determined for the reference year 2004, an interesting development for over 10 years can be drawn and interpreted (see Fig. 1). The RE-employment since then has doubled in 8 years, then rose in 4 years by a quarter. Since 2012, the gross employment subsequently declined by about 11 %. Much of the dynamic is due to the development of solar energy. The PV installations that were already increasing continuously until 2009, reached record highs of over 7 GW in annual power installations in 2010, 2011 and 2012. Since the installation capping and the reduction of feed-in tariffs, new installations have slumped. Not just the installers were negatively affected

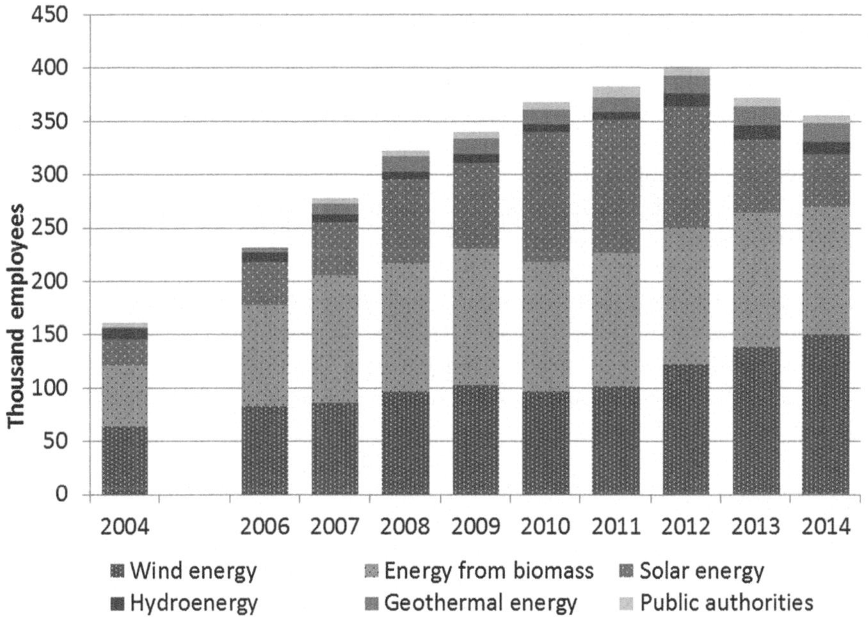

Fig. 1 Development of gross employment by energy carriers since 2004

by the drop in demand. In addition, the already-enhanced fall in prices led to a larger job-cut in domestic manufacturers. A remarkably positive influence, since the year 2012, however is the significant growth of wind energy employment. The feed-in tariff is still attractive and although offshore lags a bit behind the plan, it picks up speed. Overall, the wind industry has managed to keep its relevance ion international markets. The development in recent years has meant that new or preserved jobs are more concentrated among the coastal states, since the wind energy is strongly represented there [4].

4 Net Effects and Conclusions

Our analysis shows possible positive impacts of the expansion of RE in Germany—and the conditions and policy implication for a positive development.

Including negative effects, net employment in 2015 has been around 30,000 people. For the future development, net employment strongly hinges on the position of German producers and service companies on international markets. Increasingly, investment into RE takes place in emerging economies such as China, India or South Africa—or even in developing countries, for instance in the MENA region. Positive net employment effects strongly depend on further growth of global markets and German RE exports. Lehr et al. [5] show positive net effects in the order of 100,000–170,000 people by 2030.

The discussion of net impacts of RES has also to be related to the primary target of RE: the reduction of fossil fuel use and related externalities such as global warming, local air pollution or damages of fossil fuel extraction. If (national) accounting systems take fully account of these additional benefits of RE compared to fossil fuels, as proposed by the Commission on the Measurement of Economic Performance and social progress [6] or the OECD [7] green growth strategy, the evaluation of RE will become even more positive.

The issue of economic impacts of the expansion of RE will be part of the sustainability discussion for the time to come. On the one hand, increasing installation will bring down the specific costs through learning curves and scale effects. On the other hand, parity of electricity generation costs from RES will only be reached within the next 10–15 years. The German example shows how a large domestic market leads to the development of a successful industry. However, these successes are vulnerable to abrupt policy changes, as experiences with the US industry or the Spanish market show.

Acknowledgements This research has been supported by the German Federal Ministry for the Environment, Nature Conservation and Nuclear Safety and the German Federal Ministry of Economic and Social Affairs. The full analysis includes more aspects. Marlene O'Sullivan, Dietmar Edler, Christian Lutz, Peter Bickel, Sonja Simon, Tobias Naegler, Uwe Pfenning Fabian Sakowski, and Ines Thobe contributed to this work.

Annex

The economy-energy-environment model PANTA RHEI is at the core of our methodological approach. PANTA RHEI [8, 9] is an environmentally extended version of the econometric simulation and forecasting model INFORGE [10, 11]. Among others it has been used for economic evaluation of different energy scenarios that have been the basis for the German energy concept in 2010 [12, 13]. A similar model for Austria [14] has recently been applied to the case of sustainable energy development in Austria until 2020. The following description is taken from [4].

The behavioral equations reflect bounded rationality rather than optimizing behavior of agents. All parameters are estimated econometrically from time series data (1991–2008). Producer prices are the result of mark-up calculations of firms. Output decisions follow observable historic developments, including observed inefficiencies rather than optimal choices.

Structural equations are usually modeled on the 59 sector level (two digit NACE classification) of the input–output accounting framework of the official system of national accounts (SNA) and the corresponding macro variables are then endogenously calculated by explicit aggregation. In that sense the model has a bottom-up structure. The input–output part is consistently integrated into the SNA accounts, which fully reflect the circular flow of generation, distribution, redistribution, and use of income.

The core of PANTA RHEI is the economic module, which calculates final demand (consumption, investment, exports) and intermediate demand (domestic and imported) for goods, capital stocks, employment, wages, unit costs and producer as well as consumer prices in deep disaggregation of 59 industries. The disaggregated system also calculates taxes on goods and taxes on production. The corresponding equations are integrated into the balance equations of the input–output system.

Value added of the different branches is aggregated and gives the base for the SNA system that calculates distribution and redistribution of income, use of disposable income, capital account and financial account for financial enterprises, non financial enterprises, private households, the government and the rest of the world. Macro variables like disposable income of private households and disposable income of the government as well as demographic variables represent important determinants of sectoral final demand for goods. Another important outcome of the macro SNA system are net savings and governmental debt as its stock. Both are important indicators for the evaluation of policies. The demand side of the labor market is modeled in deep sectoral disaggregation. Wages per head are explained using Philips curve specifications. The aggregate labor supply is driven by demographic developments.

The model is empirically evaluated: The parameters of the structural equations are econometrically estimated. On the time consuming model-specification stage various sets of competing theoretical hypotheses are empirically tested. As the resulting structure is characterized by highly nonlinear and interdependent dynamics the economic core of the model has furthermore been tested in dynamic ex-post simulations. At this, the model is solved by an iterative Gauss-Seidel algorithm year by year.

The energy module captures the dependence between economic development, energy input and CO_2 emissions. It contains the full energy balance with primary energy input, transformation and final energy consumption for 20 energy consumption sectors, 27 fossil energy carriers and the satellite balance for renewable energy [15]. The energy module is fully integrated into the economic part of the model.

To fully assess the impacts from the production and operation and maintenance of renewable energy systems, input–output structures for the renewable energy sectors have been developed and integrated in the modeling framework [9]. Input–output tables provide detailed insights in the flows of goods and services between all sectors of the economy and the interdependence of the economy of a country and with the rest of the world. They are closed accounting schemes where the identity of the sum of inputs and the sum of outputs has to hold in each sector. This consistency check of course also holds true for the newly created sector “Production of systems for the use of RES”. The new sector is defined in economic terms by its input and output structure, being represented by a new column and a new row in an existing table. The input or cost structure describes the amounts of goods and services required as intermediate inputs from all other domestic sectors, the amount of imported intermediate inputs and the value added in the sector itself. The output or sales structure describes the amounts of goods and services delivered to other sectors as intermediate goods or as final goods to final demand: Wind energy offshore and onshore, PV, hydro, solar thermal heat generation, biomass electricity generation, biomass heat generation, geothermal electricity generation, heat pumps, and biogas generation.

To account for the variety of technologies involved in RES use the newly created sector is build up in a bottom up process based on ten subsectors each of which represents a defined RES technology. Compared to a previous study [9] the larger number of technologies distinguished improves the homogeneity of subsectors which is beneficial for empirical quality of representing technologies in an input–output framework. Figure 2 shows examples of the cost structures in percent of gross production volume (GPV) derived for the production of systems for the use of wind energy and photovoltaic systems in comparison to systems for the use of fossil fuels. The difference in structures shows the importance of creating a new sector in the system, since the intermediate inputs used for the respective production processes come from very different sectors of the economy so that the overall and sectoral impacts are depending on a reliable empirical representation of the new RES technologies in the applied analytical tool.

To examine the economic effects of increasing shares of renewable energy in Germany our analysis applies PANTA RHEI to a set of scenarios and compares the resulting economic outcomes.

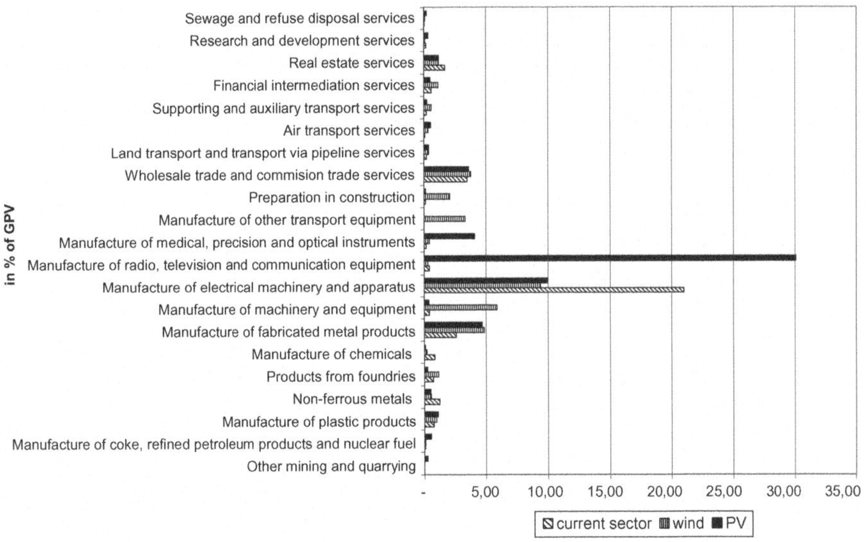

Fig. 2 Cost structure for the production of systems for the use of different energy types, in percent of Gross Production Value (GPV)

References

1. BMU, BMWi (2010) Energy concept. Federal Ministry for the Environment, Nature Conservation and Nuclear Safety (BMU), Federal Ministry of Economics and Technology (BMWi), Berlin
2. Breitschopf B, Nathani C, Resch G (2012) Methodological guidelines for estimating the employment impacts of renewable energy use in electricity generation Study commissioned by IEA-RETD, Paris
3. O'Sullivan M, Edler D, Nieder T, Rüther T, Lehr U, Frank P (2015) Gross employment from renewable energy in Germany in 2011 – a first estimate. Federal Ministry for the Environment, Nature Conservation and Nuclear Safety, Bonn
4. Lehr U, Lutz C, Edler D (2012) Green jobs? Econ Impacts Renew Energy Germany 47:358–364. doi:10.1016/j.enpol.2012.04.076
5. Lehr U, O'Sullivan M, Edler D, Lutz C, Bickel P, Simon S, Naegler T, Pfenning U, Sakowski F, Thobe I (2015) Beschäftigung durch den Ausbau erneurbarer Energien, Ausbau und Betrieb – heute und morgen. Study commissioned by the German Federal Ministry for the Environment, Nature Conservation and Nuclear Safety and the German Federal Ministry of Economic and Social Affairs
6. Stiglitz JE, Sen A, Fitoussi J-P (2009) Report by the commission on the measurement of economic performance and social progress. Commission on the Measurement of Economic Performance and Social Progress, Paris
7. OECD (2011) Towards green growth. OECD, Paris, http://dx.doi.org/10.1787/9789264111318-en
8. Lutz C, Meyer B, Nathani C, Schleich J (2005) Endogenous technological change and emissions: the case of the German steel industry. Energy Policy 33:1143–1154
9. Lehr U, Nitsch J, Kratzat M, Lutz C, Edler D (2008) Renewable energy and employment in Germany. Energy Policy 36:108–117
10. Ahlert G, Distelkamp M, Lutz C, Meyer B, Mönnig A, Wolter MI (2009) Das IAB/INFORGE-Modell. In: Schnur P, Zika G (eds) Das IAB/INFORGE-Modell. Ein sektorales makroökonometrisches Projektions- und Simulationsmodell zur Vorausschätzung des längerfristigen Arbeitskräftebedarfs. IAB-Bibliothek 318, Nürnberg, pp. 15–175

11. Meyer B, Lutz C, Schnur P, Zika G (2007) Economic policy simulations with global interdependencies: a sensitivity analysis for Germany. Econ Syst Res 19:37–55
12. Lindenberger D, Lutz C, Schlesinger M (2010) Szenarien für ein Energiekonzept der Bundesregierung. Energiewirtsch Tagesfragen 60:32–35
13. Nagl S, Fürsch M, Paulus M, Richter J, Trüby J, Lindenberger D (2011) Energy policy scenarios to reach challenging climate protection targets in the German electricity sector until 2050. Util Policy 19:185–192
14. Stocker A, Großmann A, Madlener R, Wolter MI (2011) Sustainable energy development in Austria until 2020: Insights from applying the integrated model "e3.at". Energy Policy 39:6082–6099
15. AGEB (2010) Energiebilanzen für die Bundesrepublik Deutschland, Satellitenbilanz "Erneuerbare Energien". Arbeitsgemeinschaft Energiebilanzen e.V, Berlin

The Potentials and the Benefits of Intensified RES Cooperation Between the European Union and Its Neighbours

Gustav Resch, Marijke Welisch, Gerhard Totschnig, and Andre Ortner

1 Introduction

At European level, Directive 2009/28/EC [1] subsequently named as RES Directive sets binding national targets for all EU Member States to reach an overall RES contribution of 20 % in EU's gross final energy consumption by 2020. These national 2020 RES targets are defined in a way that does not explicitly reflect the national resource availability. In order to allow for cross-border support of renewable energy in a more cost-efficient manner, articles 6–11 of that Directive introduce cooperation mechanisms, providing Member States as well as third countries with an option to agree on cross-border support of RES. Thereby, one country can partly make use of the more cost-efficient RES potentials of another country—within the EU and in neighbouring regions. By joining forces, countries may exploit potentials which otherwise would have remained untapped.

This chapter summarises key outcomes of a comprehensive model-based integrated assessment related to the potentials and the benefits of intensified RES cooperation between the European Union and its neighbours, see [2]. This assessment was a key element of the work conducted in a European joint initiative named "Bringing Europe and Third Countries closer together through Renewable Energies (BETTER)". The BETTER project can be characterised as collaborative action of several European research institutions, policy consultants and stakeholders. The overall aim of this project was to analyse in detail the role of RES cooperation between the EU and its neighbours (i.e. North Africa, Turkey and Balkan countries).

G. Resch (✉) • M. Welisch • G. Totschnig • A. Ortner
Energy Economics Group, Institute of Energy Systems and Electrical Drives, Technische Universität Wien (TU Wien), Gusshausstrasse 25/370-3, 1040 Vienna, Austria
e-mail: resch@eeg.tuwien.ac.at

T.S. Uyar (ed.), *Towards 100% Renewable Energy*, Springer Proceedings in Energy, DOI 10.1007/978-3-319-45659-1_29

This initiative could be established thanks to the financial and intellectual support offered by the Intelligent Energy—Europe (IEE) Programme of the European Commission, operated by the Executive Agency for Small and Medium Enterprises. For more details on the BETTER project, see www.better-project.net.

2 Method of Approach

2.1 The BETTER Modelling System

The approach used for the integrated model-based assessment in the BETTER project combines different dimensions:

- A comprehensive scenario-based assessment of prospects for RES cooperation from the integrated (top-down) perspective has been undertaken with TU Wien's Green-X model. This *techno-economic policy analysis* acted as key basis for our overall evaluation of prospects for RES cooperation in the enlarged geographical context (EU plus third countries). It allowed for identifying monetary savings associated with enhanced RES cooperation as well as resulting changes in costs, expenditures and benefits by region that come alongside with the changes in installed RES capacities and generation across the assessed regions. With Green-X the overall modelling of future RES developments in the EU and its neighbours was done for all energy sectors (i.e. electricity, heating and cooling and biofuels in transport) whereas our detailed assessment of enhanced cross-border RES cooperation was limited to the electricity sector.
- Complementary to above and specifically for the electricity sector, grid and transmission needs or constraints, respectively, together with the physical integration possibilities were evaluated from a technical perspective in a *power-system analysis*, done by use of TU Wien's HiREPS model.

Figure 1 gives an overview on the interplay of both models. Both models were operated with the same set of general input parameters, however in different spatial and temporal resolution. Green-X delivers a first picture of renewables deployment and related costs, expenditures and benefits by country on a yearly basis (2010–2040). The output of Green-X in terms of country- and technology-specific RES capacities and generation in the electricity sector for selected years (2030, 2040) served as input for the power-system analysis done with HiREPS. Subsequently, the HiREPS model analysed the interplay between supply, demand and storage in the electricity sector on an hourly basis for the given years. The output of HiREPS was then fed back into the RES investment model Green-X. In particular the feedback comprised the amount of RES that can be integrated into the grids, the electricity prices and corresponding market revenues (i.e. market values of the produced electricity of variable and dispatchable renewable electricity (RES-E)) of all assessed RES-E technologies for each assessed country.

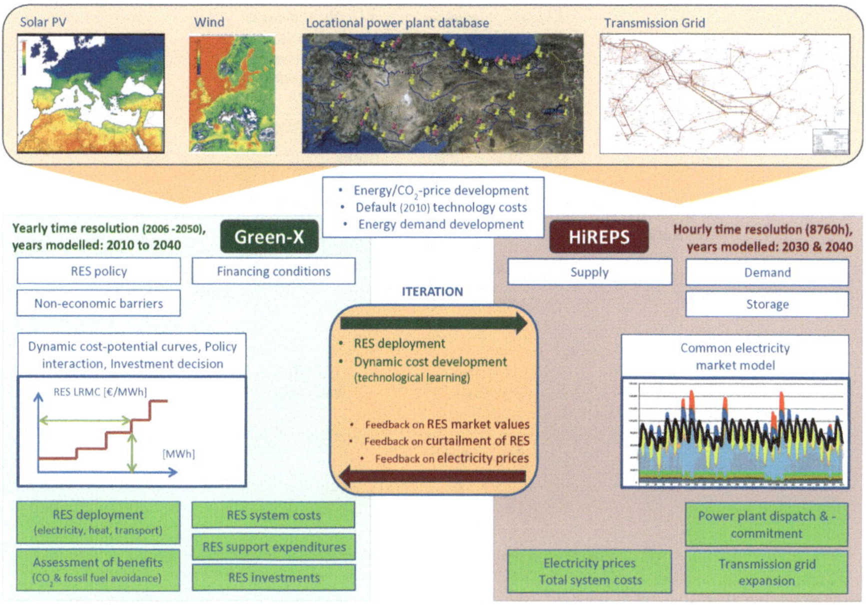

Fig. 1 Model coupling between Green-X (*left*) and HiREPS (*right*) in the integrated assessment of the BETTER project

2.2 Overview on (Selected) Assessed Cases

The integrated assessment conducted in the BETTER project served as an overarching top-down approach to identify opportunities for RES cooperation under varying policy pathways in pan-EU scenarios. These scenarios include the EU28 and North Africa, Turkey and Western Balkan regions as cooperation partners. The future scenarios are defined along two dimensions; firstly, with respect to the EU RES target ambitions, and secondly, related to the choice of cooperation strategy; and presented in Fig. 2.

For the ambition level related to the future RES expansion in Europe, two distinct RES pathways are presented in this Roadmap for 2030 (and beyond), one following a strong RES target for 2030 (i.e. 32.5 % as RES share in gross final energy demand at EU level), and one reflecting the current policy thinking, aiming for a 2030 RES share of 27 %.

Concerning the second dimension, two scenarios are distinguishable:

Reference cases (also referred to as *EU only* scenarios): RES cooperation only within the EU, i.e. no cooperation between EU Member States and the neighbouring regions.

Default cases of full RES cooperation (also referred to as *EU plus* scenarios): these scenarios assume full RES cooperation and across the EU as well as all three case regions (North Africa, Western Balkans and Turkey).

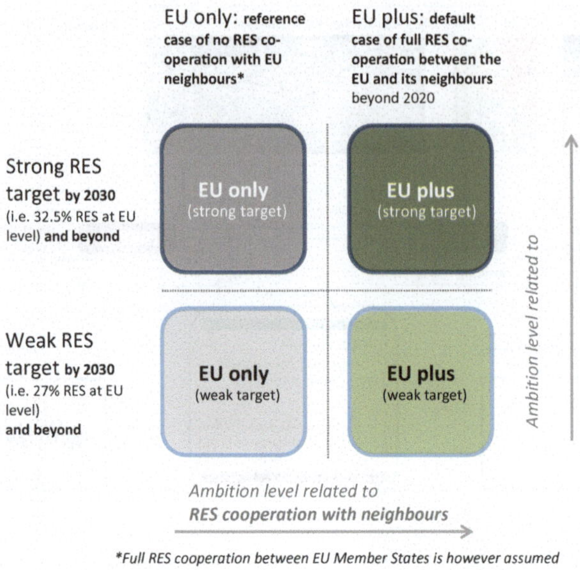

Fig. 2 Possible futures for RES cooperation: overview on (selected) assessed cases (Green-X scenarios))

3 Key Results

3.1 Total RES Deployment by Region

The previously presented quadrant (cf. Fig. 2) shows some exemplary futures that could evolve from the wide range of scenarios assessed throughout the integrated assessment of RES cooperation between the EU and its neighbours. The integrated assessment paints a big picture scenario and provides valuable policy implications for future cooperation between the EU28 Member States and their neighbouring countries. Some key outcomes related to renewables deployment are discussed below. Thus, the analysis starts by looking at the aggregated picture, taking a closer look at the expected region-specific (i.e. EU28, North Africa, Turkey and Western Balkans) deployment of RES in total, comprising energy production that stems from renewable sources in the various energy sectors (including electricity, heating and cooling and transport).

In Fig. 3 a comparison is undertaken between the reference case of RES cooperation limited to EU Member States (EU only) and the full cooperation case (EU plus) where the EU is assumed to extend the geographical scope of RES cooperation towards its neighbouring countries and regions. This exercise is done twice—i.e. once for a weak future RES target (in accordance with the agreed minimum target of 27 % RES by 2030 at EU level) and once for a strong RES target (32.5 % RES by

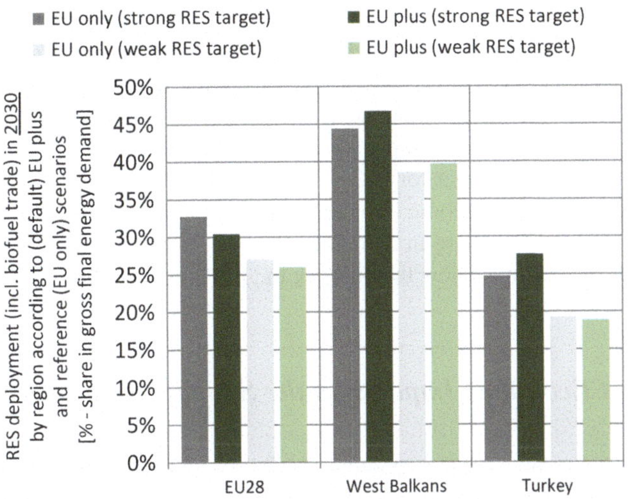

Fig. 3 RES share in gross final energy demand in 2030 by region according to EU plus (full cooperation) and reference (EU only—i.e. cooperation only within the EU) scenarios

2030). Thereby, with the exception of biofuels in transport where consumption matters, RES deployment is accounted for in the country/region of production. Since only RES developments in the electricity sector are analysed for North Africa it is left out from this discussion.

For the EU domestic RES-E deployment is substituted to a certain extent by RES-E imports when cooperation is made possible, irrespective of RES ambitions. In the time frame to 2030, the share of domestic RES generation in total energy consumption is 1 % point higher in the EU plus scenario compared to the EU only scenario. In the time frame to 2040, the difference is 3 % points.

The EU only scenario predicts a linearly increasing RES deployment in the West Balkans up to 2040 that ranges between roughly 54 and 61 % of in gross final energy demand, depending on the overall ambition level concerning RES, i.e. the RES target set for 2030 and beyond. In the EU plus scenarios, this share is a bit lower for the weak target scenario in the final period, e.g. the order of 51 % instead of 54 % by 2040 while in earlier years the opposite trend is applicable—i.e. RES deployment is higher in the case of full cooperation than under reference conditions. This implies that, following a conservative pathway for RES, the West Balkan Contracting Parties of the Energy Community offer attractive opportunities for RES investments in the short- to mid-term while other neighbours, in particular North African countries, offer a more viable long-term perspective. In the case of a strong target the attractiveness of investing more than needed domestically is applicable throughout the whole assessment period—in other words, West Balkans would act as virtual exporter to the EU28 until 2040.

In the case of a weak target Turkey would become an importer under a full cooperation (EU plus scenario) since domestic RES deployment is then lower compared to the reference (EU only) case of no cooperation with the EU28. Thus, North Africa would act under these circumstances as key host, exporting physically to the EU28 and the Energy Community (presumably including Turkey). Contrarily, assuming a strong RES target for 2030 and beyond and full cooperation (EU plus scenario), Turkey would become a major host for investments in RES, achieving a consistently higher RES share in comparison to the reference (EU only) case, that lies roughly 2–4 % points below the ones for the EU plus scenario.

3.2 *Direct Economic Impacts on the Support for Renewables*

To enable conclusions on the economic feasibility of the different levels of RES deployment, they have to be quantified, i.e. expressed in economic terms. Figure 4 shows in which ways the price for green certificates could evolve given different RES cooperation scenarios. One can see that considerable savings are possible. Comparing the reference cases (EU only) to the full cooperation scenarios (EU plus), one can observe substantial differences. Thus, extending the geographical scope of the RES market allows for significant savings related to the support for RES. This already gives a first indication how the EU, North Africa, the Western Balkans and Turkey could benefit from cooperation within a joint market for RES, in particular for renewable electricity.

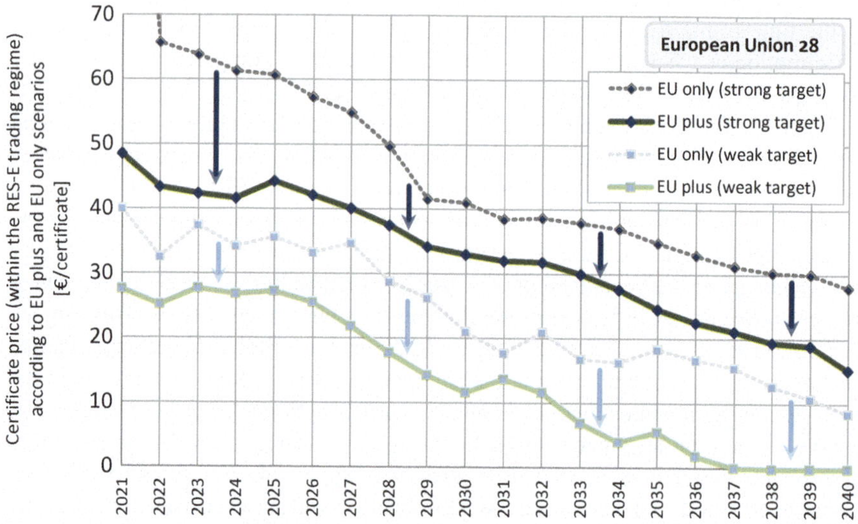

Fig. 4 Economic impacts—savings in support expenditures, depicted by the change in the certificate prices between 2021 and 2040 when comparing reference (EU only) and full cooperation (EU plus) scenarios

3.3 *Impacts on Investments and Benefits*

Through intensified cooperation investments in renewables are redirected, mainly from the EU28 to North Africa, Turkey and West Balkans. The monetary volumes appear impressive under a strong RES target: annual capital expenditures for RES-E decline by roughly 19 billion € at EU28 level on average by year in the time frame 2021–2040. In turn, they increase in North Africa (8.2 billion €), Turkey (1.4 billion €) and West Balkans (0.28 billion €). In total, it can be seen that the decline in capital expenditures within the EU is larger than the increase of investments in its neighbouring regions. The reasons for that are, on the one hand, that investments in wind and solar are redirected to more efficient sites, and, on the other hand, that increased cooperation partly replaces domestic RES-E generation in the host region, that would have been used domestically otherwise. This unintended "substitution effect" is applicable solely in North Africa and has an impact also on the benefits where avoidance of fossil fuels and carbon emissions declines in comparison to reference (of having no cooperation with the EU).

A closer look at the assessed benefits (avoidance of fossil fuels and CO_2 emissions) and their respective changes shows that at EU level impacts are smaller compared to other cost or expenditure categories. One can see an ambiguous picture for avoided fossil fuels and a negligible change to the positive side for carbon emission avoidance. In North Africa one can see more fossil fuels and CO_2 emissions due to the unintended "substitution effect" as discussed above. Contrarily, in the West Balkans region and in Turkey a clear and, in particular in the case of a strong target, significant reductions in fossil fuel and CO_2 emissions can be identified. The key reason for this positive trend is that for both regions (since assumed being part of the Energy Community post 2020) only virtual exchange of RES volumes with the EU is assumed as underlying principle in the joint RES trading regime. This lets both regions benefit locally from the increased RES expansion under a joint RES market with the EU.

4 Conclusions

The techno-economic analysis of the BETTER project confirms the results from previous studies that showed that there are significant RES-E export opportunities from the neighbouring regions to the EU and that the associated benefits for both importer and exporter are large.

In the 2020 timeframe, there is however no real basis for exploiting Article 9 cooperation, because there is no demand for imports from Member State side and the neighbouring regions have no capacities available for exports as (in most cases) their electricity demand is rising too fast.

The EUs 2030 RES policy framework as well as post 2030 RES policy framework will play a key role in determining the need for and attractiveness of RES-E exports to the EU. A weak target and/or governance framework may leave the large cost-savings potential of international RES-E cooperation unused. Thus, robust business opportunities for RES-E exports/imports will depend on the EU's seriousness on achieving its 2050 decarbonisation goals, on the political willingness to partially base its supply on RES-E imports from neighbouring regions, and on the exporters' willingness to export RES-E to the EU.

References

1. European Commission, Directive 2009/28/EC of the European Parliament and of the Council of 23 April 2009 on the promotion of the use of energy from renewable sources and amending and repealing Directives 2001/77/EC and 2003/30/EC
2. Resch G, Welisch M, Ortner A, Totschnig G, Türk A, Steiner D (2015) Long term prospects for RES cooperation with Third countries. A report compiled within the IEE project BETTER. Vienna, Austria, 2015. www.better-project.net

Gustav Resch is working as senior researcher at Vienna University of Technology, Energy Economics Group. He holds a degree in electrical engineering (Energy Technology) and a Ph.D. in energy economics at Vienna University of Technology. He is responsible for research, project acquisition and scientific coordination in the area of energy policy and energy economics with a focus on renewable energy technologies—with proven expertise on international level. His fields of activity include techno-economic assessments of (renewable) energy technologies, evaluation and design of energy policy instruments, and energy modelling (focusing on policy interactions and technology dynamics). He has contributed to several EU studies and policy assessments in the light of "20 % Renewable Energy by 2020" and the 2030 RES policy debate.

Update of World Geothermal Development (2013)

Colin C. Harvey

1 Introduction

Although geothermal projects require high upfront capital investments, the typically low operating cost (including zero expense for fuel) combined with high load factor, means that geothermal generation can be cost-competitive with fossil fuel generation. Other benefits include energy security and ability to provide reliable base load power. A 1999 report from the U.S. based Geothermal Energy Association (GEA) [1] identified 39 countries, now with a combined 800 million people, whose geothermal resources could meet 100 % of their electricity needs. Table 1 lists the current top nine geothermal generating countries.

- The United States is the current the World leader having about 30 % of the world's generating capacity (3100 MWe). Most of this development took place in the late twentieth century but right now US interest in geothermal is booming with many new projects underway. Significant growth is proposed for the period 2011–2015.
- The Philippines ranks second in the world, with 1900 MWe, providing 17 % of its electricity from geothermal. After a period of rapid development in the 1980s and 1990s the Philippines is again planning for expanded geothermal development with a target of 2500 MWe by 2020.

C.C. Harvey (✉)
Harvey Consultants Ltd, Warkworth, New Zealand
e-mail: colincharvey@gmail.com

T.S. Uyar (ed.), *Towards 100% Renewable Energy*, Springer Proceedings in Energy, DOI 10.1007/978-3-319-45659-1_30

Table 1 World leaders in geothermal power generation (2010 data)

Country	2005, MWe	2005, GWh	2010, MWe	2010, GWh	% growth over 5 years
USA	2564	16,840	3060	14,533	20
Philippines	1930	9253	1904	10,311	0
Indonesia	797	6085	1197	9600	50
Mexico	953	6282	958	7047	1
Italy	791	5340	843	5520	7
New Zealand	435	2774	780	4055	80
Iceland	202	1483	575	4597	183
Japan	535	4367	536	3064	<1

- Indonesia is currently in third place in World generation with almost 1200 MWe but it has the most ambitious goals with plans to more than triple geothermal generation by 2015. By 2025, Indonesia aims to generate 9500 MWe of geothermal energy.
- In Latin America, Mexico's 958 MWe of installed geothermal power capacity make it number four on the World list. Mexico's geothermal capacity currently exceeds that of all other countries in the region combined, but there is enormous potential in many other Latin American countries.
- Italy was the first country to generate electricity from geothermal steam in the early part of the twentieth century and holds fifth place with over 800 MWe of generation.
- Italy is close followed by New Zealand which has shown a rapid growth in generation of over 20 % per year over the past 4 years(2006–2010).
- Iceland also has experienced rapid growth since 2005 with 575 MWe installed providing 25 % of Iceland's electricity.
- Japan has enormous geothermal potential (many thousands of MWe). An early pioneer in geothermal development, its 540 MWe has not increased for almost the past 20 years. However, the recent nuclear crisis may lead to a renewed commitment to geothermal development.

A recent research report [4] projects that even without new pro-geothermal policies, global investment in this energy source will more than double from $3 billion in 2010 to $6.8 billion in 2020 while the number of countries using geothermal power is expected to jump from 24 at present to 46 in 2015.

2 Focus on the Western Pacific Sector

The regional growth of geothermal in the Western Pacific Region is currently led by Indonesia and New Zealand. Figure 1 [5] illustrates this stepped development in Indonesia which was impacted significantly by the Asian financial crises during the late twentieth Century.

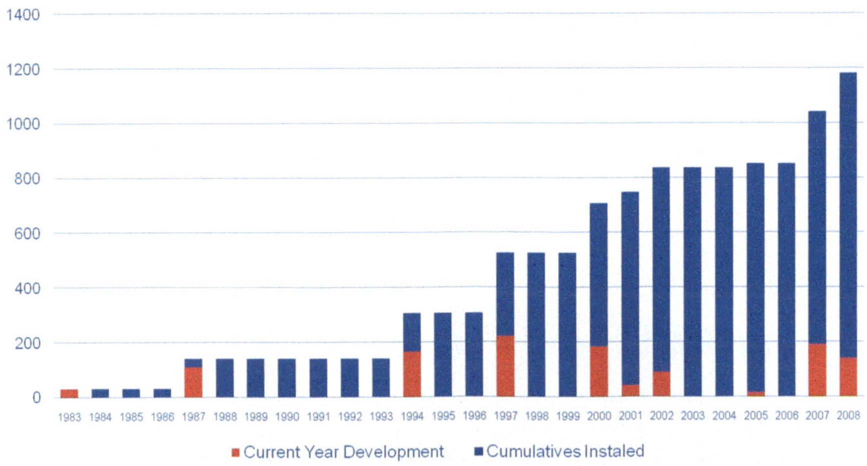

Fig. 1 Growth in geothermal generation in Indonesia 1983–2008 [5]

Table 2 Geothermal resource potential of Indonesia [6]

Location	Number of potential resources	Estimated capacity, MWe	2011 Installed capacity, MWe
Sumatra	84	13,379	12
Java	71	9996	1117
Bali	5	311	0
Nusa Tenggara	22	1461	0
Kalimantan	3	45	0
Sulawesi	55	2347	60
Maluku	22	929	0
Papua	3	75	0
Total	265	28,543	1189

The increases between 2005 and 2010 (Fig. 1) confirm a 50 % growth in generation over this period.

In mid-2011, we understood that over 2500 MWe of as yet undeveloped geothermal resources were under offer under a tender process and that projects proposing over 1000 MWe of generation had completed the tender process. The speed at which these tenders can be brought to generation is an interesting question. Many resources are in relatively remote locations where appropriate infrastructure has yet to be developed. Resources that are reasonably explored and scheduled for development include Sarulla (planned 330 MWe), Ulu Belu (220 MWe), Lumut Balai (110 MWe) Hululais (165 MWe), Sorik Merapi (55 MWe), Sungai Penuh (55 MW), Patuha (180 MWe), Karaha Bodas (140 MWe) and perhaps others. Table 2, developed in consultation with industry provides a summary of undeveloped potential resources in Indonesia.

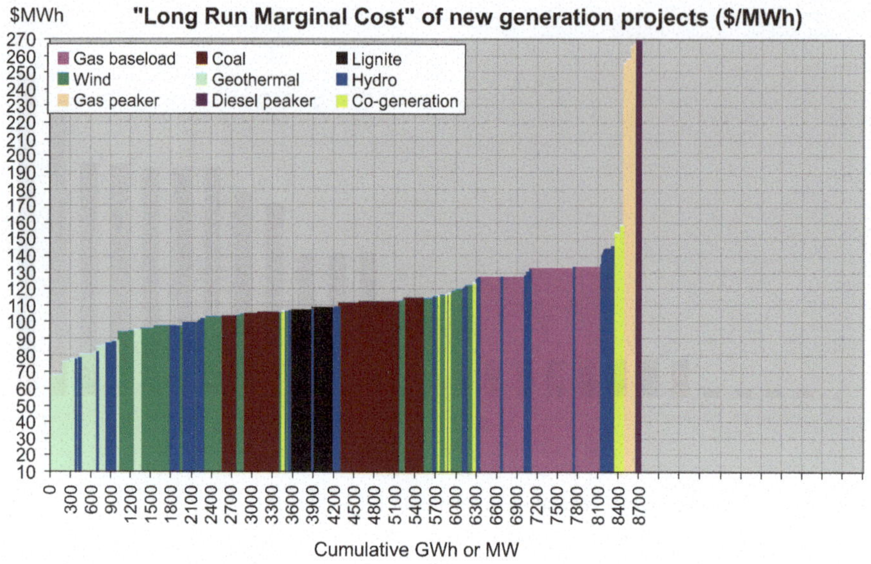

Fig. 2 Long run costs for a range of new generation options (from the New Zealand Government Website 2010)

New Zealand has an electricity generation system that is already dominated by renewable resources. A Government Energy Strategy (October 2007) [7] targeted an increase of renewable electricity contribution from around 70 to 90 % by 2025. New Zealand has a high proportion of hydro generation but relatively little storage capacity which makes both operation and prices sensitive to weather. Geothermal energy offers base load energy security. It is independent of the weather, with resources well located, close to major load centres. In New Zealand geothermal generation has grown at 20 % per year from just 430 MWe in 2006 to 780 MWe in 2010. Stimulation for the current growth has been due to declining indigenous gas reserves; concerns around the future cost and supply of imported fossil fuels; ready availability of premium geothermal resources that can be fully cost competitive with costs of all other generation alternatives; expansion opportunities on existing operations (attractive economics); readiness of developers to invest in resources through strongly established staff bases and experience, coupled with availability of skills, climate change concerns (Kyoto Protocol) and reduced greenhouse gas emissions and Government reforms to simplify consenting procedures. Figure 2 [7] is taken from the New Zealand Government Energy Strategy and shows quantities and long run costs (NZ$) associated with various energy sources.

Geothermal is currently around NZ$7/GJ and is forecast to remain below NZ$9/GJ for the next 1000 MWe. In a national context, the quantities of renewable energy options available are sufficient to satisfy 20 or so years of demand growth, unless electric vehicles develop at an unprecedented rate. At the time of compiling this chapter (March 2013) the New Zealand dollar was worth approximately 85 US cents. A wide range of new geothermal generation projects have been announced with principal sources of announced projects being Mighty River Power (a public generator-retailer) and Contact Energy (a private generator-retailer). Both of these companies have announced billion dollar investment programmes.

3 Direct Use of Geothermal

Direct use of geothermal can be a very efficient utilisation of both high temperature and low temperature resources. Iceland is the World leader in direct use with its extensive use of geothermal for space heating and other applications. Geothermal heat pumps (whether ground-source or water-source) are also making a significant contribution to direct use internationally, but are not considered further in this report.

In many countries of the World geothermal resources are often found in close proximity to forestry. This is indeed the case in the Western Pacific and specifically in Indonesia, New Zealand and the Philippines. In Indonesia it is said that possibly 70 % of its geothermal resources are in close proximity to forests, with many forestry areas in a protected category. This co-location of geothermal and forests may offer sound economic benefits for co-development. In New Zealand the growth in direct use of geothermal has virtually paralleled the growth in geothermal generation (Fig. 3 [8, 9]).

The major developments of direct use of higher temperature geothermal resources are found in the Taupo Volcanic Zone. They include greenhouse heating, prawn farming, glasshouse heating, kiln drying and for tourism. Direct use in New Zealand is dominated by steam supply to the Norske Skog mill at Kawerau which accounts

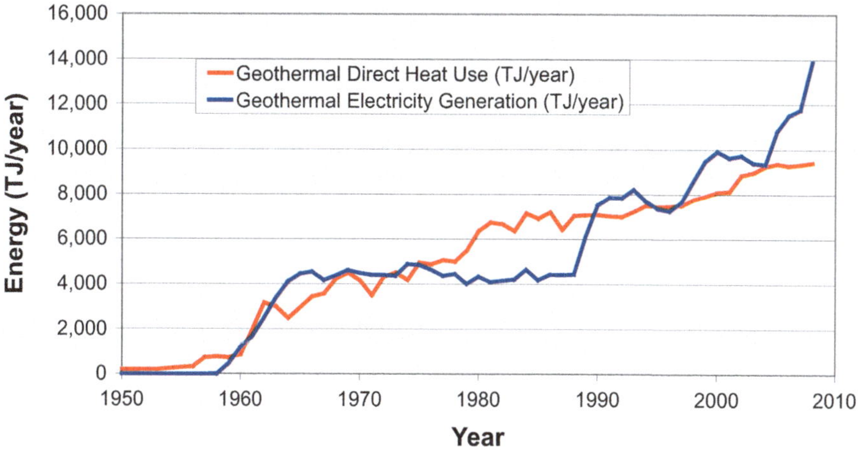

Fig. 3 Historical comparison of electricity generation and direct use in New Zealand

for 56 % of national geothermal direct energy use. This remains the largest single geothermal industrial use in the world. Direct use growth is closely associated with power generation at all major geothermal fields at Kawerau, Wairakei, Mokai and Tauhara with timber drying kilns and greenhouse heating the key areas of expansion [8, 9]. For industries located close to a geothermal resource, direct use may avoid exposure to future higher fossil fuel prices and the associated cost of emissions.

4 Barriers for Geothermal Development

Barriers or perceived barriers for development are discussed under the subheadings of policy, exploration and development risk, manpower, infrastructure and cost.

Government policy is obviously a key factor in achieving a successful development programme. In New Zealand a strongly focused national renewable energy policy was implemented before 2007 and this (coupled with significant in market gas prices) has been reflected in the high rate of growth (over 20 % annually) for the past 4 years. Resource allocation and the management of effects of development in New Zealand are handled through the Resource Management Act 1991 by local Government or the Environment Court. Over recent years efforts have been made to improve the quality of decisions and associated processes, and reduce the time required to obtain consents. In Indonesia more recent Government policies are strongly supportive of geothermal energy development which is evident by the targets for future generation detailed in this chapter. However sensitive environmental issues and the co-location of many potential resources within or adjacent to protected forests or designated national parks will require consideration for potential competing merits and uses.

In regard to exploration and investment risk, New Zealand [3] has benefited from a Government-funded extensive scientific programme (including exploration drilling) in the period 1950s–1980s which defined the resource potential of many of the potential geothermal resources. This has enabled six relatively low risk investments to be made with most developments still operating at close to their installed capacities. In Indonesia many of the resources on Java are well understood but for many of the more remote resources, access and infrastructure has constrained the level and detail of exploration. For many Indonesian resources it will be essential that detailed exploration programmes be undertaken by qualified teams of scientists before resource certainty can reach a sufficiently high level to attract both local and international investment.

In regard to generation costs, the recent major New Zealand developments in the Taupo Volcanic Zone have the advantage of established infrastructure and geographic factors. Geothermal generation costs can be fully competitive with both fossil fuel and alternative renewable energy options. However, in situations where low cost coal (as in Indonesia), gas or undeveloped hydro resources are readily available, the economics of geothermal may be less favourable. In addition, even large high temperature geothermal resources in remote locations, far from established infrastructure and markets, may be more difficult to justify development.

5 Combining Geothermal with Other Renewable Generation Technologies

Probably the most successful hybrid generation facility is the Stillwater Generation facility in Nevada , USA which was developed by ENEL of Italy and commissioned in 2011. The plant generates 33 MWe of geothermal through a binary cycle Ormat plant and 26 MWe of Solar PV. The geothermal component provides 24 h base load while the solar unit operates in a peaking role during daylight hours. In El Salvador an experimental hybrid unit uses solar energy to boost the temperature of the separated water fraction from a geothermal separator enabling extra energy to be extracted from it.

6 Geothermal in Turkey

In 2010 the installed capacity of geothermal generation in Turkey was 100 MWe while direct use installations totalled approximately 800 MWth. There are approximately 61,000 residences currently heated by geothermal fluids in Turkey. A total of 665 MWt is utilised for space heating of residential, public and private property, and 565,000 m^2 for greenhouses [2]. The proven geothermal heating capacity, according to data from existing geothermal wells and natural discharges is 3132 MWt. Present applications have shown that geothermal energy is clean and much cheaper when compared with other fossil and renewable energy sources in Turkey. It is projected that by 2015 there may be in excess of 500 MWe of generation in Turkey.

In March 2013, the IFC division of the World Bank launched a Best practice Manual for Geothermal Exploration at a seminar in Istanbul. This is part of a Worldwide initiative by the World Bank to promote greater utilisation of renewable energy.

7 Conclusions

On the World scene, geothermal development is currently accelerating both for power generation and direct use. The current World generation of over 10,000 MWe [2] is projected to double by 2016. Power generation from high temperatures geothermal resources can be fully cost competitive with other renewable energies and has the unique advantage in being able to provide sustainable base load electricity.

Direct use of both high temperature and low temperature geothermal energy is receiving increasing attention worldwide and is growing at a rate which could be comparable or exceed the growth rate of geothermal power generation. In countries where geothermal resources are co-located with forestry or other industries, the direct use of geothermal is an attractive option.

On the rim of the Pacific Ocean, the so-called Rim of Fire, there is a concentration of developable geothermal resources with Indonesia possessing what is recognised to be the World's largest number of developable geothermal resources. New Zealand has proven that relatively rapid development of resources can occur from a well proven resource base. The proposed development programme for Indonesia could move it to be the World leader in geothermal generation if the perceived or real barriers to development can be overcome. Such a rapid development programme will however require international co-operation, investment and assistance from the World's geothermal community.

References

1. Gawell K, Reed M, Wright PM (1999) Preliminary report: geothermal energy, the potential for clean power from the earth. Geothermal Energy Agency, Washington, DC
2. Bertani R (2010) Geothermal power generation in the world 2005–2010 update report. Proceedings world geothermal conference, Bali, 25–29 Apr 2010
3. Harvey CC, White BR, Lawless JV, Dunstall MG (2010) New Zealand country update 2005–2010. Proceedings world geothermal conference, Bali, 25–29 Apr 2010
4. Pike Research: geothermal power: renewable energy generation from conventional, enhanced geothermal systems, and co-produced resources: market analysis and forecasts
5. Darma S et al (2010) Geothermal energy update: geothermal energy development and utilization in Indonesia. Proceedings world geothermal conference, Bali, 25–29 Apr 2010
6. Sumiarso L (2010) Geothermal development in Indonesia, Directorate of New renewable energy and energy conservation presentation, Jakarta, November 2010
7. New Zealand Government (2007) New Zealand energy strategy to 2050 – powering our future: towards a sustainable low emissions energy system
8. East Harbour Management Services and GNS Science: assessment of possible renewable energy targets – direct use: geothermal, report prepared for Energy Efficiency and Conservation Authority (2007)
9. White B (2010) An updated assessment of geothermal direct use in New Zealand. Report prepared for New Zealand Geothermal Association, Wellington

A New Evaluation Method Applying Sustainability and Climate Change Concepts: The Case of Planning New York City 2030

Yosef Jabareen

1 Introduction

Climate change poses new risks and uncertainties that often lie outside our range of experience and that have the potential to affect the social, economic, ecological, and physical systems of any given city ([1, 2]:719). In this way, climate change and its resulting uncertainties challenge the concepts, procedures, and scope of conventional approaches to city planning, creating a need to rethink and revise current approaches [2, 3]. Decisively, it demands from us to situate the energy issue in the central when planning urban spaces. Yet a striking weakness of the scholarship on the subject is its lack of multifaceted theorizing and the fact that it typically overlooks the multidisciplinary and complex nature of urban energy planning. Moreover, the literature is vague in the context of urban energy, and there is a lack of a theoretical framework that conceptualizes urban energy planning. Therefore, the aim of this paper is to propose a new multifaceted conceptual framework for theorizing urban energy planning based on multidisciplinary literature (see [3]).

2 Methodology

A *conceptual analysis method* was used to build the conceptual framework [4]. This method is a grounded theory technique that aims "to generate, identify, and trace a phenomenon's major concepts, which together constitute its theoretical framework" (Jabareen 2009; see [2, 5, 6]). Each concept possesses its own attributes, characteristics, assumptions, limitations, distinct perspectives, and specific function within

Y. Jabareen (✉)
Technion – Israel Institute of Technology, Haifa, Israel
e-mail: jabareen@technion.ac.il

T.S. Uyar (ed.), *Towards 100% Renewable Energy*, Springer Proceedings in Energy, DOI 10.1007/978-3-319-45659-1_31

the conceptual framework. The methodology delineates the following stages in conceptual framework building: (a) mapping selected data sources; (b) reviewing the literature and categorizing the selected data; (c) identifying and naming the concepts; (d) deconstructing and categorizing the concepts; (e) integrating the concepts; (f) synthesis, resynthesis, and making it all make sense; (g) validating the conceptual framework; and (h) rethinking the conceptual framework [4].

3 Results

3.1 The Concepts of the Conceptual Framework

The conceptual framework is composed of eight concepts (see [7]), as Fig. 1 shows. These concepts are:

3.1.1 Utopian Vision

This concept is concerned with a plan's future vision. Usually, urban planning seeks to bring about a different and more desirable future. Theoretically, the power of visionary or utopian thinking lies in its inherent ability to envision the future in terms of radically new forms and values [8]. An urban vision incorporating climate

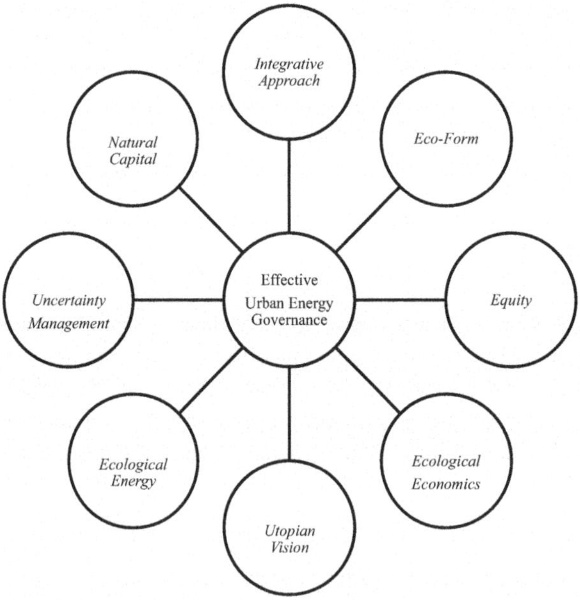

Fig. 1 Conceptual framework for urban energy

change as a central theme is of the utmost importance to practitioners, decision makers, and the public. Visionary frames are important in climate change, as they serve to identify problematic conditions and the need for change, to propose future alternatives, and to urge all stakeholders to act in concert to affect change. Climate change planning visions must provide people with an interpretive framework that enables them to understand how the issue is related to their own lives in the present and future, and to the world at large ([9, 10]:614). This concept addresses the visionary and utopian aspects regarding future urban life, the city's potential role in climate change mitigation, and the city vision regarding energy production and consumption as well.

3.1.2 Equity

Equity is a key concept in evaluating climate change policies [11]. The impacts of climate change and climate change mitigation policies are "socially differentiated," and are therefore matters of local and international distributional equity and justice ([12]:929 [13, 14]). Some argue that inequality leads to greater environmental degradation and that a more equitable distribution of power and resources would result in improved environmental quality [15–18]. Moreover, there are individuals and groups within all societies who are more vulnerable than others and lack the capacity to adapt to climate change ([1]:719). A society's vulnerability is influenced by its development path, physical exposure, resource distribution, social networks, government institutions, and technological development ([1]:719–720). The concept of equity addresses social aspects, including: environmental justice; public participation; and methods of addressing each community's vulnerability to climate change (urban vulnerability matrix).

3.1.3 Uncertainty Management

Uncertainty "is a perceived lack of knowledge, by an individual or group, which is relevant to the purpose or action being undertaken and its outcomes" ([19]:503). The new urban uncertainties posed by climate change challenge the concepts, procedures, and scope of planning. In order to cope with the new challenges, planners must develop a greater awareness and place mitigation and policies for "adaptation," or actual adjustments that might eventually enhance resilience and reduce vulnerability to expected climate changes, at the center of the planning process ([20]:720). Planners must also develop a better understanding of the risks climate change poses for infrastructure, households, and communities. To address these risks, planners have two types of uncertainty or adaptation management at their disposal: (1) Ex-ante management, or actions taken to reduce and/or prevent risky events; and (2) Ex-post management, or actions taken to recover losses after a risky event [21].

3.1.4 Natural Capital

Natural capital refers to "the stock of all environmental and natural resource assets, from oil in the ground to the quality of soil and groundwater, from the stock of fish in the ocean to the capacity of the globe to recycle and absorb carbon" ([22]:1). Maintaining constant natural capital is an important criterion for sustainability ([22]:44, [23]). The stock of natural capital should not decrease, as this could endanger the ecological system and threaten the ability of future generations to generate wealth and maintain their well-being. This concept addresses the consumption and—equally as important—the renewal of natural assets that are used for development, such as land, water, air, and open spaces.

3.1.5 Integrative Approach

Planning for climate change is more complex than the conventional approach to planning as it is undertaken in a context of great uncertainty. This context poses new challenges for collaboration among public, private, and civil institutions and organizations on all levels. Integrating the many different stakeholders and agents into planning is essential for achieving climate change objectives. The "ability of a governance system to adapt to uncertain and unpredicted conditions is a new notion" ([24]:152). Therefore, adaptive management requires new planning strategies and procedures that transcend conventional planning approaches by integrating uncertainties into the planning process and prioritizing stakeholders' expectations in an uncertain environment. Plans should also be "flexible enough to quickly adapt to our rapidly changing environment" [24].

3.1.6 Ecological Energy

The clean, renewable, and efficient use of energy is a central theme in planning for the achievement of climate change objectives. This concept evaluates how a plan addresses the energy sector and whether it proposes strategies to reduce energy consumption and to use new, alternative, and clean energy sources.

3.1.7 Ecological Economics

This concept is based on the assumption that environmentally sound economics can play a decisive role in achieving climate change objectives in a capitalist world. Cities that are committed to climate change mitigation and sustainability should stimulate markets for "green" products and services, promote environmentally friendly consumption, and contribute to urban economic development by creating a cleaner environment ([25]:11, [26]). In this spirit, the American Recovery and Reinvestment Plan, proposed by President Barack Obama, calls for spurring "job creation while making long-term investments in energy, and infrastructure," and increasing "production of alternative energy" [27].

3.1.8 Eco-Form

The physical form of a city affects its habitats and ecosystems, the everyday activities and spatial practices of its inhabitants, and, eventually, climate change. This concept evaluates spatial planning, architecture, design, and the ecologically desired form of the city and its components (such as buildings and neighborhoods). Jabareen [7] suggests the following set of nine planning typologies, or criteria of evaluation, which are helpful in evaluating plans from the perspective of eco-form as follows:

Compactness refers to urban contiguity and connectivity and suggests that future urban development should take place adjacent to existing urban structures [28]. Compact urban space can minimize the need to transport energy, materials, products, and people [29]. Intensification, a major strategy for achieving compactness, uses urban land more efficiently by increasing the density of development and activity, and involves: developing previously undeveloped urban land; redeveloping existing buildings or previously developed sites; subdivisions and conversions; and additions and extensions ([30]:243).

Sustainable Transport suggests that planning should promote sustainable modes of transportation through traffic reduction; trip reduction; the encouragement of non-motorized travel (such as walking and cycling); transit-oriented development; safety; equitable access for all; and renewable energy sources [31, 32].

Density is the ratio of people or dwelling units to land area. Density affects climate change through differences in the consumption of energy, materials, and land for housing, transportation, and urban infrastructure. High density planning can save significant amounts of energy [33–35].

Mixed Land Uses indicates the diversity of functional land uses, such as residential, commercial, industrial, institutional, and transportation. It allows planners to locate compatible land uses in close proximity to one another in order to decrease the travel distance to between activities. This encourages walking and cycling and reduces the need for car travel, as jobs, shops, and leisure facilities are located in close proximity of one another [36–39].

Diversity is "a multidimensional phenomenon" that promotes other desirable urban features, including a larger variety of housing types, building densities, household sizes, ages, cultures, and incomes ([40]:320). Diversity is vital for cities. Without it, the urban system declines as a living place [41] and the resulting homogeneity of built forms, which often produces unattractive monotonous urban landscapes, leads to increased segregation, car travel, congestion, and air pollution [28].

Passive Solar Design aims to reduce energy demands and to provide the best use of passive energy through specific planning and design measures, such as orientation, layout, landscaping, building design, urban materials, surface finish, vegetation, and bodies of water. This facilitates optimum use of solar gain and microclimatic conditions and reduces the need for the heating and cooling of buildings by means of conventional energy sources [42–44].

Greening, or bringing "nature into the city," makes positive contributions to many aspects of the urban environment, including: biodiversity; the lived-in urban environment; urban climate; economic attractiveness; community pride; and health and education [45–50].

Renewal and Utilization refers to the process of reclaiming the many sites that are no longer appropriate for their original intended use and can be reclaimed for a new purpose, such as brownfields. Cleaning, rezoning, and developing contaminated sites are key aspects of revitalizing cities and neighborhoods and contribute to their sustainability and to a healthier urban environment.

Planning Scale influences and is influenced by climate change. For this reason, desirable planning scale should be considered and integrated in plans for regional, municipal, district, neighborhood, street, site, and building levels. Planning that moves from macro to micro levels has a more holistic and positive impact on climate change.

4 Discussion and/or Conclusions

The *conceptual framework* is not a mere collection of concepts. Rather, all concepts are interrelated and interwoven with one another; each plays an important role in the framework as a whole. The conceptual framework consists of eight concepts of assessment that were identified through conceptual analyses of interdisciplinary literature on sustainability and climate change. Together, these concepts—each of which represents a distinctive aspect of urban energy planning—form the conceptual framework. Importantly, each concept contributes to the planning of urban energy in its domain. The positive contribution o all concepts together will lead to effective urban energy governance. The overlooking of one concept or more will cause various negative externalities to climate change.

References

1. IPCC – Intergovernmental Panel on Climate Change (2007) Climate change 2007: fourth assessment report of the intergovernmental panel on climate change. Cambridge University Press, Cambridge, MA
2. Jabareen Y (2015) The risk city: cities countering climate change: emerging planning theories and practices around the world. Springer, New York, NY
3. Jabareen Y (2013) Planning for countering climate change: lessons from the recent plan of New York City—PlaNYC 2030. Int Plann Stud 18(2):221–242
4. Jabareen Y (2009) Building conceptual framework: philosophy, definitions and procedure. Int J Qualitative Methods 8(4):49–62
5. Eizenberg E (2012) The changing meaning of community space: two models of NGO management of community gardens in New York City. Int J Urban Reg Res 36(1):106–120
6. Eizenberg E, Shilon M (2015) Pedagogy for the new planner: refining the qualitative toolbox. Environment and planning B: planning and design. In: Elkin T, McLaren D, Hillman M (eds) Reviving the city: towards sustainable urban development. Friends of the Earth, London
7. Jabareen Y (2006) Sustainable urban forms: their typologies, models, and concepts. J Plann Educ Res 26(1):38–52
8. de Geus M (1999) Ecological Utopias: envisioning the sustainable society. International Books, Utrecht
9. Benford RD, Snow DA (2000) Framing processes and social movements: an overview and assessment. Annu Rev Sociol 26:611–639

10. Taylor DE (2000) The rise of the environmental justice paradigm: injustice framing and the social construction of environmental discourses. Am Behav Sci 43:508–580
11. IPCC (2001) Third assessment report: climate change 2001 (TAR)
12. Adger WN (2001) Scales of governance and environmental justice for adaptation and mitigation of climate change. J Int Dev 13(7):921–931
13. O'Brien R, Leichenko U, Kelkar H, Venema G, Aandahl H, Tompkins A, Javed S, Bhadwal A, Barg LPN, West J (2004) Mapping vulnerability to multiple stressors: climate change and globalization in India. Glob Environ Chang 14:303–313
14. Paavola J, Adger WN (2006) Fair adaptation to climate change. Ecol Econ 56(4):594–609
15. Agyeman J, Bullard RD, Evans B (2002) Exploring the nexus: bringing together sustainability, environmental justice and equity. Space Polity 6(1):77–90
16. Boyce JK, Klemer AR, Templet PH, Willis CE (1999) Power distribution, the environment, and public health: a state-level analysis. Ecol Econ 29(1):127–140
17. Solow R (1991) Sustainability: an economist's perspective. the eighteenth J. Seward Johnsonlecture. Woods Hole Oceanographic Institution, Woods Hole, MA
18. Stymne S, Jackson T (2000) Intra-generational equity and sustainable welfare: a time series analysis for the UK and Sweden. Ecol Econ 33(2):219–236
19. Abbott J (2009) Planning for complex metropolitan regions: a better future or a more certain one? J Plann Educ Res 28:503–517
20. Adger WN, Agrawala S, Mirza MMQ, Conde C, O'Brien K, Pulhin J, Pulwarty R, Smit B, Takahashi K (2007) Assessment of adaptation practices, options, constraints and capacity. In: Contribution of Working Group II to the Fourth Assessment Report of the Intergovernmental Panel on Climate Change, Parry ML, Canziani OF, Palutikof JP, van der Linden PJ, Hanson CE (eds) Climate change 2007: impacts, adaptation and vulnerability. Cambridge University Press, Cambridge, pp 717–743
21. Heltberg R, Paul BS, Steen LJ (2009) Addressing human vulnerability to climate change: toward a 'no-regrets' approach. Glob Environ Chang 19(2009):89–99
22. Pearce D, Edward B, Markandya A (1990) Sustainable development: economics and environment in the Third World. Earthscan Publications, London
23. Geldrop J, Withagen C (2000) Natural capital and sustainability. Ecol Econ 32(3):445–455
24. Mirfenderesk H, Corkill D (2009) Sustainable management of risks associated with climate change. Int J Clim Change Strat Manag 1(2):146–159
25. Hsu D (2006) Sustainable New York City. Design Trust for Public Space and the New York City Office of Environmental Coordination, New York City, NY
26. Mercer Human Resources Consulting (2004) Quality of living survey: New York City, NY. Private communication, New York City, NY
27. White House (2009) http://www.whitehouse.gov/issues/Economy, http://www.recovery.gov/Pages/home.aspx
28. Wheeler SM (2002) Constructing sustainable development/safeguarding our common future: Rethinking sustainable development. J Am Plann Assoc 68(1):110–111
29. Elkin T, McLauren D, Hillman M (1991) Reviving the city: towards sustainable urban development. Policy Studies, Institute/Friends of the Earth, London
30. Jenks M (2000) The acceptability of urban intensification. In: Williams K, Burton E, Jenks M (eds) Achieving sustainable urban form. E&FN SPON, London
31. Cervero R (2003) Coping with complexity in America's urban transport sector. The 2nd international conference on the future of urban transport, Göteborg
32. Clercq F, Bertolini L (2003) Achieving sustainable accessibility: an evaluation of policy measures in the Amsterdam area. Built Environ 29(1):36–47
33. Carl P (2000) Urban density and block metabolism. In: Koen S, Yannas S (eds) Architecture, city, environment. Proceedings of PLEA 2000. James & James, London, pp 343–347
34. Newman P, Kenworthy J (1989) Gasoline consumption and cities: a comparison of US cities with a global survey. J Am Plann Assoc 55:23–37

35. Walker L, Rees W (1997) Urban density and ecological footprints – An analysis of Canadian households. In: Roseland M (ed) Eco-city dimensions: healthy communities, healthy planet. New Society Publishers
36. Alberti M (2000) Urban form and ecosystem dynamics: empirical evidence and practical implications. In: Williams K, Burton E, Jenks M (eds) Achieving sustainable urban form. E & FN Spon, London, pp 84–96
37. Parker T (1994) The land use—air quality linkage: how land use and transportation affect air quality. California Air Resources Board, Sacramento
38. Thorne R, Filmer-Sankey W (2003) Transportation. In: Randall T (ed) Sustainable urban design. Spon Press, London, pp 25–32
39. Uyen-Phan V, Senior M (2000) The contribution of mixed land uses to sustainable travel in cities. In: Williams K, Burton E, Jenks M (eds) Achieving sustainable urban form. E & FN Spon, London, pp 139–148
40. Turner SRS, Murray MS (2001) Managing growth in a climate of urban diversity: South Florida's Eastward ho! Initiative. J Plann Educ Res 20:308–328
41. Jacobs J (1961) The death and life of great American cities. Random House, New York, NY
42. Owens S (1992) Energy, environmental sustainability and land-use planning. In: Michael B (ed) Sustainable development and urban form. Pion, London, pp 79–105
43. Thomas R (2003) Building design. In: Thomas R, Fordham M (eds) Sustainable urban design: an environmental approach. Spon Press, London, pp 46–88
44. Yannis S (1998) Living with the city: urban design and environmental sustainability. In: Maldonado E, Yannas S (eds) Environmentally friendly cities. James & James, London, pp 41–48
45. Beatley T (2000) Green urbanism: learning from European cities. Island Press, Washington, DC
46. Beer A, Delshammar T, Schildwacht P (2003) A changing understanding of the role of greenspace in high-density housing: a European perspective. Built Environ 29(2):132–143
47. Dumreicher H, Levine RS, Yanarella EJ (2000) The appropriate scale for "low energy": theory and practice at the Westbahnhof. In: Koen S, Yannas S (eds) Architecture, city, environment. Proceedings of PLEA 2000. James & James, London, pp 359–363
48. Forman RT (2002) The missing catalyst: design and planning with ecology. In: Johnson BT, Hill K (eds) Ecology and design: frameworks for learning. Island Press, Washington, DC
49. Swanwick C, Dunnett N, Woolley H (2003) Nature, role and value of green space in towns and cities: an overview. Built Environ 29(2):94–106
50. Ulrich RS (1999) Effects of gardens on health outcomes: theory and research. In: Marcus CC, Barnes M (eds) Healing gardens: therapeutic benefits and design recommendations. Whiley, New York, NY

Reliability of 100 % Renewable Electricity Supply in the Australian National Electricity Market

Ben Elliston, Mark Diesendorf, and Iain MacGill

1 Introduction

This chapter reports on simulations to identify, quantify and address the challenges of reliably supplying 100 % renewable electricity to the five states and one territory spanned by the Australian National Electricity Market (NEM). Today, the NEM produces around one-third of Australia's total greenhouse gas emissions, because the system derives around 90 % of supply from bituminous coal, lignite and natural gas, and because of high per capita electricity consumption. If Australia, currently one of the world's highest per capita greenhouse emitters, is to make its fair contribution to global emission reductions, then its emissions intensive electricity industry, and indeed its whole energy sector, must decline to net zero emissions by 2050 [1]. For a rapid transition, the only zero carbon "sources" that are commercially available and seem likely to be able to make large contributions before 2025 in the Australian context are certain renewable energy sources and demand reduction (e.g. through efficient energy use).

B. Elliston (✉)
School of Electrical Engineering and Telecommunications, UNSW Australia,
Sydney, NSW 2052, Australia
e-mail: b.elliston@unsw.edu.au

M. Diesendorf
Interdisciplinary Environmental Studies, School of Humanities & Languages, UNSW
Australia, Sydney, NSW 2052, Australia
e-mail: m.diesendorf@unsw.edu.au

I. MacGill
School of Electrical Engineering and Telecommunications, UNSW Australia,
Sydney, NSW 2052, Australia

Centre for Energy and Environmental Markets, UNSW Australia,
Sydney, NSW 2052, Australia
e-mail: i.macgill@unsw.edu.au

T.S. Uyar (ed.), *Towards 100% Renewable Energy*, Springer Proceedings
in Energy, DOI 10.1007/978-3-319-45659-1_32

Numerous scenario studies, including hourly simulations, have been published that model the potential for the USA [2, 3], Australia [4–7], several other countries and regions [8–15] and the entire world [16–18] to meet 80–100 % of end-use energy demand from renewable energy by some future date, typically mid-century. Some of the studies focus on electricity while others address the whole energy sector. These scenario studies do not typically specify a transition path, nor do they share a common methodology for analysis. However, they are valuable in showing that aggressive reduction in fossil fuel use is possible and in providing a vision of how such a future energy system might look.

Electrical power systems must match time varying and somewhat unpredictable supply with time varying and somewhat unpredictable demand at all times and all locations across the network, and there are currently limited energy storage options. The variability of weather-driven electricity generation introduces additional sources of temporal and spatial variation and uncertainty [19].

This chapter reports simulations of a 100 % renewable electricity system in the region spanned by the NEM for the year 2010, using actual demand data and weather observations for that year, all converted to hourly averages. Demand is met by electricity generation mixes based on current commercially available technology: wind power, solar photovoltaics (PV), concentrating solar thermal (CST) with thermal storage, existing hydroelectric power stations, and open-cycle gas turbines fired with biofuels. After summarising the results of simulations reported in our recent publications [4, 5, 7], this chapter extends the research to investigate the effect of including a wider geographic distribution of wind farms. By minimising the number of working assumptions and by sensitivity analyses, we aim to provide some insights into the potential contribution from different renewable sources and the reliability implications of 100 % renewable electricity for the NEM.

2 Method

The computer simulation program, developed by the lead author, is written in the Python programming language. For the initial research the entire NEM region was treated as a "copper-plate", that is, power can flow unconstrained from any generation site to any load site. Hence, demand across all NEM regions is aggregated, as is supply in our initial simulations reviewed here. However, in our most recent work (to be published), supply is disaggregated into 43 regions across the NEM [20].

In general, dispatch proceeds from the lowest operating cost plants without energy storage (e.g. wind and PV) before the dispatch of flexible plant with energy storage (e.g. hydro and gas turbines). If supply cannot meet the demand, the shortfall is recorded and the hour is marked unmet. Conversely, if available supply exceeds demand, the simulation attempts to find another generator in the system that can store the excess power (e.g. by pumping at a pumped storage hydro station). Any remaining surplus power is then recorded as having been spilled.

The original data on NEM demand were 30-min averages, publicly available from the Australian Energy Market Operator (AEMO). Satellite-derived estimates of global horizontal irradiance and direct normal irradiance in 2010 for the Australian continent were provided by the Australian Bureau of Meteorology at 5 km by 5 km spatial resolution and hourly intervals. Electricity generation data reported by all wind farms over 30 MW and operating in the NEM in 2010 are also publicly available from AEMO. These data, comprising average wind power at each wind farm over the 5-min dispatch interval, were averaged into hourly values, as was the demand data.

Almost all Australia's wind farms are in essentially the same wind regime in south-eastern Australia and so the fluctuations in total wind power output for scaled up wind penetrations can be large [21]. Therefore, for the new research reported in this chapter, the wind farm data were supplemented with hourly wind data for 2010 in two locations (Hughenden and Cooranga) in the northern state of Queensland, thousands of kilometres from the wind farms in the south and south-east of the NEM. For these sites, wind speed and direction were generated using the CSIRO's "The Air Pollution Model" (TAPM). Hourly wind speeds and directions were obtained for four different levels (10, 50, 100, 200 m) and formatted into a data file suitable for use with a utility-scale wind farm model in the US National Renewable Energy Laboratory's System Advisor Model (SAM), a performance and financial model of various renewable energy technologies. The wind farm model was used to produce hourly electrical power for a 57.6 MW wind farm comprising Vestas V100-1.8 turbines. These data were then scaled to the desired wind farm size in the simulations.

SAM was also used to model the hourly power generation of PV and CST systems in selected locations in 2010, allowing realistic generation data to be included in the simulation.

More details of the method and data sources are given elsewhere [4].

In 2010 NEM peak power demand was 33.6 GW and occurred in summer. The next highest peaks were around 30 GW and occurred in winter.

The generators in the baseline scenario, which was not an optimal mix, are in order of dispatch:

1. Wind: existing wind farm output scaled to 23.2 GW; average capacity factor 30 %; supplying about 30 % of annual electrical energy.
2. PV: 14.6 GW; average capacity factor 16 %; supplying about 10 % of annual energy. Flat-plate distributed on the rooftops of the mainland capital cities of the NEM, according to population.
3. CST: six sites in three states, 2.6 GW per site, 15.6 GW total; solar multiple 2.5 and 15 h storage; average capacity factor 60 %; supplying about 40 % of annual energy.
4. Pumped storage hydro, existing, 2.2 GW. Opportunistically charged using surplus renewable power with a round-trip efficiency of 0.8; dispatched to meet critical peak loads.
5. Hydro, existing, without pumped storage, 4.9 GW.
6. Gas turbines, biofuelled, 24.0 GW, dispatched last in merit order to meet supply shortfalls; always less than 15 % of annual energy because of limited resource of biomass residues.

It should be mentioned that our subsequent research included cost projections to 2030 from the official Australian Energy Technology Assessment and used a genetic algorithm to calculate the economic optimal mix of renewable energy technologies [5]. Capacity factors and annual energy generations in this least-cost analysis were not specified initially, but were an outcome of the optimisations. The least-cost model is also more realistic in that the copper-plate assumption was partially removed by separating the NEM into its five existing market regions and by introducing regional interconnections to the simulation framework [5]. In another paper, we compared three fossil fuel scenarios—one medium-carbon (all gas) and two low-carbon (coal and gas with carbon capture and storage)—against the costs of the previous least-cost 100 % renewable electricity scenario and performed a sensitivity analysis of the results to future carbon prices, gas prices, and CO_2 transportation and storage costs [7]. However, there is insufficient space to report here on most of the results of the scenarios of [5] and [7]. Suffice it to report that the least-cost annual energy contributions for 2010 were found to be wind 46 %, CST 22 %, PV 20 %, hydro (both pumped and not pumped) 6 % and biofuelled gas turbines 6 %. However, the remainder of the present chapter focuses on the results of sensitivity analyses done by varying parameters in the original baseline scenario [4].

3 Results

3.1 *Baseline Simulation*

The baseline scenario meets 2010 demand within the current NEM reliability standard, namely an energy shortfall (unserved energy) no greater than 0.002 %, despite 6 h on winter evenings when demand was unmet [4].

3.2 *Sensitivity Analyses*

Five sensitivity analyses examine various options for improving the reliability of the system or reducing the biofuel consumption through lower utilisation of gas turbines. The first four are reported in more detail elsewhere [4] and so graphs are not included here. The fifth is original to this chapter.

1. *Reducing demand*
   A 5 % reduction in the six demand peaks that exceed supply is sufficient to bring demand and supply into balance for every hour of the year. As all these peaks occur on winter evenings, this reduction could be readily achieved through energy efficiency measures, particularly by reducing residential heating demand, or by temporarily interrupting controllable loads. Furthermore, when demand is reduced by 19 % during unmet hours, the NEM reliability standard can still be met when the gas turbine capacity is reduced from 24 to 15 GW [4].

2. *Increasing solar thermal plant capacity*
   To overcome the decline in CST generation during winter and consequent increase in bioenergy consumption, we examined the effect of doubling the total generating capacity of CST plants, while keeping the solar multiple constant at 2.5 and the storage at 15 full-load hours. This change reduces the number of unmet hours from 6 to 2, reduces the gas turbine generation modestly, but increases total spilled energy significantly. This, and current high costs of CST technologies, suggest that increasing CST capacity could be a very expensive means of meeting peak demand in winter [4].
3. *Increasing CST solar multiple*
   Another strategy to reduce biofuel consumption was tested by increasing the solar multiple of the CST plants from 2.5 to 4.0, while keeping the CST generating capacity and storage capacity constant. In other words, the size of the solar field is increased. This change is much more effective in reducing gas turbine generation than increasing the overall CST generating capacity, but is still likely to be expensive [4].
4. *Delaying CST dispatch in winter*
   An alternative winter operating strategy for the CST plants is to delay the dispatch until the evening, so that peak generation coincides with evening peak demand. Delaying CST dispatch, while maintaining reliability fixed, permits the gas turbine generating capacity to be reduced, with the minimum capacity occurring with a 5-h delay. This confirms the result by others [22] that CST contributes more to the evening peak while PV contributes to supplying daytime demand [4].
5. *Expanding the diversity of wind regimes*
   To gauge the effect of greater wind diversity in reducing bioenergy consumption, we substitute part of existing wind energy generation with output from hypothetical wind farms at Hughenden and Cooranga in Queensland, while maintaining total annual wind generation at 66 TWh. As winds at Cooranga and Hughenden are less correlated with the existing NEM wind regime, wind power in these sites reduces the need for supply from biofuelled gas turbines, as shown in Fig. 1. Hughenden, being more distant than Cooranga, eliminates more bioenergy.

4 Conclusion

This research demonstrates that 100 % renewable electricity in the Australian NEM, at the current reliability standard, would have been technically feasible for the year 2010, based on several possible mixes of commercially available renewable energy technologies. Recent work has extended the mixes to least-cost mixes [5, 7] and hourly simulations from 1 to 6 years.

The results entail a radical twenty-first century re-conception of an electricity supply–demand system. The focus is shifted away from replacing base-load coal with alternative base-load sources. Instead, generation reliability is maintained in a system with large penetrations of variable renewable sources, wind and solar PV, by

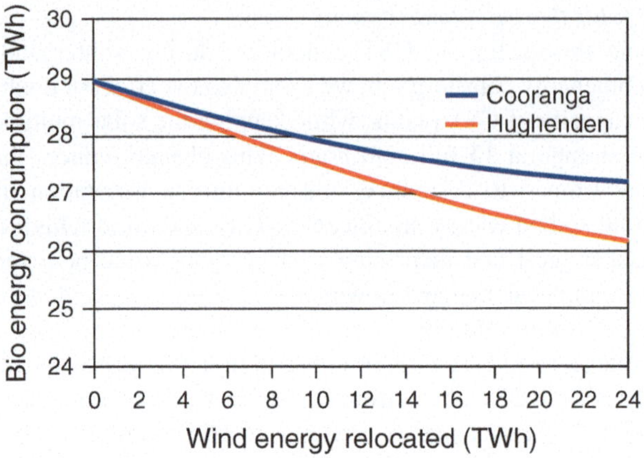

Fig. 1 Effect on biomass consumption of relocating wind energy

having as great a diversity of solar and wind sites as possible, large capacities of flexible, dispatchable peak-load generators, and thermal storage of solar energy. In a 100 % renewable electricity system based on large contributions from wind and solar, the concept of base-load power station is redundant. In particular, increasing the geographic distribution of wind farms reduces the need to operate open-cycle gas turbines to fill the infrequent, short lulls in solar plus wind generation.

In geographic regions with high levels of insolation—such as south-west USA, Mexico, North Africa, the Middle East, north-west China, north-west India and Australia—solar energy sources, both CST and PV, can together make the major contribution to electricity generation. Then the principal challenge is to generate sufficient power during the evening peak demand periods in the winter months. On winter evenings following overcast days there is insufficient energy in the CST thermal energy storage and sometimes there are also lulls in the wind. Possible solutions are to instal a high capacity of peaking plant; to increase CST generation capacity with fixed solar multiple; to increase the CST solar multiple with fixed CST generating capacity; to delay the dispatch of CST power until evenings in winter; to increase the geographic diversity of wind farms; and to reduce demand peaks, especially for the heating load on winter evenings. An economic analysis can be used to rank such options.

References

1. The Climate Institute (2016) A switch in time: enabling the electricity sector's transition to net zero emissions. The Climate Institute, Sydney, NSW, http://www.climateinstitute.org.au/articles/publications/a-switch-in-time-report-page.html
2. Mai T, Wiser R, Sandor D, Brinkman G, Heath G, Denholm P, Hostick D, Darghouth N, Schlosser A, Strzepek K (2012) Exploration of high-penetration renewable electricity futures, vol. 1 of renewable electricity futures study. Technical report TP-6A20-52409-1. National Renewable Energy Laboratory, Golden, CO

3. Jacobson MZ, Delucchi MA, Cameron MA, Frew BA (2015) A low-cost solution to the grid reliability problem with 100% penetration of intermittent wind, water, and solar for all purposes'. Proc Natl Acad Sci U S A 112(49):15060–15065
4. Elliston B, Diesendorf M, MacGill I (2012) Simulations of scenarios with 100% renewable electricity in the Australian National Electricity Market. Energy Policy 45:606–613
5. Elliston B, MacGill I, Diesendorf M (2013) Least cost 100% renewable electricity scenarios in the Australian National Electricity Market. Energy Policy 59:270–282
6. Australian Energy Market Operator (2013) 100 percent renewables study – modeling outcomes. http://www.environment.gov.au/climate-change/publications/aemo-modelling-outcomes. Accessed 2 Aug 2013
7. Elliston B, MacGill I, Diesendorf M (2014) Comparing least cost scenarios for 100% renewable electricity with low emission fossil fuel scenarios in the Australian National Electricity Market. Renew Energy 66:196–204
8. Sørensen B (2008) A renewable energy and hydrogen scenario for northern Europe. Int J Energy Res 32:471–500
9. Lehmann H (2003) Energy rich Japan. Technical report. Institute for Sustainable Solutions and Innovations. http://energyrichjapan.info
10. Lund H, Mathiesen B (2009) Energy system analysis of 100% renewable energy systems–the case of Denmark in years 2030 and 2050. Energy 34:524–531
11. Mason I, Page S, Williamson A (2010) A 100% renewable electricity generation system for New Zealand utilising hydro, wind, geothermal and biomass resources. Energy Policy 38:3973–3984
12. Connolly D, Lund H, Mathiesen B, Leahy M (2011) The first step towards a 100% renewable energy-system for Ireland. Appl Energy 88:502–507
13. German Advisory Council on the Environment (2011) Pathways towards a 100% renewable electricity system. Technical report. German Advisory Council on the Environment, Berlin, http://www.umweltrat.de/EN
14. Krajačić G, Duić N, da Graça Carvalho M (2011) How to achieve a 100% RES electricity supply for Portugal? Appl Energy 88:508–517
15. Fernandes L, Ferreira P (2014) Renewable energy scenarios in the Portuguese electricity system. Energy 69:51–57
16. Jacobson MZ, Delucchi MA (2011) Providing all global energy with wind, water, and solar power, part I: technologies, energy resources, quantities and areas of infrastructure, and materials. Energy Policy 39:1154–1169
17. Delucchi MA, Jacobson MZ (2011) Providing all global energy with wind, water, and solar power, part II: reliability, system and transmission costs, and policies. Energy Policy 39:1170–1190
18. GEA (2012) Global energy assessment – toward a sustainable energy future. Cambridge University Press, Cambridge, http://www.globalenergyassessment.org
19. MacGill I (2010) Electricity market design for facilitating the integration of wind energy: experience and prospects with the Australian National Electricity Market. Energy Policy 38:3180–3191
20. Elliston B, Riesz J, MacGill I (2016) What cost for more renewables? The incremental cost of renewable generation – an Australian National Electricity Market case study. Renew Energy 95:127–139
21. AEMO (2011) South Australian supply and demand outlook. Section 5.4. http://www.aemo.com.au/planning/SASDO2011/sasdo.html
22. Denholm P, Mehos M (2011) Enabling greater penetration of solar power via the use of CSP with thermal energy storage, Technical report TP-6A20-52978. National Renewable Energy Laboratory, Golden, CO

Managing Waste for Energy Use in Turkey

Wietze Lise

1 Rationale

Background and motivation:

- Dramatic increases in electricity consumption, volatility in commodities (fuel) markets, and fluctuations in electricity prices.
- Global awareness regarding sustainability in the energy sector.
- The development of energy recovery from municipal and industrial waste (MIW) has been less systematic and more localized in a few countries. Therefore there is a significant potential in this field.
- Integrated waste management approach is crucial for the sustainable development of international economies.
- Turkey has not achieved significant progress in the utilization and valorization of these resources so far. The economic growth has led to a significant increase of the potential for these projects.
- Waste disposal sector is the second largest GHG producer in Turkey. In 2012 the GHG emissions from waste disposal were 36.2 Mt CO_2-eq.

2 Objectives

The general objectives of this project are to perform a characterization of the availability of wastes for energy recovery.

W. Lise (✉)
Energy Markets Department, ÅF-MERCADOS EMI, ODTU-Teknokent, MET Alani,
Eskisehir Yolu 9th km, D-3 Blok No: 208, Çankaya 06520 Ankara, Turkey
e-mail: wietze.lise@afconsult.com

T.S. Uyar (ed.), *Towards 100% Renewable Energy*, Springer Proceedings
in Energy, DOI 10.1007/978-3-319-45659-1_33

- Assess the current market for supply (e.g., number, size and type of municipal landfills, LFGTP facilities).
- Assessment of type of waste, historical and projected volumes, historical and projected evolution of specific energy content, etc. and the current regulatory framework prevailing in the country.
- The overall analysis of MIW energy recovery alternatives which involves a review of the existing available waste to energy technologies (gasification, combustion alternatives, etc.). Focusing on the commercially available technologies.

3 The Market: Waste Market

The characterization of the availability of wastes for energy recovery is done in three steps, namely to discuss the current waste disposal methods for, first municipal waste, second manufacturing industrial waste and third wastewater. Current disposal methods for municipal waste in Turkey:

- In 2014, municipal solid waste (MSW) was 28 million tons (about 1 kg/capita/day).
- About 36 % of the MSW is not landfilled. This has decreased significantly from a high of 68 % in 2002/2004.
- The proportion of controlled landfill sites has significantly increased from 28 % in 2002 to 64 % in 2014 (Fig. 1).

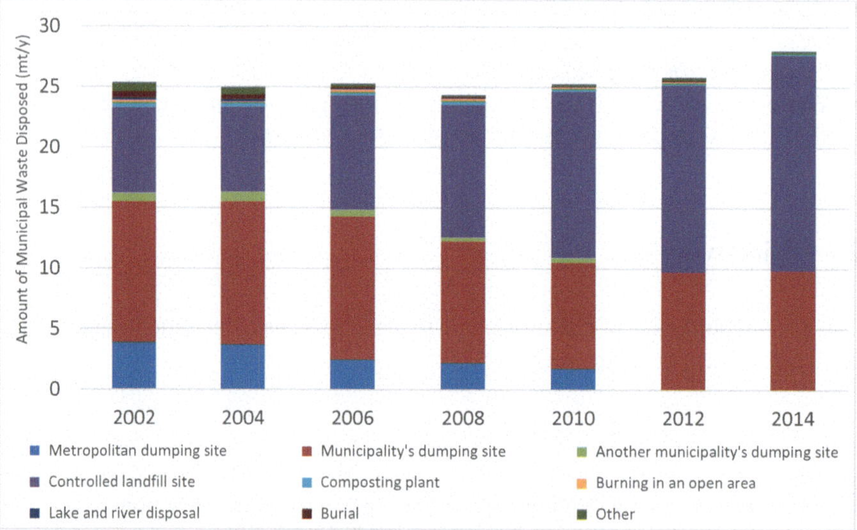

Fig. 1 Current disposal methods for municipal waste in Turkey

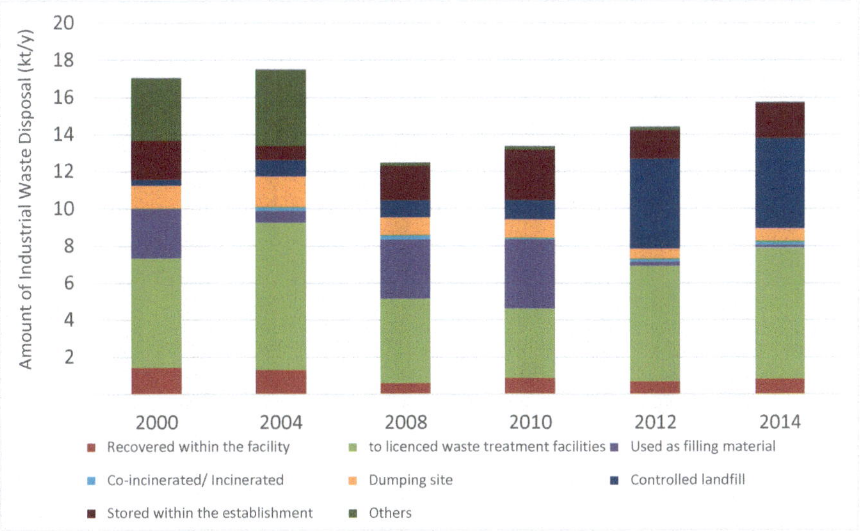

Fig. 2 Current disposal methods for manufacturing industrial waste in Turkey

Current disposal methods for manufacturing industrial waste in Turkey:

- In 2014 manufacturing industrial waste was 15.7 million tons.
- The largest energy recovery potential in the industrial sector is the utilization of the wastewater treatment sludge, covering nearly 50 % of total waste.
- Some of the manufacturing industrial waste streams are utilized in co-incineration plants.
- Other major uses are: 31 % disposed at controlled landfills and 12 % stored at the establishment.
- The total amount of waste co-incinerated was 203,000 t in 2014 (Fig. 2).

Amount of wastewater treated by wastewater treatment plants in Turkey:

- In 2014, 81 % of the total wastewater went to an either physical, biological and/or advanced wastewater treatment plant.
- 60 % of Istanbul's wastewater passes through pretreatment, 39 % through advanced biological and 1 % through biological treatment.
- Energy produced from the sludge during the waste water may be used for power production. The power production from anaerobic digestion (AD) in this plants is usually used for their own consumption (Fig. 3).

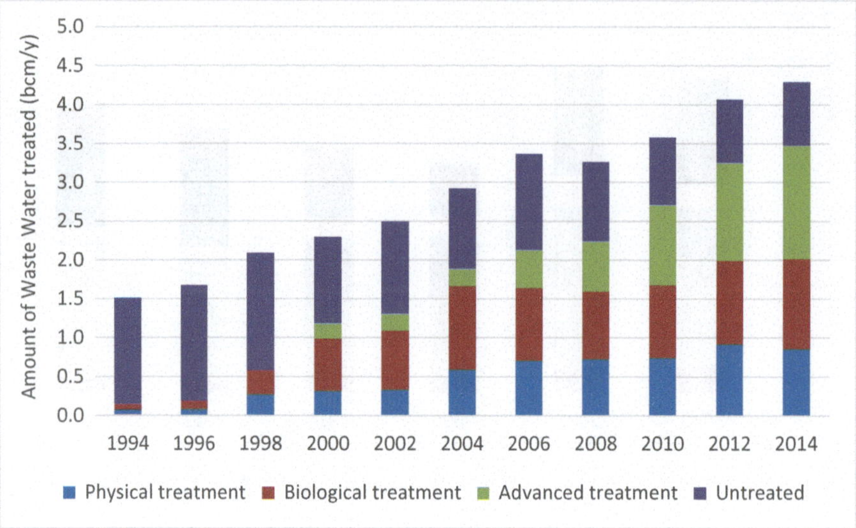

Fig. 3 Amount of wastewater treated by wastewater treatment plants (billion m³/year)

4 The Market: MISW Collection, Transport, and Disposal

1. The municipalities are responsible of the solid waste management.
2. Among the advances which may be observed in the development of waste management services should be highlighted the so-called "Union of Municipalities in Turkey." The association holds the following objectives: (a) to construct municipal waste landfill, (b) to reduce the solid waste amount, (c) to increase recycling practices, and (d) to optimize the solid waste transportation cost.

- As of end 2014, there are various operational LFGTP (*see* Fig. 4).
- Population wise: a coverage of around 50 % (38 million out of 77 million in 2014) is observed.
- There is still a market open for business, also because the full population is not yet covered in the provinces that have operating LFGTP.
- "In order to be able to plan and run satisfactory solid waste management services with a sustainable cost, service population shall not be less than 300,000."

5 Hazardous Waste

1. The collection and transport of waste from the manufacturing and health industry (including hazardous waste) is mostly developed by private companies.
2. The online platform "Environmental Information System" (Ministerial and Provincial Organizations track the production of hazardous waste) facilitates the managing activities in this field (Table 1 and Fig. 5).

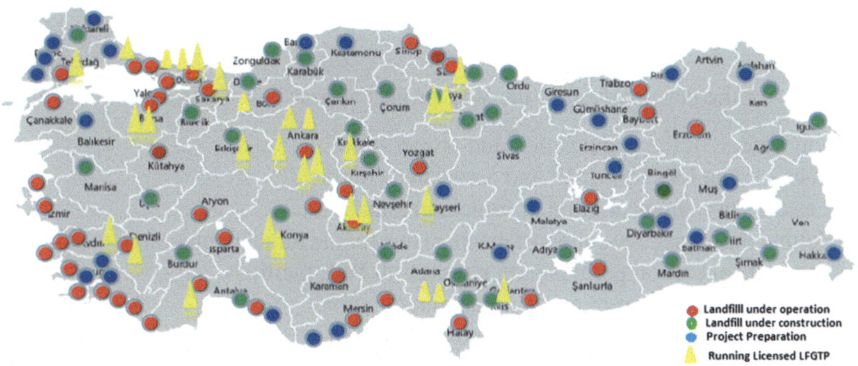

Fig. 4 Overview of major landfill sites in Turkey

Table 1 Hazardous waste treatment facilities as of late 2014

Industrialized region	Landfill	Incineration	Gasification
Trakya	ISTAC (capacity: 105 kt/year) Erdemir (capacity: 6 kt/year)	–	Istanbul (capacity: 29 kt/year)
Eastern Marmara	Kocaeli 1 unit planned (capacity: 160 kt/year)	Izaydas, Kocaeli (capacity: 35 kt/year)	–
Aegean	Manisa 1 unit planned (capacity: 240 kt/year)	PETKIM, Izmir (capacity: 17 kt/year)	–
Central Anatolia	Ankara (capacity: 200 kt/year)	Two units planned (capacity: 120 kt/year)	–
Mediterranean	ISKEN (capacity: 11 kt/year)	–	–

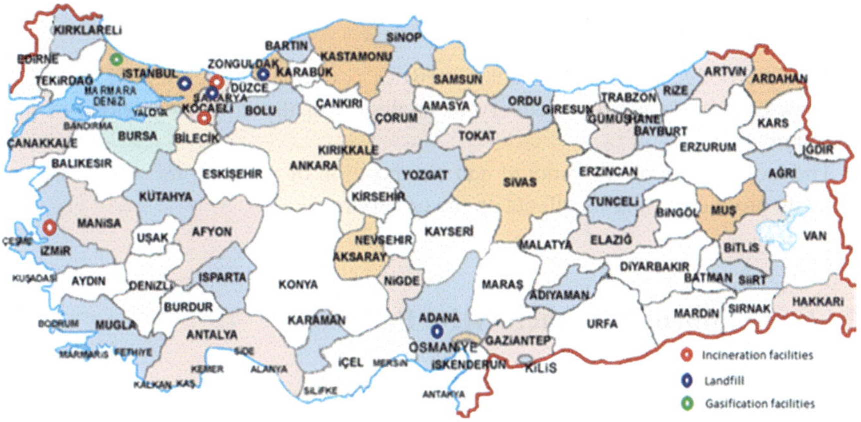

Fig. 5 Location of major hazardous waste disposal facilities in Turkey

6 Barriers

Barriers for the development of the WtE project appear in relation to:

- Knowledge of the composition and properties of the waste.
- Lack of gate fees for MSW and associated contracts.
- Lack of economics of scale in some municipalities.
- Setting the appropriate support mechanisms for the implementation of the system and to its operation permanently to cover all WtE facilities.
- Lack of enforcement of environmental regulations.
- Municipalities not recovering the cost of the service.

"The population benefiting from waste disposal services in line with the current regulation accounts for 45 %, i.e., the lack of enforcement of environmental regulations causes that less than 50 % of the population enjoy the appropriate/legal disposal services"—Union of Municipalities in Turkey.

Institutional issues:

- Municipalities do not show the required level of activity and attention in disposal of solid waste management.
- Geographic scale at which municipal waste management should be organized (30–40 km are the break-even point for introduction of waste transfer).
- Capabilities of the municipalities in dealing with the allocated responsibilities (e.g., landfill locations).

Economic and financial barriers:

- The penetration of these technologies is not natural in current markets, subsidies, Gate fees or FIT are required.
- In Turkey, there is no gate fee for MSW.
- The municipalities are not collecting the cost of waste management services; any attempt to develop WtE facilities requires external subsidies (i.e., FIT).

Informative barriers:

- The knowledge of the waste characteristics is critical to identify the different types of suitable energy recovery technologies in the different regions of the country.
- The lack of data on the amount and composition of wastes, rate of increase in waste production and its effects on the environment.

7 Environmental Impact: GHG Emissions

According to the Turkish National Inventory Report 2014 prepared under the UNFCC the waste disposal sector is the second largest GHG producer in Turkey with a share of 9 % of the total GHG emissions. In 2012, the GHG emissions from waste disposal were 36.2 Mt CO_2-eq.

The avoided emissions are 0.6 tCO_2-eq./tMSW for the LFGTP project and 0.4 tCO_2-eq./tMSW for the waste incineration project.

The environmental and social effects of the WTE projects are mainly positive. The possible negative environmental effects related to LFG and biogas utilization are air pollutant formation, noise and land usage, are easily reduced by flue gas and by silencers if necessary.

The social impacts of a WTE project are mainly positive. Both LFG and biogas utilization and waste incineration create jobs.

8 Main Sequence of Steps and Participants in LFGTP Projects

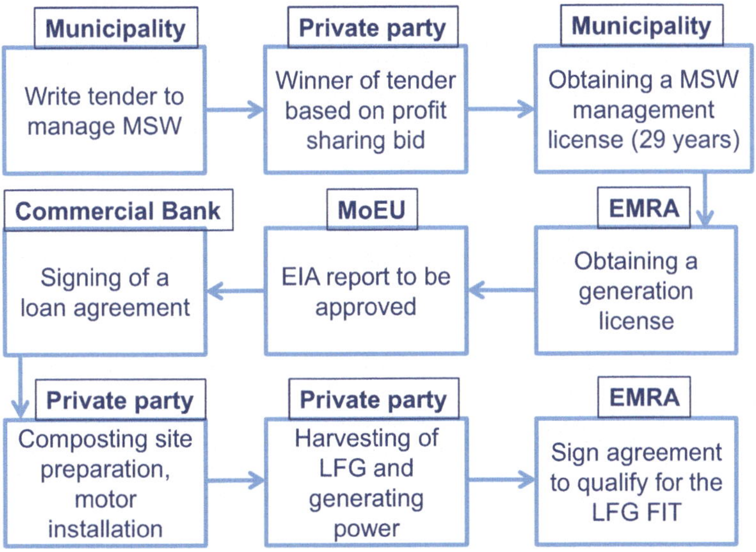

9 WTE Technologies

Qualification of various WTE technologies to Carbon Credits:

	Technology	FIT	Carbon credit	Comment	Waste
1	LFGTP	133 US $/MWh	GS+	Common	MSW
2	Co-incineration	no	VCS, VER+	Common	ISW
3	Incineration	no	VCS, VER+	Pilot	MSW
4	Gasification	133 US $/MWh	GS+	Pilot	MISW
5	Pyrolysis	133 US $/MWh	GS+	Pilot	ISW
6	Wastewater AD	133 US $/MWh	GS+	Often not sold to grid	MIWW

10 Conclusions

- There is a remarkable potential for the development of WtE facilities. Our rough estimation shows about 750 MWe of remaining potential.
- Currently LFGTP is the most profitable technology, although foreseeable boundaries such as space availability, amount of waste allowed to be landfilled, and environmental constraints, may enhance the deployment of other technologies.
- The settlement of an integrated Waste Management stream between the Municipalities is a must for scaling up the process.
- Gate fees are required to enhance the development of incineration technologies.
- The deployment of proper output based incentives for WtE projects (FIT) alone or in combination with gate fees will foster their power efficiency.
- The sustainability of the waste management services is an issue since the municipalities are not collecting the cost of the service. It seems that the principle of "polluter pays" pushed by the legislation is not fully applied in Turkey, even when affordability seems not to be an issue.
- The market for Hazardous Waste elimination seems to be short. Current gate fees for Hazardous Waste might ensure the development of incineration and gasification facilities.

Dr. Wietze Lise—Principal consultant at AF-Mercados EMI based in Ankara since February 2012 and an Energy Markets expert. His main work experience is in power and gas markets both in Europe and Turkey. However, he also has sound knowledge and experience in energy-economic modelling, economics of climate change, and resource management. During his two decades of research and consultancy experience, he worked in Holland at several research and consultancy companies who were leaders in the energy and environment field. In addition to that, he worked together with multidisciplinary teams both in Europe and Turkey, and acquired, produced and managed a myriad of projects. He has more than thirty published and widely cited scientific articles in various international journals. He holds an MSc in Applied Mathematics from Twente University, Holland and a PhD in Economics from Delhi School of Economics, India

Acknowledgements Funding for this research was provided by the European Bank or Reconstruction and Development (EBRD. However, the opinions expressed are solely mine. Collaboration with Leonardo Lupano and Jose Lopez in the course of the project are also highly appreciated. Remaining error and omissions, if any, are mine.

Evaluation of Wind–Solar Hybrid System for a Household in Northern Cyprus

Nafi Cabacaba and Serkan Abbasoğlu

1 Introduction

The energy efficiency and the solutions are very important for our growing world. The need for electricity is increasing linearly with the growing technology and population [1]. The oil prices and electricity bills are going up every day. Also, there is a significant climate change with increasing CO_2 emissions amount. Furthermore, there are several legislations that are supporting the sustainable development. Therefore, the world has a lot of natural sources and there is a real potential to invest in energy efficient technologies.

Nowadays, the new energy source is derived from nature as an alternative to the fossil fuels which is called renewable energy source (RES). Renewable energy is a growing need for a developing world due to reduction and sustainability of fossil fuel production, and increase in prices [3]. It is an energy derived from natural sources such as sun, water, and wind, which are renewable in nature. Using two different electricity generation sources together is called hybrid system [4].

Wind and solar power are the fastest growing renewable energy sources in the world. Both are common and free all around the world. Photovoltaic systems and wind turbines have been developed to convert sunlight and wind into electricity, respectively.

In this study, wind and solar PV hybrid renewable energy generation will be discussed to produce electricity for a household in North Cyprus. The system is grid connected hybrid system works in parallel with the electric utility grid. This research is focused on the analysis of the technical characteristics and economical implications of the PV–wind hybrid system for a household. HOMER software is used in

N. Cabacaba (✉) • S. Abbasoğlu
Department of Energy Systems Engineering, Institute of Graduate Studies and Research, Cyprus International University, Mersin 10, Nicosia, Turkey
e-mail: nafi@cabacaba-solar.com; sabbas@ciu.edu.tr

T.S. Uyar (ed.), *Towards 100% Renewable Energy*, Springer Proceedings in Energy, DOI 10.1007/978-3-319-45659-1_34

this study [3]. NREL (National Renewable Energy Laboratory) has developed software called HOMER (Hybrid Optimisation Model for Electric Renewable) to which is freely available and can be used to evaluate the size and cost of hybrid renewable energy systems. HOMER allows the user to incorporate several parameters such as wind, solar, and heat data and optimize their use in light of the expected electricity demand for accurate results. For a given load, HOMER software will be used to optimize the size, capacity, and the number of each of the components to find the lowest net present value for the hybrid system to produce adequate electricity for the desired load. The hybrid system will then be analyzed in HOMER to find the exact configuration for the desired demand for electricity for the household load. Environmental impacts, financial considerations and the effectiveness of the hybrid system will be analyzed.

2 Methodology

In this section, wind speed, solar irradiation household load, and energy system selection with HOMER software is evaluated. PV–Wind hybrid system will be used to generate electricity for the desired demand which is calculated by using household load analysis.

2.1 *Evaluation of Methodological Data*

Wind–solar PV hybrid system will be mounted in the capital city of Nicosia in North Cyprus where the location is 35°08′N 33°28′E. Solar irradiation, temperature and wind speed data's are collected with approved technologies to get accurate results.

Solar radiation must be inserted into HOMER software to calculate the PV performance under specific solar radiation values [5, 6]. The solar radiation data is obtained from government officials [2, 3] for a specific location, namely Nicosia, with latitude 35.8 North and longitude 33.28 east at the Eastern Europe time zone. The average daily radiation is 4.724 kWh/m^2/day. Figure 1 was obtained from HOMER and shows the solar radiation on an annual basis. The data is collected in 2010. The most radiation occurs between May and August. The highest solar radiation detected in Nicosia was in July with a 6.563 kWh/m^2/day.

Wind speed must be inserted into HOMER to calculate the wind turbine performance [7, 8]. The wind speed data is obtained from government officials. Once again it was obtained for only Nicosia, the capital city of Cyprus. The average wind speed for a year is 3.588 m/s. Furthermore, Fig. 2 was obtained from HOMER. It shows the wind speed in an annual basis. The given data is collected in year 2010. The highest wind speed was recorded in June with 4.2 m/s.

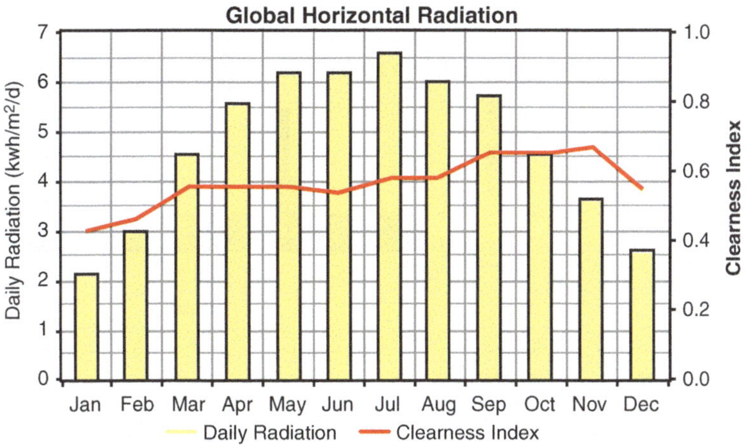

Fig. 1 Annual solar radiation

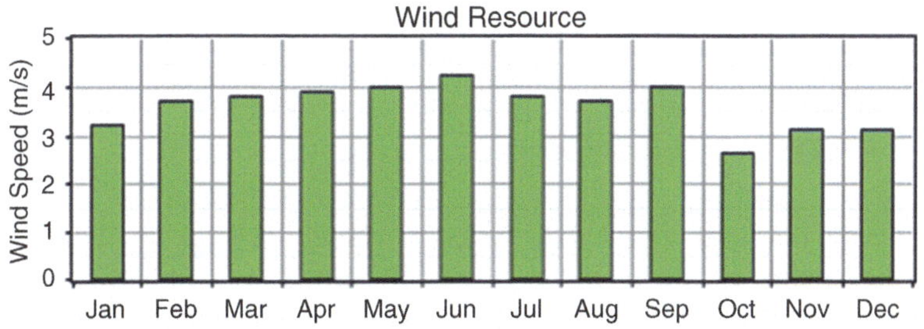

Fig. 2 Annual wind data

2.2 *Load Determination*

The house use in this study is 300 m^2 with an assumption of perfect insulation which means no heat loss. It includes three bedrooms, one living room, one kitchen, one bathroom, and a big balcony. It is assumed that, all electrical devices in the house are working with the AC load. The house includes light bulbs, a refrigerator, a freezer, a TV set, one personal computer, two air conditioners of size 12,000 BTU, one air conditioner of size 22,000 BTU, a washing machine, a dishwasher, an oven, and several other small electrical appliances. The AC average daily load is about 26 kWh to high loads of air-conditioners during both heating and cooling. The assumptions made above, leads to find the total consumption of 26 kW/day.

It is assumed that both heating and cooling is done by air conditioners within the same consumption hours. In Fig. 3, the data's inserted to the HOMER shows that the average daily electricity profile in the household. The peak load is 5.4 kW. The load factor is 0.204.

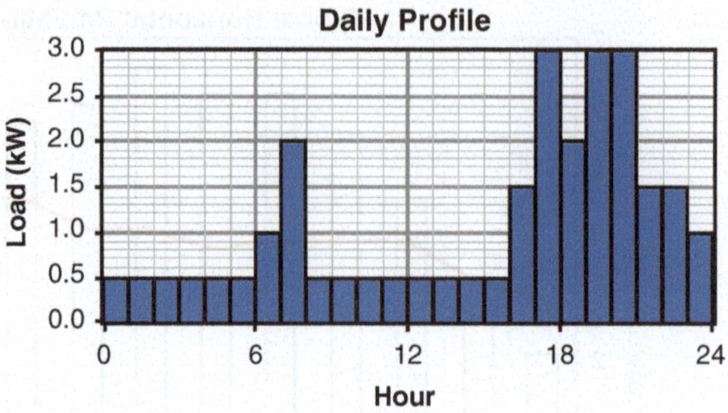

Fig. 3 Daily electricity profile

Table 1 The three ways to generate electricity

	PV	Wind	Wind–PV hybrid
Initial cost ($)	16,701	21,000	16,600
Capacity (kW)	6	12	6
Operation maintenance Cost ($/kWh)	2248	6750	1386
Total NPC ($)	45,443	54,242	34,319
Renewable energy fraction (%)	57	45	86

2.3 *Energy System Selection*

In this study, the wind–solar hybrid system is chosen for a specific household to obtain the required electricity. In this system, PV panels, wind turbine, and inverter are used. The system is based on an on-grid application. Therefore, the certain amount of electricity is transferred to the utility grid during low consumption times and the reverse situation is done during peak hours or when the excess electricity is needed.

2.4 *Calculation and Analysis*

Different solutions of wind and solar PV systems are analyzed with HOMER and the chosen solution is described.

2.4.1 System Description

The HOMER illustrated three systems: only PV, only wind, and wind–PV hybrid systems to find the best optimum way to generate electricity for the chosen household. Table 1 shows the properties of the chosen systems.

Finally, HOMER illustrated the suitable capacity for the project by using the properties stated above. PV–wind hybrid system is chosen due to its financial advantage.

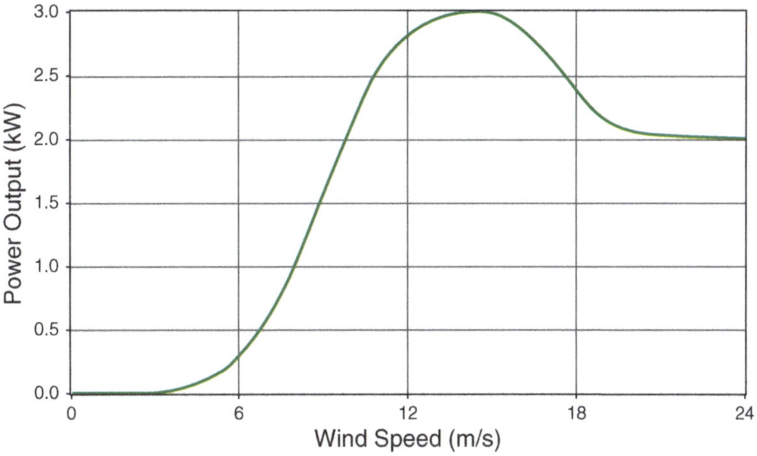

Fig. 4 The power curve of the 'Generic 3 kW' wind turbine

According to the daily profile, the household needs 3 kW peak solar photovoltaic system with a 3 kW wind turbine. The complete system for electricity generation will include wind turbine, PV arrays, and an inverter.

A converter with 99 % efficiency is selected with a maximum life time of 25 years. The average capital cost of 5 kW converter is $2500 [9]. Also, the operation and maintenance cost of 5 kW inverter is about $10 per year.

The size of the PV modules for this project is 3 kW and the capital cost is $2300 per kW [9]. This price includes the installation cost as well as the wiring, control system and shipping costs. The operation and maintenance cost is $30 per year. The life time for the PV system is assumed as 25 years. The slope assumed as 30°. There will be no tracking system.

The type of the wind turbine for this project is Generic 3 kW with rated power of 3 kW. The capital cost of the wind turbine is $7000 and the operation and maintenance cost is $150 per year [9]. The life time for the selected wind turbine is 15 years and the hub height is 12 m. The number of wind turbines for this project is one. Figure 4 shows the power curve of the selected wind turbine.

The rotor diameter of the selected wind turbine is 4.5 m. The cut-in speed is 3 m/s, maximum output is 3 kW with the wind speed of 13.5 m/s.

2.4.2 Cost Summary

The number of devices needed and the price of the system is analyzed by HOMER based on the data entered and the chosen devices. HOMER will also estimate the complete system life-cycle and installation and operation costs. In other words, Homer will be used to perform a feasibility analysis for each house and study every possible system configuration. As part of this process the software will estimate the total Net Present Cost (NPC) and the life-cycle of the system with the amount of the CO_2 production in every feasible system. The net present cost (NPC) contains all the cost occurs during systems' life time. It represents capital costs, replacement costs, operations and maintenance costs and selling price of the excess electricity. Eq. (1) shows how to calculate the net present cost. As a result, an optimal configuration of the designed system will be chosen for the house.

$$\text{NPC} = \text{Capital Cost} + \text{Replacement Cost} + \text{Operation and Maintenance Cost} \\ + \text{Fuel Cost} + \text{Salvage Cost} \tag{1}$$

3 Results

The house has 26 kWh demand per day. It needs 5.4 kW during peak hours. HOMER software provides a wealth of information including the number of components, capital cost of the simulated systems, operating cost, the net present value of the chosen system, properties of the chosen components, and working hours of the system. Also, it provides cost and electrical summary for each component for each simulated/modeled system. The optimum value is founded after the trial of three systems by the HOMER software.

Regarding to the analysis of the system for the house, the best option with the given data is using 3 kW PV arrays, 3 kW wind turbines, and a 5 kW inverter. The initial capital cost is $16,600, the operating cost is $1386 per year and the total net present cost is $34,319. The renewable energy fraction is 86 % and the utility grid electricity cost is $0.217 per kWh. Also, the project assumed to be long last for 25 years and there is no feed-in tariff.

Figure 5 shows the cash flow summary of each component for the best choice. Table 2 shows how the electrical production is split between PV arrays, wind turbine and the utility grid. The percentage of the electricity produced from PV arrays is 21 while the electricity produced from wind turbine is 65 %. Also, there is a purchase

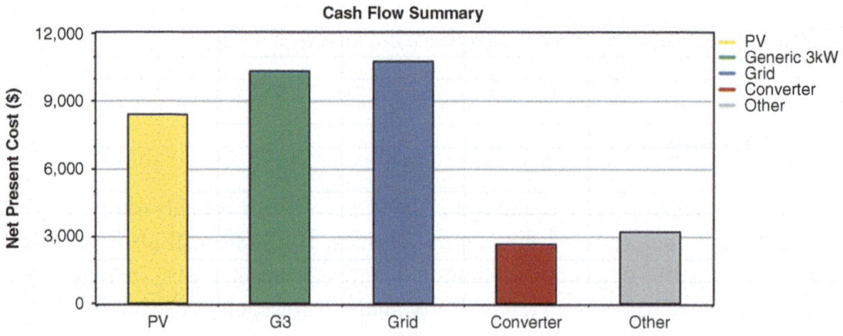

Fig. 5 The cash flow summary

Table 2 The percentage of the electricity production of each component

Production	kWh/year	Percentage (%)
PV array	4669	21
Wind turbine	14,359	65
Grid purchase	3110	14
Total	22,138	100

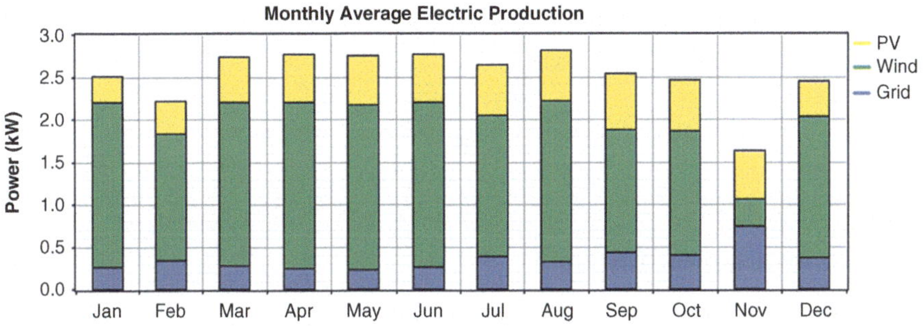

Fig. 6 Monthly average electric production by type

Table 3 The percentage of hybrid electricity consumption

Consumption	kWh/year	Percentage (%)
AC primary load	9600	44
Grid sales	12,315	56
Total	21,915	100

from utility grid which covers 14 % of total electricity consumption. Figure 6 shows the monthly average electricity production.

Table 3 shows the general information about excess electricity and general consumption of hybrid power system. The AC primary load is 9600 kWh per year which covers 44 % of the total production, and the gird sales is 12,315 kWh per year which covers 56 % of total hybrid electricity production. The renewable energy fraction is 0.859 (~86 %).

The electricity price from national grid is \$0.217 per kWh due to the high consumption level. For the chosen household, the total electricity consumption price is \$51,879 for 25 years which is derived from (2).

$$\$0.217 \times \frac{26.2\,\text{kWh}}{\text{day}} \times \frac{365\,\text{days}}{\text{year}} \times 25\,\text{year} = \$51{,}879 \tag{2}$$

The electricity derived from hybrid power system covered 86 % of the total electricity consumption. Therefore, the total consumption from grid is only 3110 kWh per year. According this information, the total electricity consumption price from grid after the hybrid system installation is dropped to \$16,871.75 for 25 years as seen in (3).

$$\$0.217 \times \frac{3110\,\text{kWh}}{\text{year}} \times 25\,\text{year} = \$16{,}871.7 \quad (3)$$

The total electricity production is 21,915 kWh per year and the capital investment is \$16,600 for the hybrid system. The payback period is only 3.5 years as the total avoided electricity payment is \$4756 each year. The (4) shows how to calculate the payback time for the installed system.

$$\frac{\$16{,}600}{\$0.217 \times \frac{21915\,\text{kWh}}{\text{year}}} = 3.5\ \text{year} \quad (4)$$

4 Environmental Impacts

Table 4 shows the avoided greenhouse gas emissions in each year by using renewable solar–wind hybrid power system. There is a 5817 kg CO_2 emission loss per year.

5 Conclusion

In conclusion, it is a fervent hope that the renewable energy systems will be turned into profitable and useful systems for a better sustainable world so the future generations can use sustainable and profitable electricity. The hybrid system uses less land and the GHG emission is zero which makes it attractive. However, installing such a system will require a very large capital investment and the life time of the hybrid system is only approximately 25 years. In summary, all the relevant parameters such as PV and wind turbine prices, labor and heat, solar and wind data all affect the complete system. These parameters and data were entered into Homer as inputs to find optimal results. The net present value of the hybrid system is investigated in this study to determine the most profitable way for meeting the needs of electricity consumption in the household in North Cyprus.

In this study, the payback period is 3.5 years. The capital investment is 16,600\$ and the annual renewable electricity production is 21,915 kWh/year. On the other hand, the electricity bought from utility grid decreased into 14 % which is 3110 kWh/

Table 4 Greenhouse gas (GHG) emissions

Pollutant	Emissions (kg/year)
Carbon dioxide	5817
Sulfur dioxide	25.2
Nitrogen oxides	12.3

year. The aim of this project was to decide which energy application was more suitable for a chosen household in North Cyprus. It is more attractive option to set up a PV–wind hybrid system as it needs less capital cost while having almost the same amount of profit with only PV system. The greenhouse gas emission is almost zero for the house. The system's life cycle is estimated to be 25 years and the net present cost includes replacement and operating costs as well. The total net present cost is $34,319 and the profit is $84,570 when compared with the utility grid system. All the prices are calculated for 25 years.

References

1. Barker PP, Bing JM (2004) Advances in solar photovoltaic technology: an applications perspective, IEEE power engineering society summer meeting, San Francisco, CA, vol 1, pp 1–6
2. NREL (2005) Getting started guide for homer version 21. National Renewable Energy Laboratory, US Department of Energy, Washington, DC
3. Cabacaba N (2011) Wind-solar hybrid system to generate electricity for a household in Cyprus and obtain heat by using heat pump system, Faculty of Engineering and Physical Sciences, University of Surrey
4. Ahmed NA et al (2008) Power fluctuation suppression of standalone hybrid generation combining solar photovoltaic/wind turbine and fuel cell systems. Energy Conver Manag 49:2711–2719
5. Redlinger RY, Andersen PD, Morthorst PE (2002) Wind energy in the 21st century economics, policy, technology and changing electricity industry. UNEP Collaborating Centre on Energy and Environment, London
6. Nelson V (2006) Wind energy renewable energy and the environment. Taylor & Francis Group, London
7. Manwell JF, Mcgowan JG, Rogers AL (2002) Wind energy explained theory, design and application. John Wiley & Sons Ltd, West Sussex
8. Lam GCK (2006) Wind energy conversion efficiency limit. Wind Eng 30(5):431
9. www.cabacaba-solar.com . Accessed 8 Jan 2013

How do The External Costs of Renewable and Fossil Fuel Energy Compare?

Rohit Mistry and Lawrie Harper-Simmonds

1 Introduction

This chapter presents the methodology and results of the Externalities of Energy project [1], which was a large international research study sponsored by the European Commission. The ExternE methodology was developed to provide a consistent and transparent framework for transforming all the non-negligible external impacts of energy production into monetary values. The five general stages in the ExternE methodology are as follows:

- Definition of the activity to be assessed and the background scenario where the activity is embedded. Definition of the important impact categories and externalities.
- Estimation of the impacts or effects of the activity (in physical units). In general, the impacts allocated to the activity are the difference between the impacts of the scenario with and without the activity.
- Monetisation of the impacts, leading to external costs.
- Assessment of uncertainties, sensitivity analysis.
- Analysis of the results, drawing of conclusions.

External costs are grouped into three categories: environmental impacts, global warming impacts and accidents. Environmental impacts, defined as being caused by substances or energy being released into environmental media (air, soil, water), are assessed using the impact pathway approach (IPA), which forms the main methodological focus of this chapter. After this is described, the methods of energy generation are

R. Mistry (✉)
EFTEC (Economics for the Environment Consultancy), 73-75 Mortimer Street, London W1W 7SQ, UK
e-mail: rohit@eftec.co.uk

L. Harper-Simmonds

T.S. Uyar (ed.), *Towards 100% Renewable Energy*, Springer Proceedings in Energy, DOI 10.1007/978-3-319-45659-1_35

individually assessed in terms of their most significant impacts, and the externalities arising from these.

In order to quantify the externalities arising from energy generation, the ExternE project adopted a fuel cycle analysis. The general stages of such a process are as follows:

- Production of construction materials.
- Transport of construction materials.
- Construction of plant.
- Exploration of fuel.
- Extraction of fuel.
- Processing of fuel.
- Transport of fuel.
- Transport of personnel.
- Treatment of flue gases.
- Generation of wastes and by-products.
- Further treatment of waste.
- Removal of plant at the end of its service lifetime.
- Restoration of sites after closure.

These stages help to define the boundaries of the analysis. A broad categorisation allows for the consistent framework required by the ExternE project, even though some stages will not be relevant for specific fuel cycles (the extraction of fuel for wind power, for example).

Furthermore, although the ExternE guidelines state that "no impact that is known or suspected to exist, but cannot be quantified, should be ignored for convenience … it should be retained for consideration alongside whatever analysis has been possible", it is not necessary to conduct a complete analysis at each stage of the fuel cycle. This is because the analysis focuses on "non-negligible" impacts; given the resource cost of quantifying and valuing an impact, it would be prohibitively expensive to conduct this for every single impact that occurs as a result of the fuel cycle. This relates to the concept of the prioritisation of impacts.

The prioritisation of impacts follows the identification of fuel cycle burdens, and the consequent impacts that may arise from them. The ExternE project defines a burden as "anything that is, or could be, capable of causing an impact of whatever type", and adopts the following broad categories:

- Solid wastes.
- Liquid wastes.
- Gaseous and particulate air pollutants.
- Accidents.
- Occupational exposure to hazardous substances.
- Noise.
- Presence of human activity.
- Others (e.g. exposure to electromagnetic fields).

Having then reported all possible impacts that could arise from the relevant burdens across the fuel cycle, these impacts can be prioritised to identify the most significant. These impacts are then quantified using the impact pathway approach.

2 Impact Pathway Approach

The impact pathway approach is a bottom up methodology which traces the chain of causal relationships from the source of a burden to subsequent impacts to relevant receptors. The practical application of the approach means that it is possible to estimate the marginal costs associated with specific changes in the emissions of certain pollutants; this is in contrast to the estimation of average costs which would occur through a top down approach.

The advantage of estimating marginal costs is that these may not be constant across different levels of the overall concentration of the pollutant within the environmental medium; that is the concentration response functions may not be linear. In such a case, the top down method would not accurately measure the external costs associated with a change in the level of emissions. These concepts are further developed below.

It should be noted that an impact pathway may be very simple or extremely complicated depending on the impact under consideration. For example, when considering the effect of the construction of wind turbines on visual amenity, the pathway of source to receptor is very direct.[1] All that is required to understand the nature of the impact are the descriptive characteristics of the turbines being constructed, their number and their location. This provides the requisite information for valuation to occur.

On the other hand, the emission of airborne pollutants presents a much more complex series of relationships until the nature of the impact can be fully described. Figure 1 illustrates the discrete stages of the impact pathway approach, as applied to airborne pollutants.

- *Emissions*: the first step is the modelling of emissions (or the burden) from the activity of interest, in this case energy generation. In general, the objective is to assess the difference in emissions arising under some proposed policy scenario (the "with" case) and a continuation of the current situation (the "without" case) which is also referred to as the "baseline". Achieving the correct specification of relevant technologies and pollutants (i.e. the amount of carbon dioxide emitted in kg per GWh of energy produced by a particular power plant at a specific site) can require detailed modelling. Reference should also be made to the type of emission source, which can be categorised as follows: a point source (single source of

[1] It should be noted that this does not necessarily imply that wind turbines represent a visual disamenity in all cases; this is dependent on the affected population and their preferences. In order to assess the effect in Turkey a study assessing this would likely be necessary. It is also true that preferences are dynamic, and can change over time.

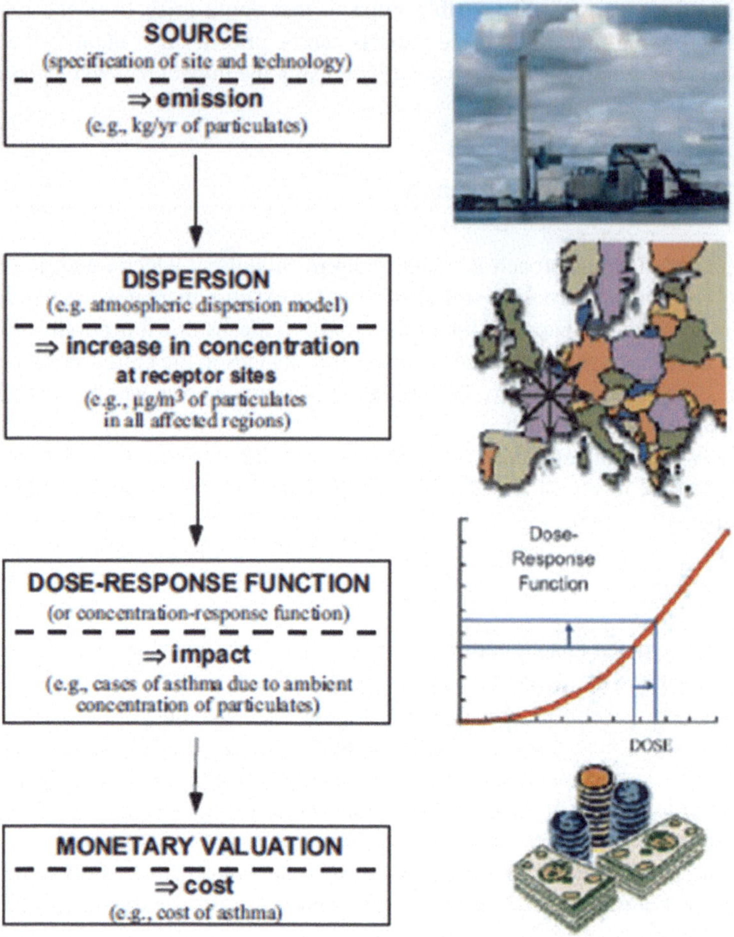

Fig. 1 The impact pathway approach. *Source*: ExternE [2]

emissions), line source (one dimensional), area source (two dimensional) and volume source (three dimensional).

- *Dispersion model*: in the second step of the pathway, atmospheric dispersion modelling predicts the concentration levels of a pollutant over a spatial area and how this changes with changes in emissions. In general, while the concentration of a pollutant resulting from an activity decreases as distance from the source of emissions increases, the total area affected increases, along with the number of potential receptors.
- *Impacts*: understanding the concentration of air pollutants across a spatial area permits "at risk" receptors to be identified. On this basis, the objective of this step in the pathway is to quantify the physical impacts on receptors that occur due to the change in air quality (pollution). This most accurately

requires epidemiological studies, which assess the effects of pollutants on real populations.

The concentration levels of pollutants in different areas, as calculated by air quality models, can be translated into impacts through concentration (or dose or exposure) response functions, which estimate the relationship between the presence of a pollutant and the scale of physical impacts. These can take a number of forms, being linear, non-linear or even subject to threshold effects, and a function is required for the assessment of each impact. Impacts may be viewed first within a group of receptors, such as human health, ecosystems and vegetation, and materials, and then at a more detailed level within each receptor; for example, within human health, rates of mortality and morbidity will have separate exposure response functions. To ensure they are as accurate as possible, these functions are mostly derived from meta-analyses of relevant studies, and are constantly being updated. Figure 2 illustrates how dose response functions can differ.

- *Valuation of impacts*: in the final step of the pathway the quantified physical impacts are valued in monetary terms. This requires the application of economic valuation methods, which were discussed in Paper 1.

This broad approach can be similarly applied to the other burdens that were previously identified.

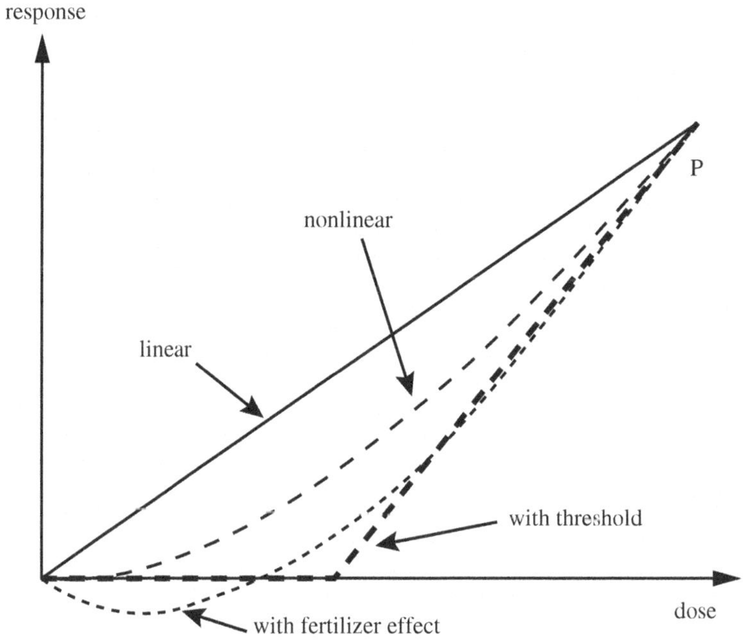

Fig. 2 Possible forms of the dose response function. *Source*: ExternE [2]

There are a great deal of uncertainties that arise when calculating the externalities from energy generation. It is necessary to make these clear to ensure that the results are taken in the right context, and that new information can be applied when it becomes available. Uncertainties can be broadly grouped into five categories [2]:

- Data uncertainty, e.g. slope of a dose-response function, cost of a day of restricted activity.
- Model uncertainty, e.g. assumptions about causal links between a pollutant and a health impact, assumptions about the form of a dose-response function (e.g. with or without threshold effects), the choice of models for atmospheric dispersion and chemistry.
- Uncertainty about policy and ethical choices e.g. the discount rate for intergenerational costs, the value of a statistical life.
- Uncertainty about the future, e.g. the potential for reducing crop losses by the development of more resistant species.
- Idiosyncrasies of the analyst, e.g. the interpretation of ambiguous or incomplete information.

Data and model uncertainty are scientific in nature, and allow the possible formation of confidence intervals to demonstrate exactly how reliable the data is. Policy and temporal uncertainty can be accounted for to some extent by the use of sensitivity analysis to test how the results change with different assumptions. The greatest uncertainties in the ExternE analysis fall under the scientific impacts and monetary valuation, but these vary with the extent of available information.

3 Priority Impacts

The original ExternE [1] analysis highlighted priority impacts for quantification across the fossil fuel, nuclear and wind and hydro generation methods. In comparison to the more recent methodology [3] that was used to generate the external cost estimates presented later in this chapter, solar as a renewable energy is missing, but the priority impacts stayed much the same. The main differences are associated with improving dose-response functions and including and updating the monetary valuation of various impacts. However, these priority impacts were not necessarily successfully quantified and incorporated into the external cost estimates.

Fossil fuels (all):

- Effects of atmospheric pollution on human health.
- Accidents affecting workers and/or the public.
- Effects of atmospheric pollution on materials.
- Effects of atmospheric pollution on crops.
- Effects of atmospheric pollution on forests.
- Effects of atmospheric pollution on freshwater fisheries.
- Effects of atmospheric pollution on unmanaged ecosystems.

- Impacts of global warming.
- Impacts of noise.

Fossil fuels (fuel cycle dependent):

- Impacts of coal and lignite mining on ground and surface waters.
- Impacts of coal mining on building and construction.
- Resettlement necessary through lignite extraction.
- Effects of accidental oil spills on marine life.
- Effects of routine emissions from exploration, development and extraction from oil and gas wells.

Nuclear:

- Radiological and non-radiological health impacts due to routine emissions.
- Radiological and non-radiological health impacts due to accidental emissions
- Occupational health impacts.

Wind:

- Accidents affecting the public and/or workers.
- Effects on visual amenity.
- Effects of noise emissions on amenity.
- Effects of atmospheric emissions related to the manufacture of turbines and construction and servicing of the site.

3.1 Valuation of Health Impacts

Although methods of economic valuation have been discussed in Paper 1, it may be useful to briefly consider how health impacts, which constitute the majority of the valued impacts, are incorporated into external cost estimates.

The valuation of health impacts considers three key impacts: (1) the potential for an increased risk of mortality, which may be viewed in terms of lowered life expectancy or deaths brought forward; (2) morbidity, which is the increased incidence of certain illnesses; and (3) resource costs which arise in relation to lost productivity (loss of earnings) and the treatment of ill-health.

3.1.1 Mortality: Adults

The estimation of values for mortality involves assessing risks which may affect overall life expectancy across a population (chronic mortality) as well as those which target a particular population, particularly the elderly or those with underlying health conditions, bringing their deaths forward by a number of months (acute mortality).

The initial progress in valuing mortality risks was made in transport and other health and safety contexts. The usual approach to estimate mortality risks associated with accidents (such as road or rail fatalities) has been to estimate the "value of a statistical life" (VOSL).[2] This is calculated by estimating individuals' WTP to secure a risk reduction arising from some defined policy, or their WTA a risk increase. WTP and WTA measures can be obtained either from revealed preference or stated preference methods depending on the context. By dividing the change in risk into the WTP for that risk reduction, the VOSL is calculated. In order to measure the aggregate benefit of the policy to society, the number of premature mortalities is estimated through the use of a concentration response function and multiplied by the VOSL. The [4] study for input to the ExternE project assessed the VOSL as implied by the median annual WTP for a five in 1000 risk reduction across the UK, Italy and France. The respective amounts were €772,000, €1,448,000 and €958,520. A general VOSL for use across the EU was calculated at €1,052,000. A calculation using a VOSL would take the form:

$$\text{Number of mortalities} \times \text{VOSL}$$

However, this approach is concerned with a policy that has immediate effects on the current population. The nature of the health impacts of some of the pollutants from energy generation are such that they are unlikely to have immediate impacts, and if they do, there will be discrepancies within the population with the greatest effect on the most vulnerable. Estimates of the VOSL are also context dependent; in particular in relation to the type of risk and level (the baseline) and the change from it. Results from empirical studies also show that individuals tend to value voluntary risks (e.g. risk of accidents from driving a car) lower than involuntary risks (e.g. health effects of air pollution). These points imply that a VOSL estimate cannot necessarily be transferred straightforwardly from one context to another (e.g. transport accidents to health impacts of air pollution).[3]

Difficulties in applying VOSL estimates to air pollution contexts lead to the development of an alternative measure, the "value of a life year lost" (VOLY). An advantage of the VOLY is that it can be estimated for different groups of individuals, which can be of key importance for policy-making. VOLY estimates were derived from the NewExt VOSL estimates stated previously; for example, to estimate the value of a life expectancy extension of a month, the respondent's WTP for the annual five in 1000 risk reduction is divided by their life expectancy extension. This generates a payment to be made every year for 10 years, which totals an implied value of a statistical life year of €55,800 (rounded to €50,000 given the uncertainties arising from the method). This is assumed to be implicit discounting. A calculation using VOLYs would take the form:

---

[2]The use of the phrase "statistical life" is intended to emphasise that VOSL is measuring a statistical change in the risk to a life, and not the value of the life of a known individual. While any given individual may value their or another's life infinitely high, the implicit valuation through their behaviour shows that individuals accept more than zero risk to their life.

[3]Relevant studies also highlight methodological issues in valuation such as challenges in conveying small changes in risks to respondents such that they fully comprehend them [4]. There are also key issues in comparing results from WTP and WTA estimates for risk reductions [5].

$$\text{Number of life years lost} \times \text{VOLY}$$

With regard to different age groups in the population, the relationship between risk and WTP could take several forms. For example, WTP may decline with age, i.e. as an individual gets older, they have fewer years to benefit from a current risk reduction. WTP could also increase because of the scarcity value of time (with less time left, each unit of time valued more) or because of saving over lifetime, the opportunity cost of spending money on the reduction of a mortality risk declines. The empirical evidence is not conclusive on this issue; for example Chilton et al. [7] find that WTP declines with age, while Johannesson and Johansson [8] find that WTP increases with age.

A further issue to consider is that an individual's WTP may change depending on the quality of life they expect to have in the time added due to reduced risk to health. This relates to the calculation of quality adjusted life years (QALYs) which weight the number of life years generated by an intervention by the quality of that life; perfect health is weighted by 1, while death is weighted by 0. Therefore, an individual gaining an extra life year in perfect health will have one extra QALY, while someone in a health state regarded half as good as perfect has half an extra QALY. There are some states worse than death, which may be given a score of less than zero. This overall weight can then be applied to the VOLY for valuation purposes. Disability adjusted life years (DALYs) is a further measure that extends this concept to include an age weighting factor and the loss of health from a given health profile that is used as a reference point. Consequently, lower weights are applied to the health of the very old and the very young.

3.1.2 Mortality: Children

Children are exposed to different risks than adults, as their daily life differs, and are more susceptible to certain health risks, such as those associated with exposure to air pollution [9]. This warrants consideration of an estimate for the value of health effects on children separate to that of adults. However, there are difficulties associated with this, particularly in terms of the appropriate methodological approach (stated preference methods are not intended to be administered to children). The usual approach is to administer WTP questions to parents, but they are likely to be much more risk averse, and, as Pearce et al. [10] suggest, when combined with their altruistic nature, may make eliciting accurate WTP values for children very difficult.

Indeed, the time scale involved with children is so large that, when combined with discounting,[4] any estimates will be subject to large confidence intervals and

[4] Discounting is the process by which costs or benefits that occur in the future are given less weight than those that occur in the present. This is for a number of reasons, including impatience (or "pure time preference" that individuals prefer benefits now and costs to be deferred), uncertainty regarding the future state of the world, and the opportunity cost of investment (i.e. £1 saved today will return £1 + i in a year's time where i is the interest rate). The process of discounting generates the present value of a given sum of money:

high sensitivity to small changes in the initial parameters. There is also a risk of double counting in aggregating benefits, and a degree of arbitrariness in deciding when values should be assigned to "children" or "adults".

3.1.3 Morbidity

As the effects of morbidity do not result in fatalities, it is inappropriate to apply either VOSLs or VOLYs as discussed above. Rather, valuation focuses on the reduction in well-being which results from impaired health states. This includes:

- Hospital admissions for respiratory and cardiovascular illnesses.
- Bronchitis.
- Asthma.
- A reduced ability to engage in normal activities.

The typical approach is to calculate the cost of illness, which comprise market costs that can be measured directly, such as: (1) the costs of illness to individuals and companies in terms of lost earnings and time incurred; and (2) actual expenditures arising from the illness; which include the costs of treatment and medicines. To estimate the cost of illness, the per day wages lost and medical expenses incurred would be multiplied by the number of days lost to the illness:

$$\text{Number of lost days} \times (\text{wage loss per day} + \text{medical expenses per day}) \\ + \text{WTP to avoid pain and suffering}$$

Stated preference methods can be used to estimate an individual's WTP to avoid a hospital admission.

4 Results

The ExternE-Pol [3] extension to the original ExternE calculates the external costs that arise from the generation of energy, according to the method of generation used. The process by which these external costs are estimates can be re-summarised as follows:

1. Calculate the emissions of the pollutant per kWh of electricity generated. The ExternE figures are shown in Table 1.

$$\text{PV} = \frac{x}{(1+r)^2}$$

Where PV is the present value of the sum of money x, r the discount rate and t the number of years into the future this sum of money is being considered.

Table 1 Emissions per unit of energy produced

kg/kWh	Coal			Oil		Natural gas		Cogeneration			Nuclear	
	Lignite	Hard coal	Hard coal PFBC	Oil	Oil CC	Gas	Gas CC	Diesel SCR 200 kWe	Gas $\lambda = 1$, 160 kWe	Gas lean burn 1 MWe	LWR	PWR centrifuge enichment
Greenhouse gases	1.23*E*+00	1.07*E*+00	7.98*E*–01	8.82*E*–01	5.26*E*–01	6.40*E*–01	4.23*E*–01	7.31*E*–01	6.27*E*–01	5.90*E*–01	7.64*E*–03	4.83*E*–03
SO_2	6.95*E*–03	3.25*E*–03	2.76*E*–04	6.61*E*–03	1.06*E*–03	2.19*E*–04	1.47*E*–04	1.04*E*–03	3.28*E*–04	2.34*E*–04	3.78*E*–05	2.31*E*–05
NO_x	1.49*E*–03	2.26*E*–03	3.86*E*–04	2.82*E*–03	5.21*E*–04	7.20*E*–04	3.29*E*–04	1.06*E*–03	3.80*E*–04	7.80*E*–04	3.94*E*–05	3.19*E*–05
PM2.5 (incl. primary nitrates and sulphates)	5.08*E*–03	2.07*E*–04	4.24*E*–05	1.45*E*–04	3.09*E*–05	1.46*E*–05	1.07*E*–05	5.68*E*–05	1.23*E*–05	1.06*E*–05	9.50*E*–06	8.26*E*–06
Heavy metals (total, unweighted)	2.37*E*–07	4.58*E*–07	2.50*E*–07	3.93*E*–06	1.65*E*–07	2.47*E*–08	4.44*E*–08	3.40*E*–07	6.67*E*–08	4.17*E*–08	1.43*E*–07	1.31*E*–07
NMVOC (total, unweighted)	4.01*E*–05	1.06*E*–04	5.45*E*–05	3.96*E*–04	2.43*E*–04	2.72*E*–04	1.81*E*–04	8.41*E*–04	3.51*E*–04	3.83*E*–04	8.53*E*–06	7.73*E*–06
Radioactive Emiss. (unweighted)	2.97*E*–10	4.33*E*–10	1.83*E*–10	4.21*E*–10	2.78*E*–10	4.96*E*–11	3.59*E*–11	3.95*E*–10	8.07*E*–11	6.02*E*–11	2.61*E*–08	2.29*E*–08
kg/kWh	*Hydropower*	*Photovoltaic*			*Wind*							
	Alpine reservoir	*Panel, mounted (South-Europe)*	*Panel, integrated (South-Europe)*	*Panel, integrated (South-Europe)*	*Onshore, 800 kW*	*Offshore, 2MW*						

(continued)

Table 1 (continued)

	Coal			Oil		Natural gas		Cogeneration			Nuclear	
kg/kWh	Lignite	Hard coal	Hard coal PFBC	Oil	Oil CC	Gas	Gas CC	Diesel SCR 200 kWe	Gas $\lambda = 1$, 160 kWe	Gas lean burn 1 MWe	LWR	PWR centrifuge enichment
Greenhouse gases	3.70*E*–03	5.38*E*–02	5.34*E*–02	3.41*E*–02	1.05*E*–02	1.34*E*–02						
SO_2	4.15*E*–06	1.49*E*–04	1.50*E*–04	1.21*E*–04	3.81*E*–05	4.44*E*–05						
NO_x	2.87*E*–05	1.84*E*–04	1.82*E*–04	1.53*E*–04	3.85*E*–05	5.62*E*–05						
PM2.5 (incl. primary nitrates and sulphates)	1.46*E*–05	1.72*E*–05	1.71*E*–05	1.46*E*–05	1.18*E*–05	1.55*E*–05						
Heavy metals (total, unweighted)	3.35*E*–08	4.45*E*–07	4.42*E*–07	4.23*E*–07	6.18*E*–07	6.37*E*–07						
NMVOC (total, unweighted)	4.34*E*–06	3.75*E*–05	3.67*E*–05	2.83*E*–05	9.03*E*–06	1.20*E*–05						
Radioactive emiss. (unweighted)	2.06*E*–11	4.55*E*–10	4.56*E*–10	3.32*E*–10	6.84*E*–11	7.73*E*–11						

Table 2 Damage cost by pollutant

Pollutant	Damage cost (€ per ton, 2000 prices)
CO_2 equivalent	19
SO_2	2939
NO_X	2908
PM_{10} $PM_{2.5}$	11723 19539
Arsenic	80,000
Cadmium	39,000
Chromium Chromium—VI Chromium—other	31,599 240,000 0
Lead	1,600,000
Nickel	3800
Formaldehyde	120
NMVOC	1124
Nitrates, primary	5862
Sulphates, primary	11,723
Radioactive emissions	50,000 (€ per DALY, 2000 prices)

Note: Disability-Adjusted Life Years (DALY), assumed equal to the unit value of chronic Years of Life Lost (YOLL)

- Emissions / kWh

2. Apply the concentration response function from the scientific literature. This defines the level of the impact (i.e. health) at a given level of emissions.
   - Impact / unit of emission
3. Apply the valuation of the impact in question (i.e. cost of increased cases of respiratory hospital admissions).
   - /impact
4. Combine steps 2 and 3 to calculate a damage cost. This is shown in Table 2.
   - (Impact / unit of emission) (€ / impact) = € / unit of emission
5. Calculate the external costs of electricity generation relating to a particular impact by combining steps 1 and 4. The aggregated version of this (across all included impacts) is shown in.
   - (Emissions / kWh) (€ / unit of emission) = € / kWh

The damage cost methodology adopted here allows for the application of these valuation methods in a very widespread fashion (Fig. 3). However, this analysis clearly omits some impacts identified in the impact pathways; this is discussed subsequently.

Fossil fuel generation methods have much higher external costs than renewables or nuclear. Externalities for coal, oil and gas are measured at approximately €0.04,

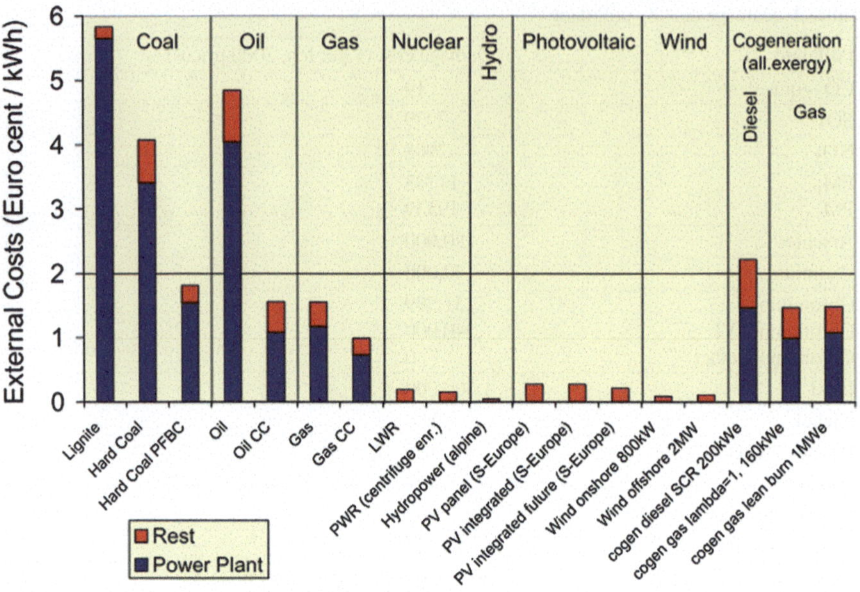

Fig. 3 External costs by generation method in 2000 prices

0.05 and 0.015 kWh^{-1} respectively, although these can vary depending on the technologies used. In contrast, the measured externalities from hydropower, photovoltaics and wind power are in the region of €0.0004, €0.002 and €0.001 kWh^{-1}, respectively. Nuclear power has measured externalities of around €0.0004 kWh^{-1}. Thus, the externalities from the least polluting renewable source (hydropower) and nuclear are around 1 % of the measured externalities from coal.

Figure 4 illustrates how the external costs are split across these different pollutants by generation method. It illustrates the relative importance of the sources in determining the external costs. Much of the external costs that arise for renewables and nuclear methods arise not during the generation phase, but rather the construction and installation of the plant. These represent fixed external costs, which means that the average external costs of the plant decrease as more energy is produced. However, as highlighted earlier, some external costs are omitted, and so the analysis does not present a full consideration of the external costs of energy generation.

The exclusion of visual impacts can likely be explained by the fact their valuation is strongly site and population specific; it depends on the preferences of the individuals affected and the current state of the environment. Decommissioning impacts are also site specific, and with regard to nuclear power, the long term impacts of nuclear waste storage are highly uncertain.

DECC (2009) [11] guidance stipulates that new renewable investments should be considered as displacing other renewable sources rather than conventional fossil fuel based generation methods. Therefore, the avoided cost (benefit) of introducing new renewable generation sources is smaller than would be expected. In the UK,

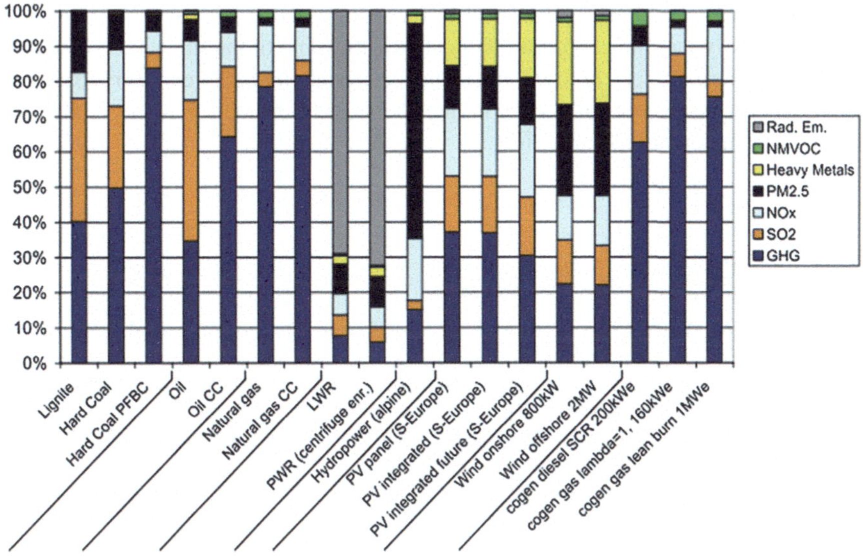

Fig. 4 Composition of external costs by pollutants

this may be the case; however, in Turkey, such renewable installations would only be replacing fossil fuels, and, as such, the benefits relating to renewable generation methods are greater.

5 Conclusion

This chapter illustrates that under the ExternE analysis, the measured external costs from renewable energy sources are significantly smaller than the measured external costs from fossil fuel energy generation. However, this is accompanied by the caveat that some external costs have been omitted, and these may have a proportionally greater effect on the external cost impact for renewables than for fossil fuel.

The omitted external costs for renewable generation methods are heavily site specific. For example, the visual amenity impacts depend on the preferences and size of the local population, which can vary significantly from site to site. This is also true for the disturbances of ecosystems that may arise from the construction of a generation plant; it is therefore extremely important to consider the baseline of the situation, and not to over-generalise these impacts. These externalities can therefore mitigated against by a careful choice of siting.

Measured external costs for nuclear are also significantly lower than those for fossil fuels. While some impacts have been omitted, to include also the preferences that some individuals may have against nuclear fuel as opposed to other sources, there are also particularly large uncertainties associated with some of the impacts. For example, the long term effects, both in human and ecosystem terms, of exposure to radioactive emissions are not fully known, and the storage of nuclear waste is also an unknown with potentially large risks.

References

1. ExternE (1995) Externalities of energy vol. 2: methodology. Directorate General XII; Science, Research and Development, Brussels
2. ExternE (2005) Externalities of energy: methodology 2005 update. European Commission, Directorate General for Research, Brussels
3. ExternE-Pol (2004) Externalities of energy: extension of accounting framework and policy applications. http://www.externe.info/exterpols.html
4. IER, Germany ARMINES / ENSMP, France PSI, Switzerland Université de Paris I, France University of Bath, United Kingdom VITO, Belgium (2004) New Elements for the Assessment of External Costs from Energy Technologies, Final Report to the European Commission, DG Research, Technological Development and Demonstration (RTD) http://www.ier.uni-stuttgart.de/forschung/projektwebsites/newext/newext_final.pdf
5. Hammitt J, Graham J (1999) Willingness to pay for health protection: inadequate sensitivity to probability? J Risk Uncertainty 18:33–62
6. Horowitz JK, McConnell KE (2002) A review of WTA/WTP studies. J Environ Econ Manag 44:426–447
7. Chilton S, Covey J, Jones-Lee M, Loomes G, Metcalf H (2004) Valuation of health benefits associated with reductions in air pollution. Department of Environment, Food and Rural Affairs, London
8. Johannesson M, Johansson P (1997) To be, or not to be, that is the question: an empirical study of the WTP for an increased life expectancy at an advanced age. J Risk Uncertainty 13(2):163–174
9. Tamburlini G (2005) Overview of the risk differences between adults and children, in economic valuation of environmental health risks to children. OECD, Paris
10. Pearce D, Atkinson G, Mourato S (2006) Cost-benefit analysis and the environment: recent developments. OECD, Paris
11. DECC 2009, UK Renewable Energy Strategy, https://www.gov.uk/government/uploads/system/uploads/attachment_data/file/228866/7686.pdf

Wind Energy Statistics in Europe: Onshore and Offshore

Klinger Friedrich and Müller Lukas

1 Statistics

During 2012, 12,744 MW of wind power was installed across Europe (Fig. 1), of which 11,895 MW was in the European Union. Of the 11,895 MW installed in the EU, 10,729 MW was onshore and 1166 MW offshore. Investment in EU wind farms was between €12.8 and €17.2 bn. Onshore wind farms attracted €9.4–€12.5 bn, while offshore wind farms accounted for €3.4–€4.7 bn. In terms of annual installations, Germany was the largest market in 2012, installing 2415 MW of new capacity, 80 MW of which (3.3 %) offshore. The UK came in second with 1897 MW, 854 MW of which (45 %) offshore, followed by Italy with 1273 MW, Spain (1122 MW), Romania (923 MW), Poland (880 MW), Sweden (845 MW), and France (757 MW). Among the emerging markets of Central and Eastern Europe, Romania and Poland both had record years—both installing around 7.5 % of the EU's total annual capacity. Both markets are now consistently in the top ten in the EU for annual installations. It is also important to note the amount of installations in the UK, Italy, and Sweden. These three markets represent, respectively, 16 %, 11 %, and 7 % of total EU installations in 2012. Offshore accounted for 10 % of total EU wind power installations in 2012, one percentage point more than in 2011. Tables 1 and 2 show an overview of the installed wind power MW in 2011 and 2012 [1].

Figure 2 shows that wind power accounted for 26.5 % of new installations in 2012, the second biggest share after solar PV (37 %) and before gas (23 %). Solar PV installed 16 GW (37 % of total capacity), followed by wind with 11.9 GW

Note: The following text and figures are a summary of the publication: "Wind in Power 2012 European Statistics" by EWEA. Other sources are appropriately marked as footnotes.

K. Friedrich (✉) • M. Lukas
INNOWIND Forschungsgesellschaft mbH in cooperation with Fraunhofer IWES, Altenkesseler Str. 17/D2, 66115 Saarbrücken, Germany
e-mail: f.klinger@wind-energy-research.de

T.S. Uyar (ed.), *Towards 100% Renewable Energy*, Springer Proceedings in Energy, DOI 10.1007/978-3-319-45659-1_36

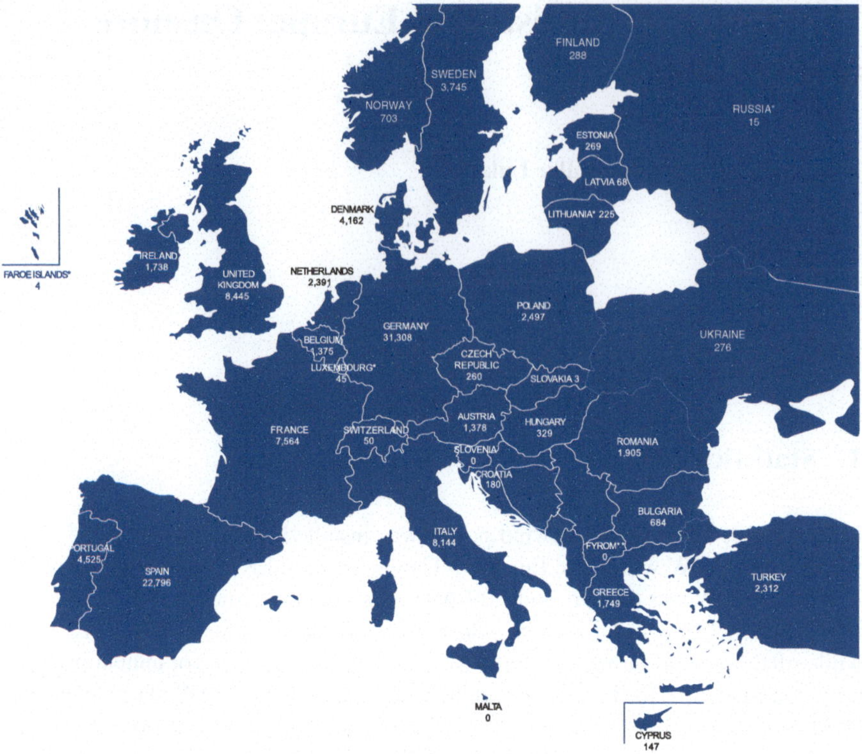

Fig. 1 Map of wind energy installed in Europe by the end of 2012 in MW (cumulative) [1]

(26.5 %), and gas with 10.5 GW (23 %). No other technologies compare to wind, PV, and gas in terms of new installations. Coal installed 3 GW (7 % of total installations), biomass 1.3 GW (3 %), CSP 833 MW (2 %), hydro 424 MW (1 %), waste 50 MW, Nuclear 22 MW, fuel oil 7 MW, ocean technologies 6 MW, and geothermal 5 MW [1].

Overall, during 2012, 44.9 GW of new power generating capacity was installed in the EU, 1.7 GW less than in 2011, which was a record year for new power capacity installations. During 2012, 5.5 GW of gas capacity was decommissioned, as were 5.4 GW of coal, 3.2 GW of fuel oil, and 1.2 GW of nuclear capacity. After 2 years of installing more capacity than it decommissioned, coal power installations reduced by almost 2.4 GW in 2012 (Fig. 3) [1].

In 2012, a total of 31 GW of renewable power capacity was installed. Almost 70 % of all new installed capacity in the EU was renewable. It was, furthermore, the fifth year running that over 55 % of all new power capacity in the EU was renewable [1] (Figs. 4 and 5).

Wind power's share of total installed power capacity has increased five-fold since 2000; from 2.2 % in 2000 to 11.4 % in 2012. Over the same period, renewable capacity increased by 51 % from 22.5 % of total power capacity in 2000 to 33.9 % in 2012 [1].

Table 1 Installed wind power of the EU-Members [1]

	Installed 2011	End 2011	Installed 2012	End 2012
EU capacity (MW)				
Austria	73	1084	296	1378
Belgium	191	1078	297	1375
Bulgaria	28	516	168	684
Cyprus	52	134	13	147
Czech Republic	2	217	44	260
Denmark	211	3956	217	4162
Estonia	35	184	86	269
Finland	2	199	89	288
France	830	6807	757	7564
Germany	2100	29,071	2415	31,308
Greece	316	1634	117	1749
Hungary	34	329	0	329
Ireland	208	1614	125	1738
Italy	1090	6878	1273	8144
Latvia	17	48	21	68
Lithuania[a]	16	179	46	225
Luxembourg[a]	1	45	0	45
Malta	0	0	0	0
Netherlands	59	2272	119	2391
Poland	436	1616	880	2497
Portugal	341	4379	145	4525
Romania	520	982	923	1905
Slovakia	0	3	0	3
Slovenia	0	0	0	0
Spain	1050	21,674	1122	22,796
Sweden	754	2899	846	3745
United Kingdom	1298	6556	1897	8445
Total EU-27	9664	94,352	11,895	106,040
Total EU-15	8524	90,145	9714	99,652
Total EU-12	1140	4207	2181	6388

[a]Provisional data or estimate

2 Wind Turbine Manufactures

The top ten global wind turbine manufactures rankings 2012 are as follows[1] (Fig. 6):

Among the worldwide top ten wind turbine manufactures are four from Europe (Fig. 7).

Siemens is the leading offshore wind turbine supplier in Europe with 58 % of total installed capacity. Vestas (28 %) is the second biggest turbine supplier, followed

[1]MAKE Consulting, 2013.

Table 2 Installed wind power of the Non-EU-Members [1]

	Installed 2011	End 2011	Installed 2012	End 2012
Candidate countries (MW)				
Croatia	52	131	48	180
FYROM[a]	0	0	0	0
Serbia	0	0	0	0
Turkey	477	1806	506	2312
Total	529	1937	554	2492
EFTA (MW)				
Iceland	0	0	0	0
Liechtenstein	0	0	0	0
Norway	99	537	166	703
Switzerland	3	46	4	50
Total	88	583	170	753
Other (MW)				
Faroe Islands[b]	0	4	0	4
Ukraine	66	151	125	276
Russia[b]	0	15	0	15
Total	66	171	125	296
Total Europe	10,361	97,043	12,744	109,581

Note: Due to previous year adjustments, 207 MW of project de-commissioning, re-powering, and rounding of figures, the total 2012 end-of-year cumulative capacity is not exactly equivalent to the sum of the 2011 end-of-year total plus the 2012 additions

[a]Former Yugoslav Republic of Macedonia

[b]Provisional data or estimate

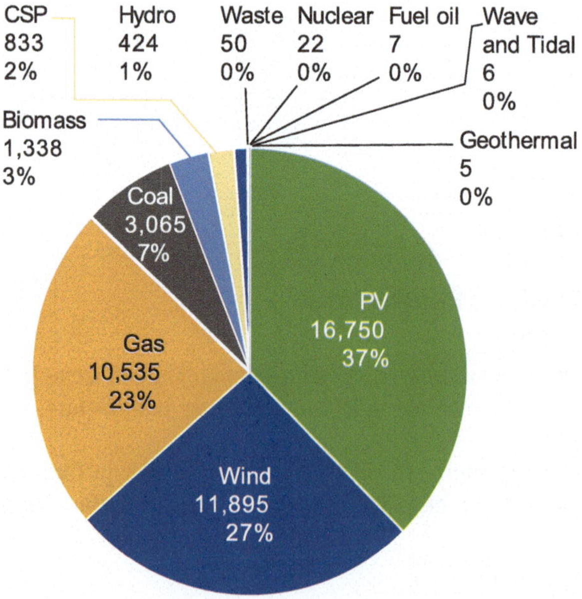

Fig. 2 Share of new power capacity installations in EU total 44,601 MW in 2012 [1]

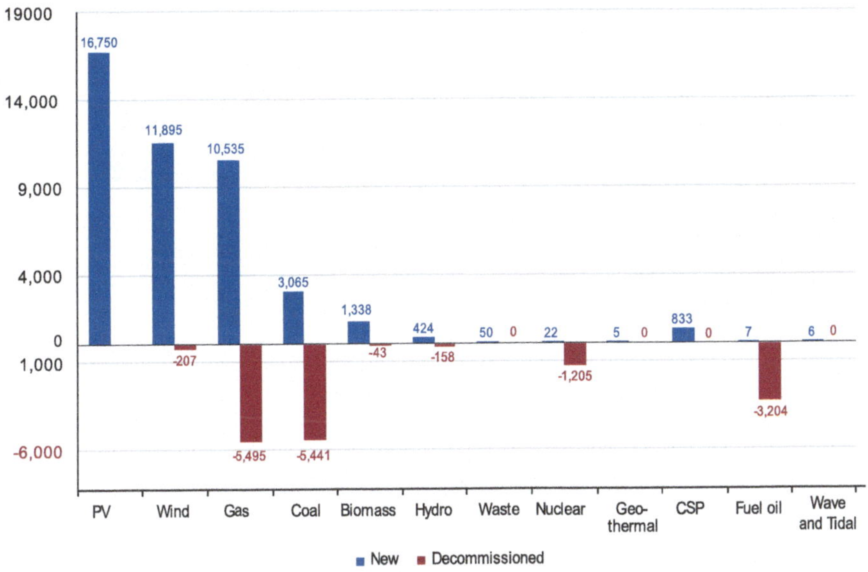

Fig. 3 New installed power capacity and decommissioned power capacity in MW during 2012 [1]

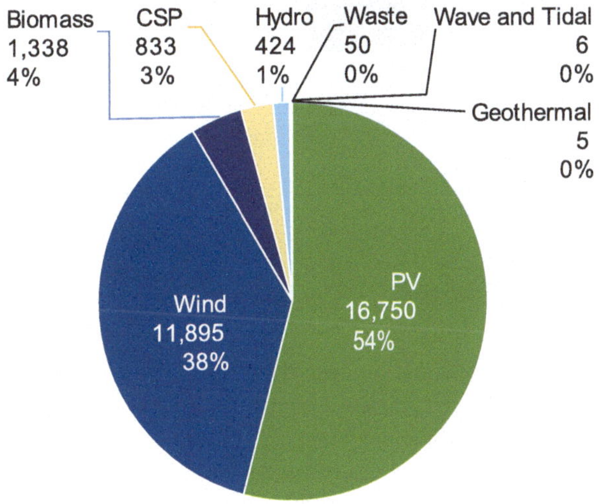

Fig. 4 2012 share of new renewable power [1]

by REpower (8 %), BARD (3 %), WinWind and GE with 1 % each, and AREVA Wind with 30 MW connected [2].

As in 2011, DONG remains the biggest operator of offshore wind power in Europe with 22 % of the cumulative installations at the end of 2012. Vattenfall (15 %), E.On (11 %), RWE (10 %), and SSE (7 %) follow (Fig. 8) [2].

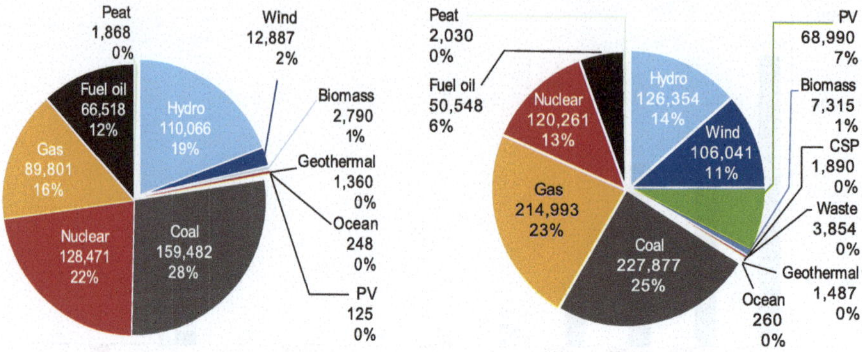

Fig. 5 EU power mix 2000 and 2012 [1]

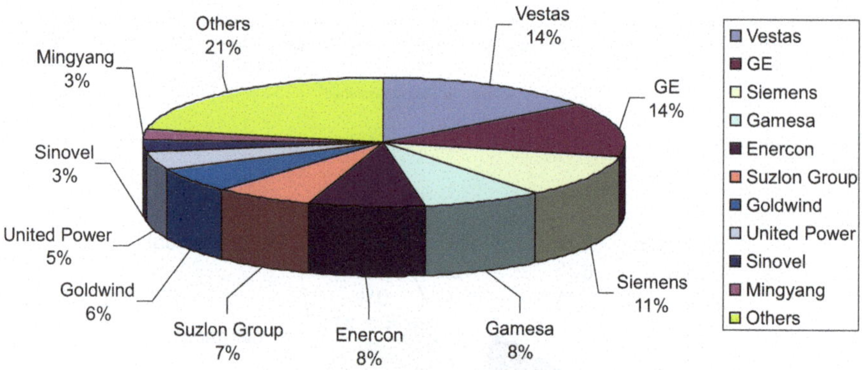

Fig. 6 Diagram top ten global wind turbine manufactures 2012 (numbers from MAKE Consulting)

3 Onshore or Offshore?

German electricity is to become greener. That's the official stance of the government since its decision 2 years ago to go for an "energy turnaround"—opting to shut down all nuclear plants by 2022 and boost electricity from renewable sources. It's a decision that has sparked a lot of controversy, as the path towards clean energy is lined with obstacles. Consumers complain about prices going through the roof, industry demands more financial support for the development of new technologies, while policymakers just seem rather lost. Everyone wants the turnaround, but no one seems to be able to agree on exactly how to achieve it. The latest quarrel in this affair is about offshore wind farms. One disadvantage of offshore installations is the geographical conditions. "Unlike Germany, many other countries are increasing their focus on offshore wind energy. But there the wind farms are closer to the coast, and this is impossible for Germany as we have the tidal flats here," says Niels

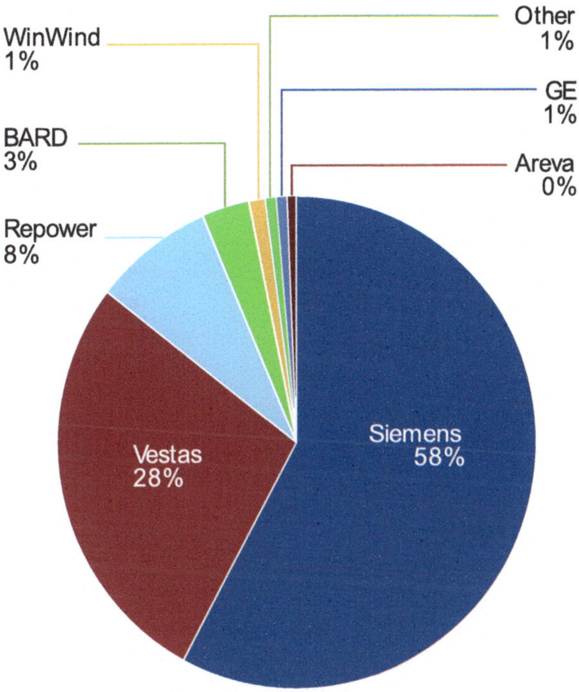

Fig. 7 Offshore wind turbine manufactures' share at the end of 2012 [2]

Schnoor of Federation of German Consumer Organizations (VZBV). Germany would therefore be the only country that has to erect its wind parks far offshore—which, from an economic point of view, makes no sense. Building offshore wind farms is "the wrong road to go down, both economically and technologically speaking," concludes an analysis by the consumer rights group the Federation of German Consumer Organizations (VZBV). The organization is opposed to the commissioning of any further offshore projects, according to Niels Schnoor. "It doesn't make sense to support this technology, as it is too expensive," said the VZBV's Schnoor. "There are more economical methods than that of offshore technology." He added that it would be in consumers' interests to bring about the energy turnaround as cost-efficiently as possible.[2]

German politicians have a broad scope for action designing policies for the regional distribution of renewables and could even save EUR 2 billion a year until 2023, a study commissioned by the Agora Energiewende[3] finds. According to the study this would require a faster expansion of wind turbines onshore and a slower expansion of offshore wind power (while maintaining the total output of power generated by wind power). Savings would be approximately the same whether the further expansion prioritized wind turbines primarily in the north ("best generation

[2] http://www.dw.de/do-german-offshore-wind-farms-have-a-future/a-16717347.

[3] http://www.germanenergyblog.de/?p=12797.

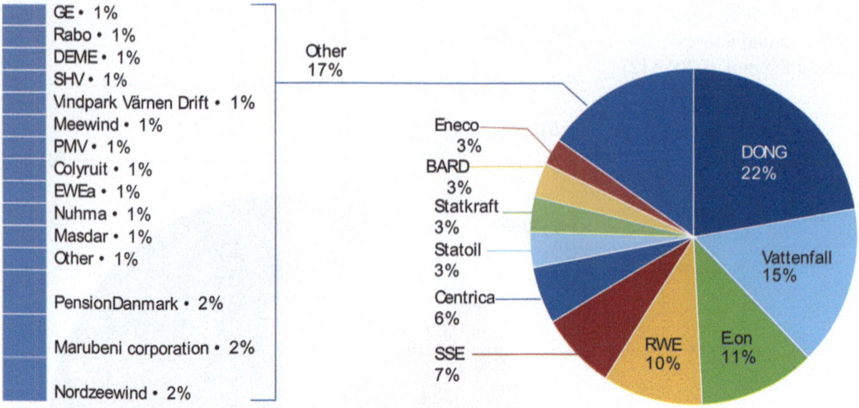

Fig. 8 Owners share of installed capacity in MW [2]

site scenario") or wind park sites in proximity to areas of high power consumption ("consumption related expansion scenario"), the study concludes (see footnote 10).

The study compares two options for expanding wind and solar power in Germany: One focuses on the best generation sites with wind power plants primarily installed in the north of Germany and solar power primarily in the south. The other scenario focuses on creating production facilities near the centers of power consumption. The results show that both expansion paths lead to approximately the same total costs. In the consumption-driven scenario, a somewhat higher number of wind- and solar power plants have to be built. As they, however, produce power at different times and closer to the consumers, they relieve the power system and have to be curtailed less often than in the resource-driven scenario.[4]

Wind power is to be part of the German "energy mix," there's no doubt about it. But it's not the wind far out at sea that will contribute to German electricity. "The energy turnaround will happen onshore, not offshore—that much is certain," says Matthias Hochstätter, adviser to the German Wind Energy Association. It's this source of energy that will provide the main part of the future energy mix. According to Hochstätter, southern Germany in particular still has a lot of potential for onshore wind farms. "We're not asking for a complete shutdown of the offshore facilities. But we want to see their expansion reduced. The current goal is to extend offshore wind farms to output of 14 gigawatts by 2020," explains Schoor. He believes that an expansion of up to 5 GW is enough to achieve the energy turnaround, and that attention should be focused instead on onshore and photovoltaic facilities.[5]

On the contrary, Sophia von Waldow, an analyst at Bloomberg New Energy Finance, says:

[4] http://www.dw.de/do-german-offshore-wind-farms-have-a-future/a-16717347.

[5] http://www.germanenergyblog.de/?p=12797.

> Offshore wind plays a special role in the future German energy mix, considering its large industrial supply chain involving turbines, power cables, vessels and ports. It's about jobs and the development of an industry Germany has the potential to lead in.[6]

The European Wind Energy Association writes in a report: "In 2012, despite the negative headlines over European finance, there was an increase in the number of transactions and in the overall amount of financing committed to the industry. A number of useful new precedents were set. The market will continue to be active in 2013 (and likely move back to Germany, following the resolution of the grid issues, which discouraged new transactions in 2012) with a number of transactions expected in the coming months." [2].

References

1. European Wind Energy Association (EWEA) (2013) Wind in power 2012 European statistics. (https://windeurope.org/wpcontent/uploads/files/about-wind/statistics/Wind_in_power_annual_statistics_2012.pdf)
2. European Wind Energy Association (EWEA) (2013) The European offshore wind industry—key trends and statistics 2012. (https://windeurope.org/wp-content/uploads/files/about-wind/statistics/European_offshore_statistics_2012.pdf)

[6] http://www.businessweek.com/news/2013-01-16/merkel-s-offshore-wind-power-dream-for-germany-stalls.

Issues in Accessing Enormous Renewable Resource in Ireland

Grattan Healy

1 Renewable Energy Resource in Ireland

The Republic of Ireland is recognised as having some of the best renewable energy resources in Europe, primarily due to the weather conditions to which it is exposed on the Eastern edge of the Atlantic Ocean. Satellite-based ocean wind mapping shows that, along with Iceland and Greenland, Ireland's Western seaboard is hugely endowed (illustrated in Appendix A).

As technology for marine energy is developing, such as wave energy converters, but in particular floating wind turbines, that position is being strengthened, as Ireland has a 200 mile Exclusive Economic Zone (EEZ) of some 410,000 km^2, or approx. 6 times its land area of some 70,000 km^2. Ireland's continental shelf area is larger again (illustrated in Appendix B).

Electricity demand in Ireland is currently approx. 25 Terawatt-hours (TWh) per year, while the Total Primary Energy Requirement (TPER) peaked at around 190 TWh in 2008 [1]. The technical onshore wind energy resource is estimated at some 2000 TWh [2] with a further 7000 TWh from floating wind turbines at sea inside the EEZ [3].

2 Accessing Ireland's Renewable Energy Resource

Ireland's share of total primary energy requirement (TPER) supplied from renewable sources has fallen dramatically to around 6 %, while the rest comes from fossil fuel sources. Dependency on external energy sources has remained at around 90 % since the start of the millennium [1]. Under the EU's Renewables Directive, the Republic

G. Healy (✉)
Ierne Limited, Teach Mór Thiar, Inverin, County Galway, Ireland
e-mail: info@ierne.ie

T.S. Uyar (ed.), *Towards 100% Renewable Energy*, Springer Proceedings in Energy, DOI 10.1007/978-3-319-45659-1_37

has an obligation to source 16 % of its energy demand from renewable sources by 2020 [4], which would reverse those trends. That target has been translated by the Irish authorities into a transport target of 10 %, heating/cooling target of 12 % and an electricity target of 40 % coming from renewable sources, primarily wind energy [5].

At the same time, discussions between the Irish and UK Governments have been considering an arrangement for the export of electricity generated from renewable sources from Ireland to the British mainland, given Ireland's rich endowment, and the UK's impeding needs. However, the targets under the Renewables Directive are legally binding, while the provisions relating to exports merely facilitate it. Consequently, the European Commission is unlikely to facilitate such exports where an exporter cannot demonstrate its ability to meet its own national targets. There is growing concern about the Republic's ability to meet its binding targets by 2020 (illustrated in Appendix C).

3 Issues in Accessing Ireland's Renewable Energy Resource

Due to the strength of the winds, Ireland's most economical energy source is electricity produced from onshore wind [6], and it is therefore the primary focus of renewable energy policy today. At the same time, the two electrical interconnection links between the island of Ireland and elsewhere use direct current (DC) rather than alternating current (AC) technology, which means that the combined electrical system of the Republic and Northern Ireland forms a relatively isolated network of modest scale. The primary challenge then is in reliably accommodating vast amounts of electricity generated from variable renewable sources on a relatively small and isolated electricity network with variable demand. Plans are in place in the two jurisdictions to reach an average renewable generation of 40 % of all electricity demand on the combined island system by 2020, which in turn implies that at some points in time, the level of wind generation on the total system will reach up to 75 % of electricity demand. This is very ambitious, even in world terms, and means that the island's authorities are trying to break new ground in the management of the network without a diminution in the level of reliability (illustrated in Appendix D).

3.1 Operation of the Electricity System

The growing disparity between variable sources and variable demand creates an obstacle to renewables, because of the need to match supply and demand at all times. The availability of a single 292 Megawatt (MW) pumped storage station at Turlough Hill and a few hundred MW of large hydro stations helps, and the interconnectors at Moyle in Northern Ireland and the East-West Interconnector (EWIC) from Wales to the Dublin area allow some electricity trading to assist system balancing. But in the end, the all-island system operator, Eirgrid-SONI, is relying on dispatchable generation plant that is almost entirely based on fossil fuels to balance the system.

Ireland is reaching another obstacle related to the dynamic stability of the relatively isolated all-island network, long before most other places, due to the degree of isolation and the demanding penetration levels of variable renewable generation. The Republic's

System Operator, Eirgrid, presented a significant report in 2010, known as the 'Facilitation of Renewables Study', which dealt with the various issues as regards the dynamic stability of the all-island system [7], such as voltage and frequency control, inertia and fault response. The study separated the equipment on the system into synchronous and non-synchronous, the latter category being made up primarily of non-synchronous renewable generators and DC interconnectors. The authors considered that the non-synchronous equipment could not provide the requisite inertia in an instantaneous passive manner and could not be relied upon to respond adequately to frequency excursions and faults. The implications are twofold. Firstly, that the system has to be considered isolated for dynamic stability analysis purposes, since both interconnectors are DC, not AC. Secondly, within the island network, a limit of 50 % should be imposed on the instantaneous level of System Non-Synchronous Penetration (SNSP) and that level would rise to 75 % depending on certain aspects of the system being improved. In other words, the rather ambitious target of having an average of 40 % of all electricity coming from renewable sources by 2020, which in turn means an instantaneous level of some 75 % at times, relies very heavily on the implementation of that set of measures. Their implementation is currently being handled through a programme known as DS3 [8].

This is not without some controversy, since both the interconnector and wind turbine manufacturers consider that their systems provide a type of inertia, known as 'proto-inertia', which can assist system stability. Ireland will continue to operate under this limiting regime for the moment, but as the limits start to prevent wind development, the industry will inevitably force a review, which should take full account of the extensive capabilities of these new generation and transmission technologies, as well as further new technologies, like synchronous wind turbines.[1]

Both of the above issues have encouraged the System Operator to require quite a high level of fossil plant minimum generation (MinGen) and quite often renewable generators are curtailed to facilitate this MinGen, well before the SNSP limit is reached.

3.2 Access to the Electricity System

Another significant limitation has been imposed on renewables development in Ireland by the very long delays incurred by projects in connecting to the grid. From around 2000, as the quantity of wind projects with planning and grid applications increased, and the spare capacity on the transmission system reduced, projects were considered to 'interact' with one another very significantly, causing increasing delays. Additionally, there was a widely held view amongst the authorities that the network could only accommodate some few hundred MW of wind, which resulted in a moratorium on connections being declared by the Republic's regulator, the Commission for Energy Regulation (CER), on 3rd December 2003. After a hiatus of almost 1 year, a new grid application system for renewables was introduced, known as 'Group Processing', which handled applications from adjacent renewable projects in groups. However, in the meantime, a huge queue of grid applications built up out of concern for the increasing delay, which persists to this day. Many projects started to apply for grid without having secured planning permission. The CER then began a series of

[1] For example, see Dewind's synchronous version of their wind turbines, www.dewindco.com.

'Gates' in which a defined part of the Republic's grid application queue was allowed to receive a grid connection offer, while all other projects were made to wait for a subsequent Gate. The processing of these Gates became increasingly complicated, in terms of allocating grid connection costs between projects, estimation of the impact of network limitations on project output (local 'constraints' and national 'curtailment)', timing of the availability of firm access (where projects receive the market price for output lost due to network issues), and more recently, updated lists of Associated Transmission Reinforcements (ATRs). Some projects in Gate 3 applied for grid connection in 2004, and still do not have complete grid connection offers, they might connect in 2016 and get firm access to the network by 2019, or even later.

During this development, it became clear that such lengthy delays for grid access had extended beyond the planning permission time limit, which was initially 5 years, but was extended by the authorities to as much as 10 years. This meant that planning permission could not be a requirement for receiving a connection offer, since otherwise the planning would have expired by the time the grid connection was delivered. Therefore, these extensive delays in grid connection now mean that projects must apply for connection before they apply for planning permission, which is the opposite of a normal development model; what this author terms a 'backward development model'. A whole new set of rules has had to be defined to cope with projects that have a grid connection but are without planning, which permits the splitting, merging and relocation of grid. This 'backward development model' effectively causes the chance of project success to fall from some 30 % to as low as 5 %, while considerably raising costs (illustrated in Appendix E). In the ground-breaking German 1990 law, the 'Einspeizegesetz', a forward model was inherent, in that projects with planning were pretty much guaranteed grid access and price support [9]. The island's authorities need to put a time limit on grid delivery of say 3 years, and then make planning a condition for grid access again, reverting to a similar 'forward development model'.

3.3 Legal Issues as Regards Grid

Apart from setting binding primary energy targets, the EU's Renewables Directive strengthens the rules within the electricity market in favour of renewable generators. Article 16 sets out extensive operation and access rules, subject only to the reliability and safety of the network.

When dispatching generators, renewables must be given priority, and this rule has been implemented, without reference to cost, by the Single Electricity Market Committee (SEMC), which is the joint regulatory authority for the electricity market on the whole island of Ireland.

However, adequate measures are not being taken by the authorities to 'guarantee transmission' and prevent the inexorable growth in the levels of constraint and curtailment, which are completely undermining the financing of renewable energy generation projects.

The authorities argue that they need not build energy storage or undertake similarly costly measures to guarantee transmission, relying on the excuse that the obligation is subject to safety and reliability of the grid. Were it correct, this interpretation would render that obligation void, which was clearly not intended by the EU. The correct interpretation is that they must adopt measures to guarantee transmission while maintaining reliability and safety, which of course those very measures would do.

In the Irish Government's 'National Renewable Energy Action Plan' (NREAP), dated July 2010, Ireland opted for priority access to the grid under Article 16 of the Directive. But the opposite has been the case, since fossil fuel plant that applied after renewable projects have connected and been given firm access first, even where that firm access allocation was revised in December 2012 (illustrated in Appendix F). The CER has justified its decisions on the grounds of 'security of supply', but this is not the qualification on priority access provided for in the Directive.

To circumvent these issues, two enormous onshore wind projects sought direct connection to the UK in order to export surplus power under the UK support scheme, while avoiding the local grid constraints. The authorities on the island would of course be anxious to incorporate these 'interconnectors' into the local network, arguing overall economic benefits.

3.4 Other Obstacles

The proportion of projects achieving planning in the Republic has increased to quite a high level, despite increasing planning requirements, many of which come from the EU. However, it should be noted that the total number of planning permissions being achieved has peaked, and has been falling due to increased requirements and costs, larger scale of projects and emerging social acceptance issues (illustrated in Appendix G).

The Renewable Energy Feed In Tariff (REFIT) is the support scheme in the Republic of Ireland, managed by the Department of Communications, Energy and Natural Resources (DCENR). It takes the form of a guaranteed floor price for electricity produced from specific renewable energy sources, where the price depends on the technology. It does not cover solar, offshore wind or marine energies. Its primary weakness is that it is paid on metered output only. Losses of output caused by grid issues lead to an uncertain revenue stream, which directly contradicts the whole point of a guaranteed price scheme. Grid connection costs account for an increasing proportion of project capital cost, reducing the attractiveness of the REFIT. Indeed, charging projects for grid connections, which they must then gift to the network owner (the incumbent ESB) is an economically inefficient way of funding grid connections (KHSK Economic Consultants (2010) "Efficient funding of transmission network connection costs." For NOW Ireland, unpublished). A preliminary analysis of the cost benefits of measures dealing with these two issues suggests a lower REFIT price and a consequent reduction in cost to the consumer (illustrated in Appendix H).

There have also been periods in time when there was no support scheme at all on offer to wind projects, such as between December 2009 and March 2012, while issues of state aid were resolved with the European Commission. The current REFIT 2 scheme for wind closed for applications at the end of 2015 and plant must be built by 2017, placing further doubt over targets.

4 Conclusion

Despite a truly enormous renewable energy resource, large enough for exports many times local demand, Ireland faces significant difficulties in achieving relatively modest binding renewable energy targets. Grid is central to these difficulties, because the vast bulk of the available renewable energy is to be delivered in the form of electricity, while the local network is relatively isolated. Rather conservative analysis for the grid authorities has led them to impose quite stringent limits on the progress of variable renewable generation. Enormous connection delays and costs, as well as uncompensated constraint and curtailment further undermine the development of the sector. There is no possibility of a 100 % renewables system under these conditions, and exports simply cannot happen without much more significant progress towards EU targets. A forward development model is essential, along with payment of REFIT supports on the available output of renewable energy projects, proper funding of grid connections, and most importantly, investment in various network assets to allow higher renewable penetrations, with a view to a 100 % renewable energy future.

Acknowledgments The author would like to acknowledge the wind resource assessment work of Brian Hurley.

Appendix A

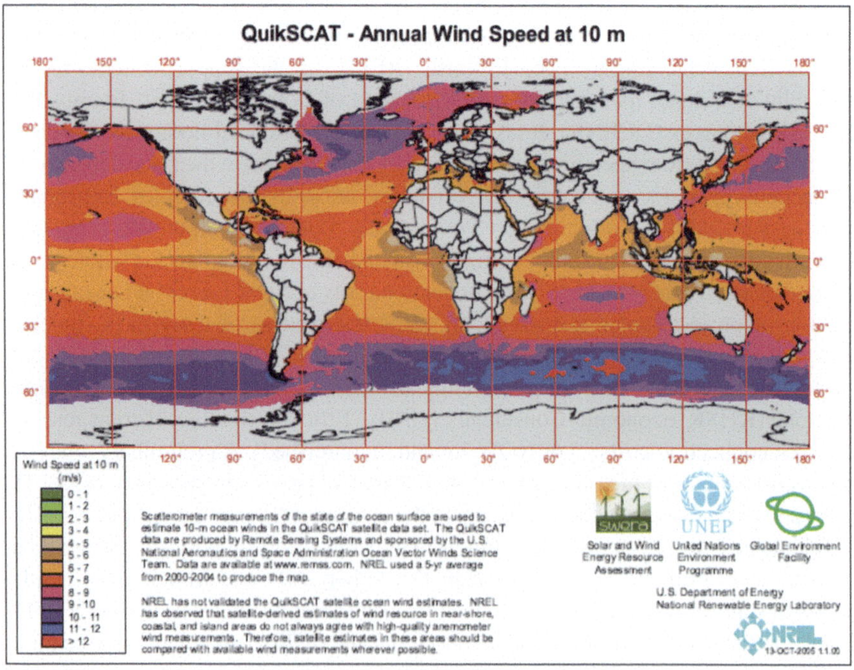

Global annual wind speeds as measured by QuikSCAT satellite, presented by NREL (based on World monthly offshore wind maps, posted at www.retscreen.net).

Appendix B

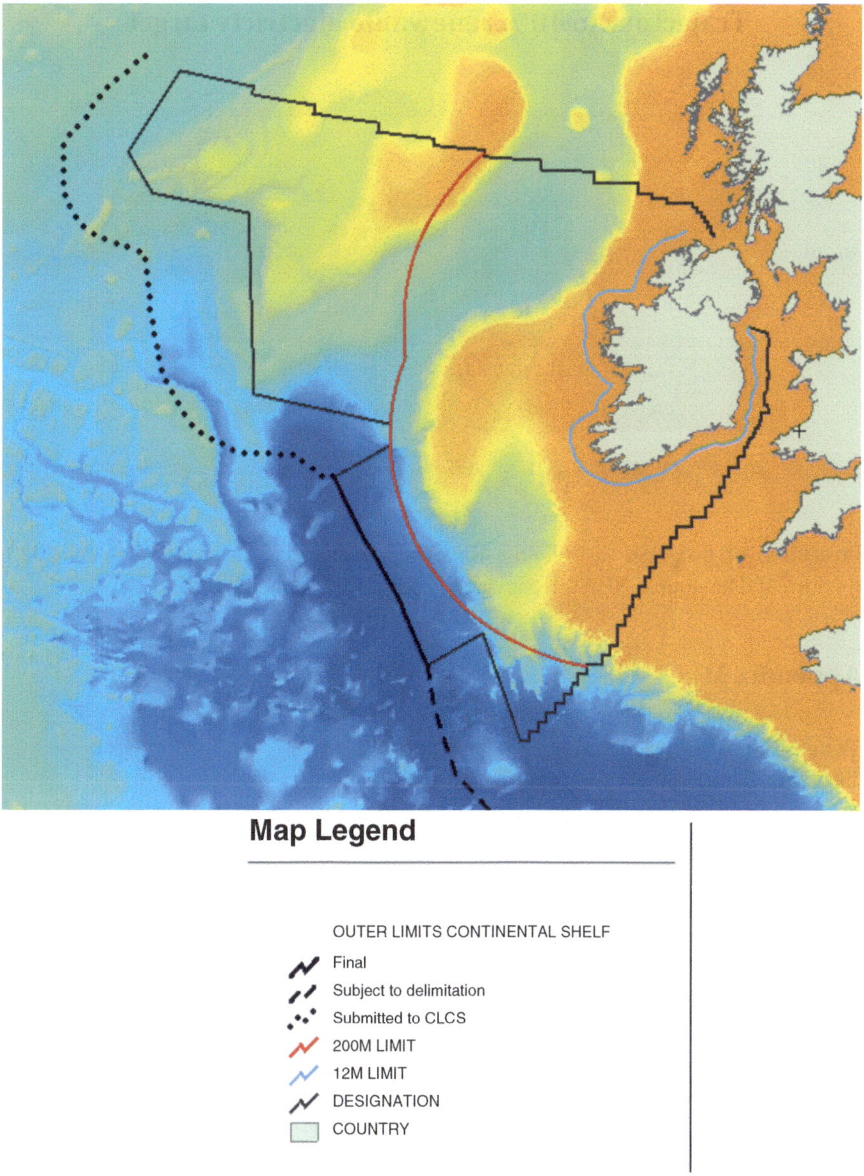

Ireland's Nautical Territorial Boundaries.

Source: Department of Communications, Energy and Natural Resources (DCENR), Ireland; Petroleum Affairs Division.

Appendix C

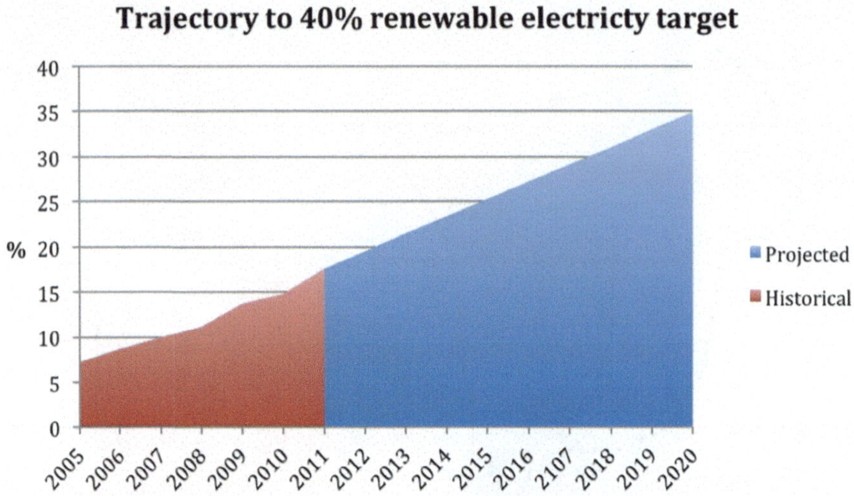

Projection of progress to Ireland's Renewable elecricity target of 40 % by 2020 (historical data source SEAI, Renewable Energy in Ireland 2011, 2012).

Appendix D

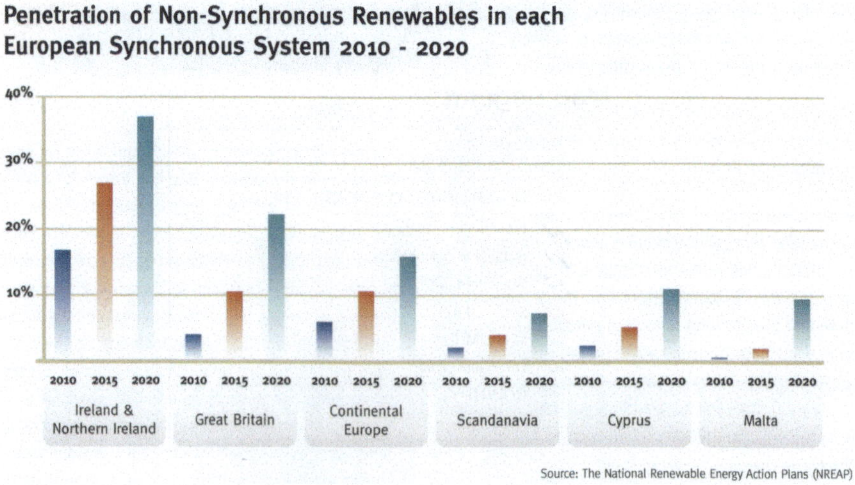

Projected non-syncrhronous penetrations on various electrical systems (*Source*: Eirgrid Group, Annual renewable report 2012, www.eirgrid.com).

Appendix E

	Approx. probability (%)	Approximate timing (years)	Rough cost estimate
Backward development model			
Get into a gate	50	3	€7 k + €20–80 k
Get grid offer	100	3–6	€300 k/MW
Get planning	30	1–5	€250 k+
Get REFIT	50	1–2	€0
Get finance	70	1–2	€100–200 k
Overall	5	15–20	€1 million+
Forward development model			
Get planning	30	1–5	€250 k+
Get grid offer	100	0.5	€7 k + €20–80 k
Get grid	100	3	€50 k/MW
Get REFIT	100	0	€0
Get finance	90	1	€100 k
Overall	27	5–8	€ half million

Estimated risk and cost comparison between 'forward' and 'backward' development models for renewables

Appendix F

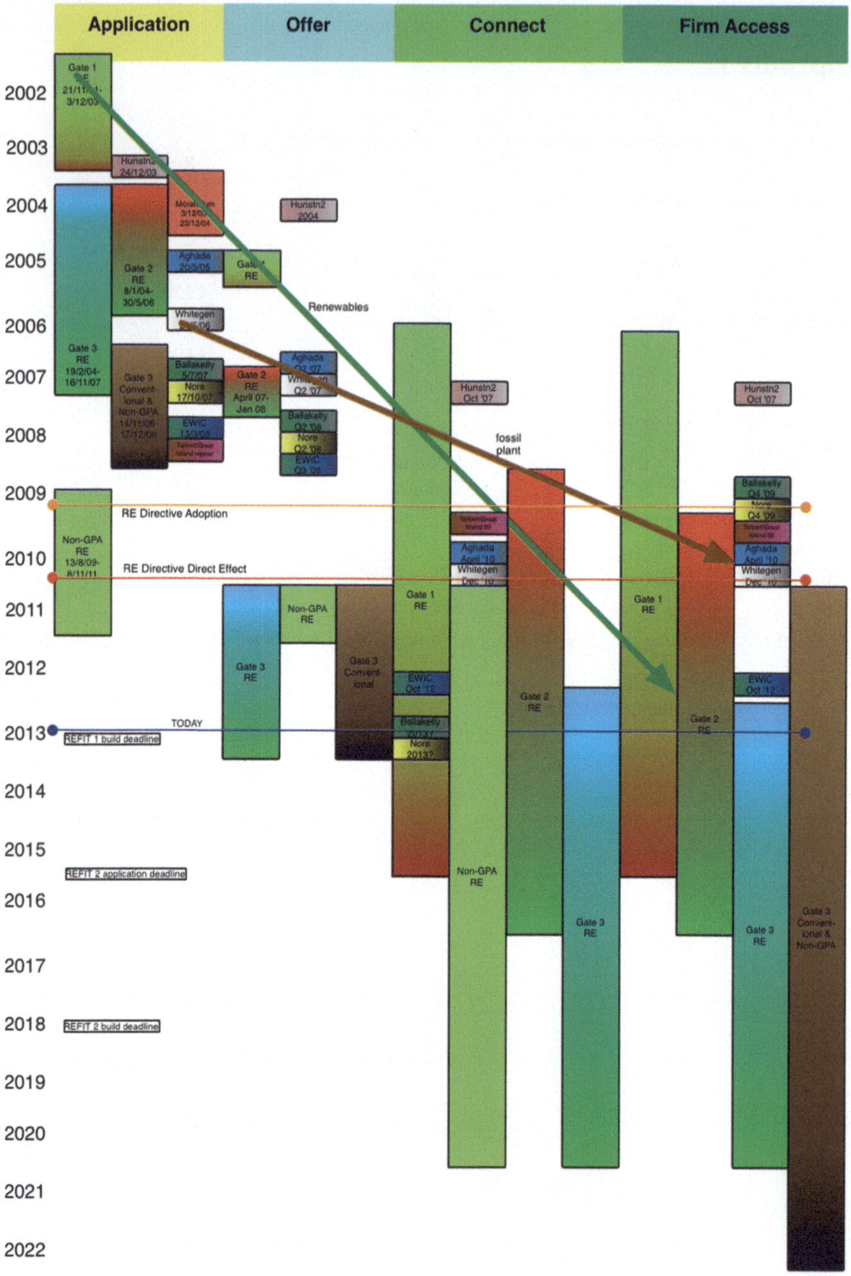

Renewable and fossil plant timelines in Republic of Ireland since 2002

Appendix G

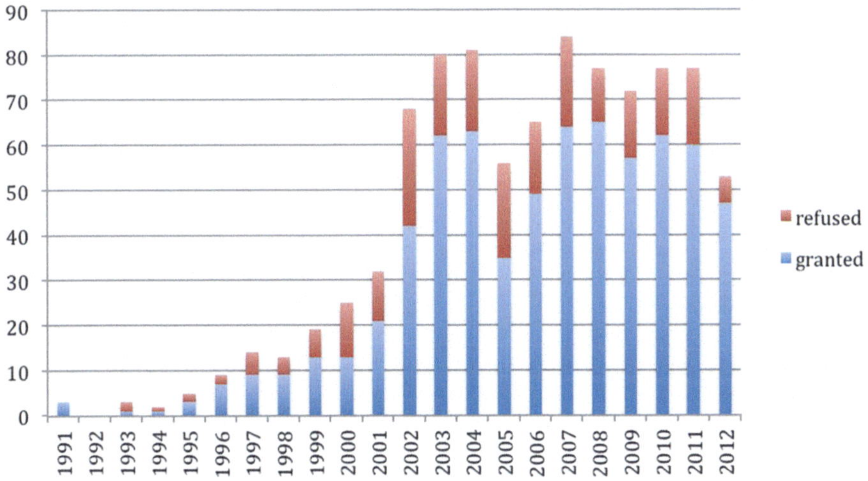

Total grants/refusals of planning for wind farms (including appeals) 1991–2012, Republic of Ireland (data source SEAI)

Appendix H

10 MW project; constant 20 year Equity IRR	Existing model	Constraint and curtailment paid at REFIT	Underground cable connection paid	Both
REFIT start price (unadjusted) (€c/kWh)	6.808	6.808	6.808	6.808
REFIT start price (adjusted) (€c/kWh)	6.808	6.121	5.794	5.224
CAPEX (€)	14,347,826	14,347,826	11,739,130	11,739,130
Annualised interest rate (4 loan payments/ year) (%)	6.00	5.50	6.00	5.50
Gearing (%)	80.00	85.00	80.00	85.00
Unpaid constraint and curtailment (%)	6–9	0	6–9	0
Average annual paid production (kWh/year)	26,961,513	29,290,074	26,961,513	29,290,074
20 year Equity IRR (unadjusted) (%)	10.61	16.88	19.53	29.63
20 year Equity IRR (adjusted) (%)	10.61	10.61	10.61	10.61

(continued)

(continued)

10 MW project; constant 20 year Equity IRR	Existing model	Constraint and curtailment paid at REFIT	Underground cable connection paid	Both
PSO cost (incl. 0.5c balancing; SMP at 5.8c) (€/year)	406,580	240,472	133,190	0
Connection cost to consumer (financed by ESB at 4 % over 50 years) (€/year)	0	0	121,435	121,435
Consumer cost (PSO + connection) (€/year)	406,580	240,472	254,625	121,435

Estimate of cost benefit of paying for constraint and grid (based on author's proprietary wind project financial model)

References

1. Sustainable Energy Authority of Ireland. Energy in Ireland 1990–2011. www.seai.ie
2. ESB International, Future Energy Solutions, Energy Research Group (2004) UCD, updating the renewable energy resource in Ireland. Final report. Sustainable Energy Ireland. www.sei.ie
3. Halligan B, Hurley B, Healy G, O'Rorke B. Europe's energy exporter (paper unpublished)
4. Directive 2009/28/EC of the European Parliament and of the Council of 23 April 2009 on the promotion of the use of energy from renewable sources…, OJEU L140/16, 5 June 2009
5. Government of Ireland (2010) National renewable energy action plan. www.dcenr.ie
6. Meitheal na Gaoithe (2013) Embedding sustainability, the business case for small wind. www.mnag.ie
7. Bömer J, Burges K, Nabe C (Ecofys), Pöller, M. (DIgSILENT) (2010) All island TSO facilitation of renewables studies. Final report for work package 3. www.eirgrid.com
8. Eirgrid plc. Delivering a secure sustainable electrical system. www.eirgrid.com
9. Germany. Electricity In-feed law, as of 7 Dec 1990 (English translation: http://ierne.ie/Ierne_Wind_Energy_Consulting_Ltd/Reference.html)

Integration of Large Rooftop Photovoltaic Plants in Industrial or Commercial Areas

Andrea Bodenhagen, Claire Guesdon, Andy Parr, Patrick Clough, and Maarten van Cleef

1 Introduction

With an anticipated GDP growth rate for the next 15 years of 6.7 % Turkey is likely to be the fastest growing economy among OECD countries. While the current reliable electricity generation is balanced with the demand, by 2015 the energy demand is forecasted to exceed reliable generation capacities leading to import demands or greater efforts to increase local renewable energy generation. Today 64.5 % of the Turkish energy demand is satisfied through conventional fossil resources, hydro accounts for 32.7 %, and geothermal as well as wind contribute another 2.8 %. Due to a lack of financial incentives PV has only played a minor role in the Turkish energy mix, albeit the fact that with 1500 kWh/kWp its average PV energy yield exceeds that of Spain by 10 % [1].

When looking at Turkey's solar potential numbers up to 500 GW are mentioned, focussing mainly on ground mounted installations. However, with its textile, food processing, auto electronic and paper industries, Turkey provides also a large base of rooftop space that can be converted into clean energy generating plants, profiting all, country, businesses, and environment.

2 Solar Technology: A Brief Overview

A solar or photovoltaic module is the main component of a grid-connected photovoltaic system. Other components are mounting structure, wiring, inverter to convert direct current into alternating current, and options like energy meter, monitoring system,

A. Bodenhagen (✉) • C. Guesdon • A. Parr • P. Clough • M. van Cleef
Marketing Department, United Solar Ovonic Europe GmbH,
Robert-Koch-Straße 50, 55129 Mainz, Germany
e-mail: andrea.bodenhagen@gmx.de

T.S. Uyar (ed.), *Towards 100% Renewable Energy*, Springer Proceedings in Energy, DOI 10.1007/978-3-319-45659-1_38

etc. The main differentiator of PV module technology is whether the photovoltaic cells are made from crystalline silicon or in a thin-film coating process.

Crystalline silicon photovoltaic cells are extremely fragile and therefore need to be protected against breakage and humidity by encapsulating the cells behind glass. These PV modules are usually installed tilted on a substructure for optimal energy generation.

Thin-film solar cells or modules vary according to their composition and substrate material. They are coated with amorphous silicon or other photovoltaic material. The range of their physical characteristics and performance depends on the material. Due to the thin layer of active material coated onto a substrate, the material input required is much lower than for crystalline cells.

Deposition of thin-film materials can be performed onto rigid or flexible substrates. By an intelligent selection of the encapsulation material the flexibility of the solar cells can be maintained.

3 Traditional PV Systems

3.1 Weight and Area Usage

Due to the weight of the glass and the substructure of traditional PV systems the total weight of such PV systems often reaches 25 kg/m^2 or more. Because the system is propped up, additional wind loads need to be taken into consideration, which can increase the total accountable weight load by more than 10 times the module weight.

Crystalline silicon PV systems need to be optimally aligned to face the sun (south facing with a 30° tilt angle) and will suffer significant reduction in performance when these optimal conditions cannot be achieved. It therefore becomes necessary to introduce adequate spacing between the module rows to avoid self-shading which significantly (by more than 30 %) the available area for locating the solar panels, this low area usage is often ignored at the time of selection.

3.2 Efficiency Versus Energy Output

Efficiency and nominal power (Pnom) of a PV module are measured under standard test conditions (STC) (1000 W/m^2, AM 1.5, 25 °C Cell Temperature). Experts agree that real world operating conditions vary greatly from standard test conditions, but have a much greater impact on the final energy output produced. Hence the expressiveness of the efficiency and nominal power to judge the performance of a solar system are misleading. The efficiency determines only the amount of space required to install a certain amount of nominal power. The nominal power is one single point of the matrix of power production conditions under real environmental conditions. The brightness of the sun, latitude, and season, time of the day, air mass, cloud cover, and pollution are factors influencing the energy yield of a photovoltaic system. Additionally, increasing cell temperature, impacted by ambient temperature, insolation, and wind, influences energy production. As a result, a module with a 14 % theoretical efficiency will not produce twice as many kWh per kW than a 7 % efficiency module. As all photovoltaic modules are influenced by real world outdoor conditions, the energy yield, i.e., kWh/kWp, is the main indicator of the suitability of a particular solar system at a certain location.

4 The UNI-SOLAR Approach

The UNI-SOLAR approach for large low-load bearing roofs is based on durable, easy to install, high energy yield rendering, lightweight, building integrated PV solutions.

Durable materials guarantee the long-term availability, low maintenance, and related down-times of the PV system, even in severe conditions. PV modules normally come with a performance guarantee of 20–25 years. This implies that all material related to them and the roof construction should have a minimum lifetime of this time frame. The roofing substrate materials approved by UNI-SOLAR as a base layer to its PV laminates have been chosen among other requirements due to their proven long-life time under all weather conditions, while our thin-film amorphous PV technology has been installed since more than 10 years and has been tested under severe conditions ranging from space, ocean to desert applications.

Easy to install systems imply that no specific tools or equipment is required to apply the modules to the roof.

High energy yields are possible due to the multi-junction technology of the amorphous thin-film solar cells employed. Multi-junction means the thin-film layer consists of various semiconductor layers with each layer collecting a specific range of the light spectrum. Blue light is dominant in the morning and afternoon of the day as well as when it is cloudy or overcast. Hence, these PV cells collect light over a longer time period of the day. Furthermore the multi-junction technology makes these PV cells less sensitive to orientation and tilt angle—allowing them to be installed directly onto the roof. Because the amorphous silicon cells have also a very low temperature coefficient, they are less affected by rising module temperatures, do not need ventilation or cooling and are hence ideally suited for building integration. In addition, when the module temperature reaches around 80 °C, a phenomenon called "thermal annealing," which occurs only with thin-film technology and positively affects the module output during the summer when the most sunlight is available.

Lightweight PV modules allow the integration of solar energy systems into large building designs with limited additional load bearing capacities and hence increase the area and number of buildings available to clean energy generation.

Considering cultural aspects when planning a solar system into the design of a building usually means not to place the PV system in the focus of the viewer, but concentrating on the structural elements of the building. By integrating the solar system into the building it becomes a part and melds into the appearance of the building. This is a particularly important aspect in countries with a large touristic industry like Turkey—allowing clean energy generation while maintaining the unobstructed view on cultural landmarks.

5 Main Features of Rooftop Electricity Generation

Rooftop electricity generation presents numerous advantages over either ground mounted PV electricity generation or traditional centralized electrical generation. First, the space used for a rooftop PV plant would not have been used to fulfill another purpose, as opposed to ground mounted PV, which sometime compete with agriculture or forestry uses. Second, the electricity produced on the rooftop of a plant or a warehouse is generally consumed in the plant or in the other buildings of the industrial area, thus avoiding the electricity losses during transportation making this a much more efficient use of solar energy. Third, in hot climates countries like Turkey, one of the peaks of electricity consumption occurs during the early afternoon, when the hot temperatures lead to a heavy consumption of air-conditioning systems. This peak corresponds with the peak of PV production: as a result, rooftop electricity generation can help soften the afternoon peak of consumption and therefore stabilize the local electricity grid. Even if a rooftop PV producer does not consume the PV energy it produces, and would rather sell it to the national operator, it would benefit from the presence of PV plants in its area.

Furthermore, a rooftop PV plant has very few drawbacks. Since there are no moving parts, the noise is inexistent. There is no exhaust gazes of liquids, and, in case of a BIPV plant, no visual impact. A rooftop BIPV plant can produce energy without being detected by passers-by. Contrary to many other types of plants, it will not decrease the value of the building that supports it or that is near to it, quite the contrary. The maintenance is recommended twice a year, requiring no more than allowing access to the roof. In the worst of case, the space required inside the plant will not exceed one room on the plant floor for the inverters; and actually these devices are often located in areas not used by the normal activities, like under the roof.

6 Technical and Organizational Challenges

The construction of large rooftop PV plants has some unique features. Building Integrated (BIPV) solutions are integrated in some way into the waterproofing material, such that the requirements for waterproofing are added to the typical issues of PV. In many countries, BIPV solutions must undergo a certification made by the same organizations that are responsible for the safety of construction materials. If the chosen solution is to apply the PV solution to the roof (BAPV), then the presence of additional weight and wind loading as a result of the inclined surfaces require careful calculations.

From an organizational point of view, the construction of a PV plant integrated to a roof—which, in most cases, covers a warehouse or a production line in activity—presents some unique issues to tackle. Indeed, the combination of working at height and with electrical material requires a special training for installer and electrician. The ability to conduct the construction work without disturbing the activity underneath the roof or the transit is crucial throughout all the construction process. The success of the future rooftop plant depends primarily on the correct interaction of very different actors: providers, investors, clients, etc.

7 Application Examples

7.1 Perfetti Van Melle (Mentos and Smint), Istanbul: 31 kWp by Our Partner Centrosolar

The installation by our partner Centrosolar consists of 108 TF Plates Professional each with two UNI-SOLAR PV laminates and produces approx. 38 MWh electricity per year while saving 25 tons of CO_2 emission.

One of the major reasons for choosing this thin-film product was its low temperature coefficient—meaning lower reduction of efficiency at higher module temperatures—in fact the UNI-SOLAR PV laminates even generate higher energy yields at high module temperatures due to the thermal annealing of the so-called Staebler Wronsky effect. With the Centrosolar's TF Plate Professional, design for installation on trapezoidal profiles it perfectly matches the existent roof structure [2].

7.2 Coca-Cola, Italy: 3.3 MWp

The choice for the PowerTilt and PowerPlate product solutions on four buildings PV plants belonging to Coca Cola HBC Italia lay in their ability to easily integrate into the existing roofs without the need for strengthening of the roof support. The key for success of this project lay in the partnership with another energy sector company, Contour Global, who have a strong position in combined heat and power plants and were able to integrate the PV-system into an overall energy efficiency program [3].

7.3 *Diesel Jeans Headquarter, Italy: 200 kWp*

The membrane solution by UNI-SOLAR's sister company Solar Integrated was chosen for this application to utilize key product characteristics of being very flexible, lightweight, durable, and reliable. The system blends with the building's unique architecture—a bold contemporary design, built with a very good energy rating and the top goal to reduce energy consumption.

7.4 Interporto Distribution Center, Italy: 25 MWp

The installation on CIS—Interporto Campano, the largest logistic site of the Southern Mediterranean, located in Nola features up to 25 MWp roof-top PV on 660,000 m^2 roof area divided on a total of 60 warehouse buildings and was realized by our partners General Membrane and Ondulit. The installation produces approx. 33 GWh/year, enough to satisfy the consumption of 13,000 families, avoiding CO_2 emissions of more than 21 million tons. It allows CIS and Interporto to realize their primary objective of achieving a neutral energy balance. This ENEL installation will be able to produce 90 % of the energy consumed by CIS and Interporto [4].

8 Conclusion

To fulfill its growing energy requirements Turkey has rightfully identified that renewable energies need to play a significant part in the energy mix. Photovoltaics should be a natural part of this, in addition to hydro, geothermal, and wind due to the high insolation levels available across the year and good match between solar generation and peak energy consumption.

The paper shows that in addition to ground mounted systems, rooftop installations on large industrial and commercial buildings can contribute considerably to the Turkish national objectives. Furthermore BIPV solutions use existing infrastructure with minimal visual impact and generate electricity close to the point of consumption. UNI-SOLAR's lightweight and flexible PV laminates offer a viable alternative to traditional PV systems of glass and metal framed modules to make low-load bearing roofs available for photovoltaic installations.

Examples from Turkey and across Europe show the broad range of application fields, roof types, and product solution, which can turn existing unused roof space into clean energy generating assets. These help to meet national goals, reduce entities' electricity costs, and benefit the environment by reducing emissions.

References

1. Oktik S (2011) How and when Turkish industries shall join "International PV Industry Value Chain". In: Proceedings 1st solar business forum for the Balkans 2011, 14–15 Apr 2011, Sofia
2. Centrosolar PR (2010) Größte Dünnschicht-Solaranlage der Türkei mit CENTROSOLAR-Modulen ans Netz gegangen, Hamburg
3. Guerreri A (2011) Realization of a multiple site PV-project on Coca-Cola bottling plants in Italy by cooperating with partners from the energy sector. In: Proceedings of the Conferenza dell'Industria Solare—Italia 2011, 25–29 Feb 2011. Rome, p 24
4. Amyuni T (2011) Case study: realization of an investor owned 25 MW roof-top PV power plant. In: Proceedings PV power plants conference 2011—Europe, 10–11 Mar 2011. Paris

Part IV
Future Directions

Archetypes of 100 % Renewable Energies Scenarios by 2050

Harry Lehmann and Mark Nowakowski

1 Introduction

Before the Fukushima disaster, the idea of a carbon-neutral society, with an electricity supply based entirely on renewable energies, was considered one of the several possible future scenarios in Germany. Since 11 March 2011 this has changed once and for all. The German government's decision to "transform the energy system" foresees Germany phasing out nuclear energy completely by 2022 and significantly increasing renewable energies' share in coming decades. So it is no longer a question of whether or not the change to a carbon-neutral, renewable-based society will happen. In a long-term perspective, there will not be any alternative to that change, not only because of the obvious threats posed by nuclear energy, but above all for climate protection reasons. Industrialized countries must reduce their CO_2 emissions by 80–95 % by 2050 if the rise in global temperature is to be limited to 2 °C above the temperature in pre-industrial times. This is crucial for keeping the impacts of climate change within tolerable limits. The electricity generation sector can make a major contribution to the desired emission reductions. Geared to using fossil energy sources, this sector is responsible for 40 % of all energy-related emissions in Germany. On the other hand, nuclear energy, supposedly clean, will no longer be available as of 2022. These facts, and the crucial role which a reliable energy supply plays for a highly developed industrialized country such as Germany, make it obvious that the electricity sector should be given priority attention.

H. Lehmann (✉)
Division of Environmental Planning and Sustainable Strategies, German Federal Environment Agency, Wörlitzer Platz 1, 06844 Dessau-Roßlau, Germany
e-mail: harry.lehmann@uba.de

M. Nowakowski
Section of Energy Strategies and Scenarios, German Federal Environment Agency, Wörlitzer Platz 1, 06844 Dessau-Roßlau, Germany
e-mail: mark.nowakowski@uba.de

T.S. Uyar (ed.), *Towards 100% Renewable Energy*, Springer Proceedings in Energy, DOI 10.1007/978-3-319-45659-1_39

2 Archetypes of a Future Renewable-Based Power Generation

With this objective at hand the Federal Environment Agency (Umweltbundesamt, UBA) has been analyzing several "archetypes" of a future renewable-based power generation. Within this framework three radically different scenarios have been developed in order to study the technical and ecological feasibility of Germany switching to an electricity supply based entirely on renewable sources by 2050. Apart from different generation structures, the studies assume different degrees of connection between individual regions in Germany and between Germany and other countries within a pan-European network. In the "Local Energy Autarky" scenario (not yet published), small-scale decentralized energy systems use locally available renewable energy sources to satisfy their own power demand without being connected with each other or with outside suppliers, i.e., without electricity imports. In the "Regions Network" scenario (published in 2010), electricity is exchanged throughout Germany and only a small part of the load is covered by electricity imports from neighboring countries. In the "International Large Scale" scenario, Germany's electricity supply is based on all renewable energy potentials in Germany, Europe, and its vicinity which can readily be tapped by large-scale technology projects and storage power plants. In this scenario, Germany imports much of its electricity demand via a well-developed intercontinental transmission grid.

These three "archetypical" scenarios represent extreme points of a solution space for a renewable electricity supply in Germany by 2050. The Federal Environment Agency hopes to show in this way that there is not just one technically and ecologically feasible path towards attaining this goal but a whole range of possible variants depending on political and social priorities. A real-life future energy supply system for Germany will most likely include characteristics of all three scenarios, however.

In July 2010, UBA published the "Energy Target 2050: 100 % Renewable Electricity Supply" study, which presents the results of the Regions Network scenario.[1] This scenario assumes that all regions in Germany make extensive use of their renewable energy potentials. Efficient energy use in all sectors compensates for the rise in consumption caused by moderate economic growth and increased e-mobility and use of heat pumps. The introduction of large-scale storage of electricity and the use of demand side management potentials make a

[1] English version: http://www.umweltdaten.de/publikationen/weitere_infos/3997-0.pdf. Other reputable institutions have also examined the feasibility of an electricity supply entirely based on renewable sources and published studies. For example, the publication "Pathways towards a 100 % renewable electricity system" by the German Advisory Council on the Environment SRU or "The Energy Report" by the World Wildlife Fund WWF presents scenarios which show how a 100 % renewables-based electricity supply can be realized. The core findings are that such a system can provide sufficient power at any time of year. The same applies to the study "Klimaschutz Plan B 2050" by Greenpeace. Other relevant studies are: "Long-Term Integration of Renewable Energy Sources into the European Energy System" by LTI Research Group (1998), or H. Lehmann et al. (2003), "Energy Rich Japan—A Vision for the Future."

substantial contribution to the balancing of load and production. A well-developed national electricity transmission grid ensures large-scale balancing across the various regions.

The simulations show that under the given assumptions for 2050 the power requirement can be covered at any moment at today's level of supply security and in an ecologically compatible way. Fluctuations in renewable energy feed-in can be offset and sufficient balancing power can be provided at the same time. Technology leaps are not assumed in this scenario—these outcomes are achievable with today's best available technologies. The potentials identified are sufficient to even cover the considerable additional power demand from e-mobility, the wide use of heat pumps for heating and hot water provision, and air conditioning. Therefore, an electricity supply system based completely on renewable energies as simulated in the Regions Network scenario does not compromise supply security. Nor does it increase the dependency on energy imports. However, essential prerequisites are a significant increase in reserve capacities for electricity generation and storage, the exploitation of the large load management potential as well as expanding and modifying existing infrastructure to facilitate this as well as electricity transport, including an ambitious expansion of the grid. The political decisions needed to make such a development happen must be taken in the years to come. The study devotes a separate chapter to corresponding policy recommendations.

Figure 1 shows the feed-in from renewable energy sources against the load in the 2050 Regions Network scenario for two different seasons. On the top chart we see a typical situation for a winter month with a major share of wind power while photovoltaic does not contribute substantially to cover the load. In combination with geothermal and hydro power the wind feed-in even exceeds the load at many times thus charging the storage facilities. Between days 23 and 26 there is no sufficient feed-in to meet the load, however, which means that stored energy is being utilized. The bottom chart of Fig. 1 reflects a typical summer supply situation with a significant feed-in of photovoltaic electricity generation and a clearly lower input from both onshore and offshore wind. While the photovoltaic midday spikes go together well with the load peaks (as a result of load management), storage capacities need to be tapped during night hours when the load is higher than the supply. The crucial part of storing away excess electric energy becomes obvious here. Short-term storage facilities such as pump storage power plants can balance out deviations between feed-in and load for one to several days. Similarly, demand side management functions as virtual short-term storage. Long-term storage systems such as chemical energy binding can balance feed-in fluctuations for any time period from days to years. Currently, the benefits of methanation in particular are subject to further investigations.

Figure 2 shows a basic concept for the transformation of renewable excess electricity via hydrogen to methane with reconversion in combined heat and power plants, gas turbines, or combined cycle power plants. Methane can be stored in underground spaces, the so-called cavern and pore storage facilities. As required, the chemically bound energy can be reconverted into electricity or used for other purposes. Chemical storage thus allows for innovative load and generation management. Storage limits the need for additional production capacities and

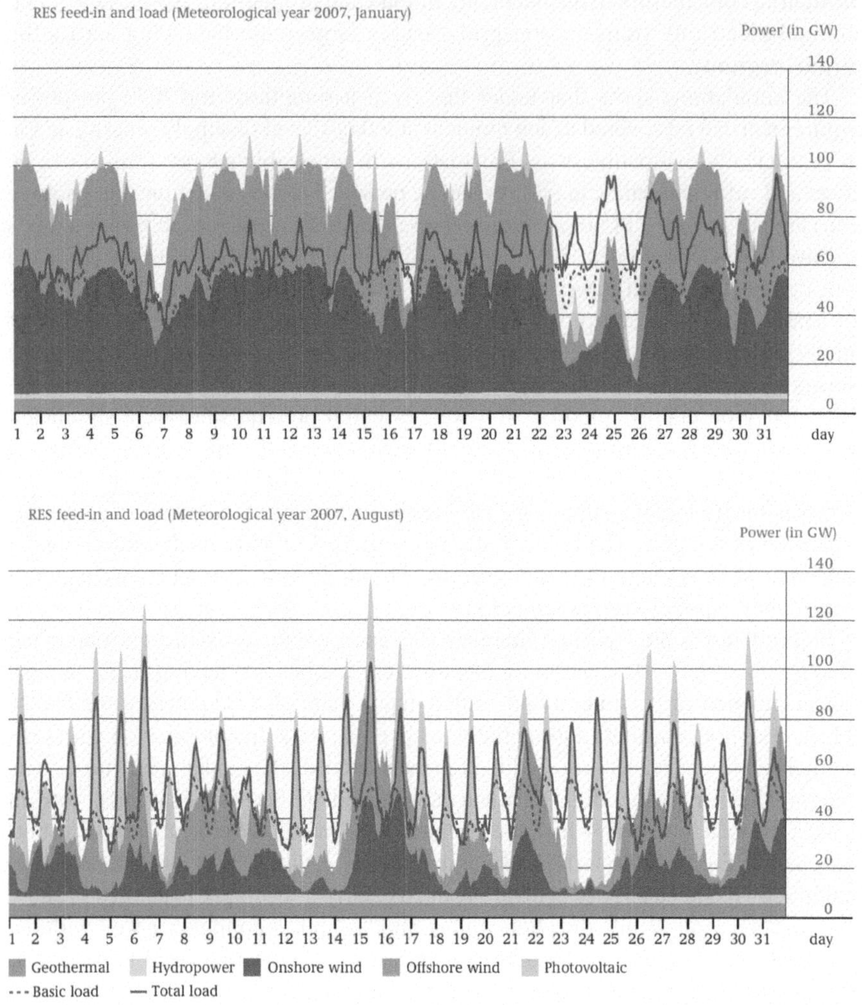

Fig. 1 Examples of feed-in from renewable energy sources in the year 2050 for the Regions Network scenario, based on the meteorological year 2007 (*top*: January, *bottom*: August)

reduces stress on the transmission grid, e.g., when methane is transported through the existing natural gas grid instead. Therefore much of the electricity from renewable energy sources which otherwise could not be absorbed by the grid can be utilized. Beyond that both methane and hydrogen can be used in the chemical industry or as alternatives to fossil fuels in the transport sector.

Unlike the Regions Network scenario, the Local Energy Autarky scenario does not consider the whole of Germany. Instead, the simulations were applied exemplarily to a rural community and a town district, each with and without trade/industry and each at a location in northern and in southern Germany, which are

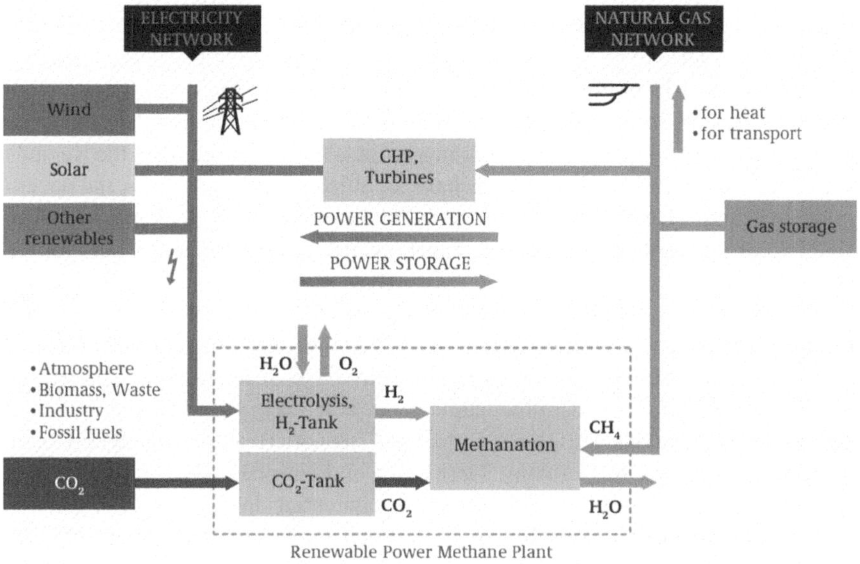

Fig. 2 To methane with reconversion in combined heat and power plants, gas turbines or combined cycle power plants

representative of settlement structures in terms of population and building density. In these simulations, the communities utilize locally available constrained renewable electricity potentials almost fully and are not part of an interconnected network, i.e., they are self-sufficient "energy islands."

The object of the simulations was to ascertain whether in 2050 the local electricity requirement can be covered year round from locally available wind and solar power generation potential using storage systems. The use of biomass or waste biomass for electricity production was excluded since it was assumed that in 2050 it will be used primarily in industry and transport to replace fossil fuels. Electricity production from deep geothermal energy or hydropower was not considered either, since in Germany the availability of these sources is the exception rather than the rule and since this would greatly improve the supply situation.

The results in this scenario show that local energy autarky, or "island solutions," may be technically feasible in individual cases, such as when geothermal and/or hydropower potentials can be harnessed. If such potentials are absent, however, which is normally the case in Germany, an electricity supply based completely on renewable energies can hardly be modeled for the rural and not at all for the urban structures considered here. This is especially true if the community (both rural and urban) has local jobs in industry and trade, assumed as a percentage of the local population (whose electricity requirement would otherwise have to be covered by other communities). The rural community simulated here, and more still the urban settlement structure, is thus reliant on additional, external generation potential.

This makes the national electricity transport network an important component of any strategy for achieving a 100 % renewable energy supply in Germany. Potentials will have to be tapped where they exist. Regions with "excess potential" will have to cater to regions with a "shortfall." Finally, harnessing offshore wind potentials appears to be meaningful only in the context of a nationwide interconnected grid. The high degree of interconnectedness within Germany in the Regions Network scenario thus shows itself to have significant advantages over the decentralized "island" electricity supply sketched in the Local Energy Autarky scenario, which according to these findings presents no viable alternative, at least not for the whole of Germany.[2]

With the International Large Scale scenario currently in preparation the question will be whether and to which extent a wider network reaching across Germany's borders can be beneficial in terms of the optimal use of renewable energies, the large-scale balancing between fluctuating renewable energy feed-in and load, and using storage potentials. This third "archetypal" scenario features the highest share of electricity imports to Germany. Germany's and Europe's electricity supply is based here on all renewable energy and storage potentials in Germany, Europe, and its vicinity which can readily be tapped through large-scale technology projects (e.g., solar thermal energy from northern Africa, geothermal energy from southern Europe and Turkey, pump storage power plants in Scandinavia, and wind power from the British Isles). A well-developed intercontinental transmission grid connects generation centers with the centers of electricity consumption, allowing fluctuations in feed-in of renewable electricity to be balanced out on a Europe wide scale. Hence, relative feed-in peaks could be reduced and the contribution which renewable energies, particularly wind energy, make to guaranteed capacity could be increased. This could also help to considerably reduce the need for storage and power plant reserve capacities. In this scenario, Germany imports a significant proportion of its power requirement.

Work is currently underway to study the technical and ecological feasibility of the International Large Scale scenario and to pinpoint the differences to other scenarios, including benefits as well as limitations of this approach. The project's central object is to model and simulate Germany's integration into a European grid and to find out whether an electricity supply based 100 % on renewable energy would be possible in 2050 at today's high level of supply security. In particular, it aims to determine how supply and demand can be balanced at any moment and what technologies are needed to achieve this.

[2] Still, further studies are currently being carried out examining other cases with high degrees of decentralized local electricity generation from renewable sources aiming at the benefits which may result from this archetype.

3 Conclusion

The studies presented here show that a renewable electricity supply in Germany by 2050 is technically and ecologically feasible and that there is even a variety of potential pathways towards such a future depending on the chosen framework. Hence, it is now high time to think and talk about the questions where we want to end up and how to get there—regarding electricity generation as well as storing technologies, demand side management, and the expansion of national and international transmission grids in Germany and Europe. Now is the time to make decisions based on society's and policy's preferences and finally take action concerning our future energy mix.

The Energy Report: 100 % Renewable Energy by 2050

Stephan Singer, Jean-Philippe Denruyter, and Deniz Yener

1 Introduction

A world powered by 100 % renewable energy. That's the WWF's vision for the middle of this century. Achieving it will mean avoiding catastrophic climate change, less pollution, increased energy security, and improved health for people worldwide.

But is it possible? We called upon respected energy consultancy Ecofys to investigate. The result is the most ambitious science-based examination ever of a renewable and clean energy future on a global scale. The study shows that it is technically possible to achieve almost 100 % renewable energy sources within the next four decades.

The Ecofys scenario raises many issues and challenges, outlined in this briefing and discussed in more detail in *The Energy Report*. Meeting the energy needs of current and future generations is one of the most important, difficult, and urgent political tasks for every government. This paper sets out our recommendations in light of these challenges (Fig. 1).

2 Why the World Needs a 100 % Renewable Energy Future

Switching to renewable energy isn't just the best choice. It's our only option. The way the world produces and uses energy today is not sustainable.

- A fifth of the world's population has no access to reliable electricity [2]. More than 2.7 billion people are dependent on traditional bioenergy (such as wood and charcoal) for cooking and heating [2]—with serious economic, environmental, and health impacts.

S. Singer (✉) • J.-P. Denruyter • D. Yener
CAN International, Brussels, Belgium
e-mail: stephansinger@outlook.com; jpdenruyter@wwf.panda.org

T.S. Uyar (ed.), *Towards 100% Renewable Energy*, Springer Proceedings in Energy, DOI 10.1007/978-3-319-45659-1_40

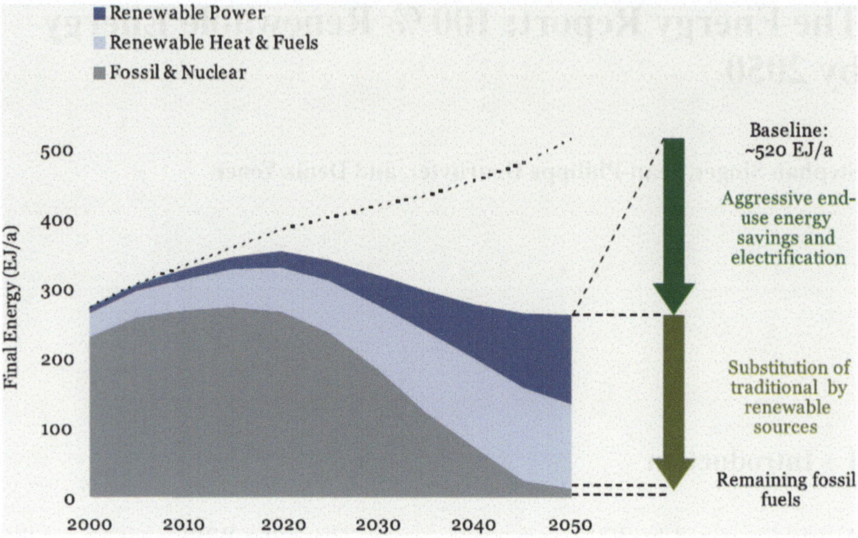

Fig. 1 Evolution of energy supply in the Energy Scenario, showing the key developments. *Source*: The Ecofys Energy Scenario, December 2010 [1]

- Production from known oil and gas reserves will fall by around 40–60 % by 2030, according to the International Energy Agency [3]. Continuing to depend on fossil fuels will mean substantially higher and more volatile energy costs, driven by the increasing scarcity of oil and gas and a move to unconventional—and increasingly environmentally damaging—sources. Supply disruptions, accidents, and disputes over energy resources will continue to challenge energy security.
- The global energy sector is responsible for around two-thirds of global greenhouse gas emissions. And its emissions are increasing at a faster rate than any other sector. "Business-as-usual" scenarios show an increase in emissions that would lead to very dangerous levels of warming, far above the threshold agreed by governments of 2 °C above pre-industrial levels.
- Nuclear is a risky and expensive option, producing dangerous waste that remains highly toxic for thousands of years. It could also potentially contribute to political instability and insecurity.

A fully sustainable renewable energy system is the only way to secure energy for everyone and avoid environmental catastrophe.

3 The Ecofys Scenario in a Nutshell

Ecofys envisages a world in 2050 where energy demand is 15 % lower than in 2005. Although population, industrial output, economic activity, passenger travel, and freight transport continue to rise as predicted, ambitious energy-saving measures allow people to do more with less. Industry uses more recycled and energy-efficient

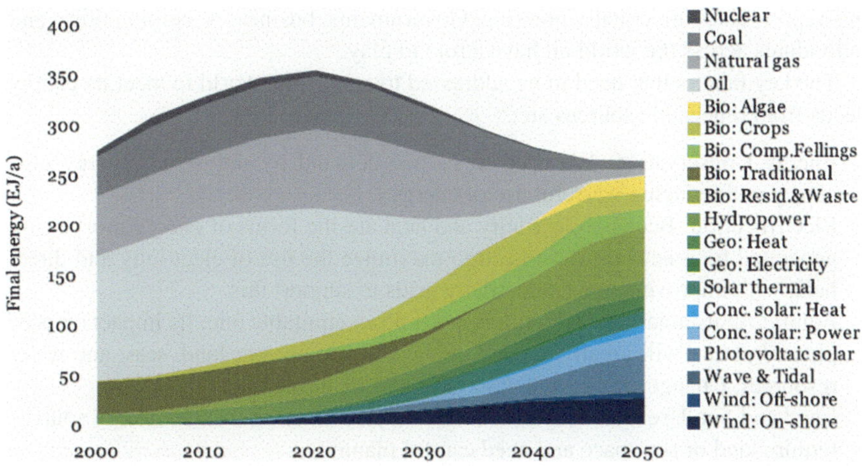

Fig. 2 World energy supply by source. *Source*: Ecofys Energy Scenario December 2010 [3]

materials, buildings are constructed or upgraded to need minimal energy for heating and cooling, and there is a shift to more efficient forms of transport.

As far as possible, the world uses electrical energy rather than solid and liquid fuels. Wind, solar, biomass, and hydropower are the main sources of electricity, with solar and geothermal sources, as well as heat pumps, providing a large share of heat for buildings and industry. Because supplies of wind and solar power vary, "smart" electricity grids have been developed to store and deliver energy more efficiently. All of this is delivered using already proven technology and processes.

Bioenergy (liquid biofuels and solid biomass) is used as a last resort where other renewable energy sources are not viable—mainly to provide fuels for aeroplanes, ships, and trucks, and in industrial processes that require very high temperatures.

By 2050, the world is saving nearly €4 trillion per year through energy efficiency and reduced fuel costs compared to a "business-as-usual" scenario where everyone carries on using energy in the same way as at present. But big increases in capital expenditure are needed first—to install renewable energy-generating capacity on a massive scale, modernize electricity grids, transform goods and public transport, and improve the energy efficiency of our existing buildings. Investments begin to pay off around 2040, when the savings start to outweigh the costs. If oil prices rise faster than predicted (the scenario uses a conservative estimate of $87 per barrel in 2030 and $142 in 2050), and factoring in the costs of climate change and the impact of fossil fuels on public health, the pay-off happens much earlier (Fig. 2).

4 Critical Issues and Recommendations

The Ecofys analysis shows that the world can technically meet its energy needs from renewable sources by 2050. But it throws up some difficult challenges—and not just technical ones. The social, environmental, economic, and political issues this report raises are equally pressing. Governments, businesses, communities, and individuals across the world all have a role to play.

The key factors that need to be addressed to enable the world to meet its energy needs from renewable sources are:

- Energy Conservation: We need to reduce demand by improving energy efficiency and reducing wasteful use of energy.
- Electrification: Because electricity and heat are the forms of energy most easily generated by renewables, we need to maximize the use of electricity and direct heat, with improvements to electricity grids to support this.
- Equity: A sustainable energy future must be an equitable one. Its impact on people and nature will greatly depend on the way we use our land, seas, and water resources. Changes in lifestyle also have a critical role to play.
- Land and Sea Use: Not only bioenergy but also other renewable energy sources require land or sea space and need careful planning.
- Lifestyle: Travel modes and meat consumption are two examples for changes that will happen.
- Finance: Moving to a renewable future will mean rethinking our current finance systems.
- Innovation: Global expenditure on R&D for renewables and efficiency must double over the next 10 years.
- Governance: Local, national, and regional governance will need to be greatly strengthened to secure an equitable energy future. We need international cooperation and collaboration on an unprecedented level to bridge the gap between the energy-rich and energy-poor, both within and between countries.

5 Conclusion

That the world faces an energy crisis is beyond doubt. There's a pressing need to secure a sustainable energy supply as demand for fossil fuels outstrip environmentally and economically sustainable supplies. A lack of access to energy is one of the main causes of poverty. On top of this, the world needs to start drastically reducing CO_2 emissions within the next few years if we're to have the best chance of avoiding catastrophic climate change.

We—individuals, communities, businesses, investors, politicians—must act immediately, and boldly. Half-hearted solutions are not enough. We must aim for a fully renewable energy supply by the earliest possible date.

It is possible. The Ecofys scenario lays out, in unprecedented detail, one way that we can do this. It isn't the definitive solution, and it isn't perfect. As we've seen, it raises many challenges and difficult questions. But it shows that solutions are at hand. We are putting it forward to catalyze debate and to spur action.

We now need to respond to the issues it raises. We need to take it further. But most of all, we need to act on it—each and every one of us. Starting today.

References

1. Singer S, Denruyter JP (eds) (2011) The energy report: 100% renewable energy by 2050. WWF-International, Gland, www.wwf.panda.org/energyreport
2. International Energy Agency (IEA) (2010) World energy outlook. OECD/IEA, Paris
3. International Energy Agency (IEA) (2009) World energy outlook. OECD/IEA, Paris

Offshore Wind Energy: Key to 100 % Renewable Energy

Klinger Friedrich and Müller Lukas

Nomenclature

CAPEX	CAPital EXpenditure
EEZ	Exclusive Economic Zone
EPC	Engineering-Procurement-Construction
EU	European Union
EUR	Euro
GBF	Gravity-Based Foundations
HVDC	High-voltage direct current
km	Kilometer
kW	Kilowatt
kWh	Kilowatt Hours
LCoE	Levelized Cost of Energy
m	Meter
MW	Megawatt
O&M	Operation & Maintenance
R&D	Research & Development
TWD	Taiwan Dollar
USD	US-Dollar
WEO	World Energy Outlook 2006
WTG	Wind Turbine Generator

K. Friedrich (✉) • M. Lukas
INNOWIND Forschungsgesellschaft mbH, in cooperation with Fraunhofer IWES,
Altenkesseler Str. 17/D2, 66115 Saarbrücken, Germany
e-mail: f.klinger@wind-energy-research.de

T.S. Uyar (ed.), *Towards 100% Renewable Energy*, Springer Proceedings
in Energy, DOI 10.1007/978-3-319-45659-1_41

1 Introduction

The major challenge for offshore wind development today is to continue to bring down costs. Selection of sites in deeper waters, further from shore, with more difficult bottom conditions and higher waves, have all contributed to driving the costs up faster than the improvements in the technology has been able to drive them down. However, technology cost reductions continue to be achieved, and this is one of the main reasons for confidence regarding offshore wind. It is expected that the cost of energy from offshore wind will come down substantially as the mass roll-out of the next generation of offshore wind turbines begins to take place. More than 90 % of the world's offshore wind power is currently installed off northern Europe, in the North, Baltic and Irish Seas, and the English Channel. Most of the rest is in two "demonstration" projects off China's east coast.

Offshore wind is an essential component of Europe's binding target to source 20 % of final energy consumption from renewables, and China has set itself a target of 30 GW of installations off its coast by 2020. The USA has excellent wind resources offshore, and many projects are under development, but there is no offshore wind power installed yet [1].

2 Statistics

2.1 In the EU

1567 MW of new offshore wind power capacity were connected to the electricity grid during 2013 in Europe, 34 % more capacity than in the previous year. 47 % of all new capacity was installed in the UK (733 MW). The share of total capacity installed in the UK was significantly less than in 2012 (73 %). The second largest amount of installations were in Denmark (350 MW or 22 %), followed by Germany (240 MW, 15 %) and Belgium (192 MW, 12 %). Of the total 1567 MW installed in European waters, 72 % were located in the North Sea, 22 % in the Baltic Sea, and the remaining 6 % in the Atlantic Ocean. Siemens continues to be the top offshore turbine supplier in terms of annual installations. With 1082 MW of new capacity connected, Siemens accounts for 69 % of the market. Bard (240 MW, 15 %), Vestas (123 MW, 8 %), and Senvion (REpower) (111 MW, 7 %) are the other three turbine manufacturers which had turbines grid connected in full-scale wind farms during 2013. Alstom and Gamesa both installed their first demonstration turbines and Siemens installed the first two of its 6 MW turbines. A total of 2080 wind turbines are now installed and connected to the electricity grid in 69 offshore wind farms in 11 countries across Europe. The total installed capacity at the end of 2013 reached 6562 MW, producing 24 TWh in a normal wind year, enough to cover 0.7 % of the EU's total electricity consumption [2].

2.2 *Rest of the World*

Chinese offshore wind made slow progress in 2013, and only Dongfang, Envision, and United Power had installation in the intertidal zone projects. 21 offshore wind turbines were installed in China during 2013, a decrease of 54 % compared to 2012, with a new installation of 39 MW, 69 % less than the previous year. At the end of 2013, there are 428.6 MW of installed offshore wind power capacity in China. The 39 MW installed during 2013 were located in intertidal zones, where the cumulative installed capacity reached 300.5 MW by the end of 2013, while the remaining 128.1 MW were installed in near shore waters. Sinovel, Goldwind, and Siemens account for the major share in offshore wind turbine manufacturers in China. Most of the installed turbines were rated 2 MW and above. Among them, the turbines from Sinovel and Siemens were mainly installed in offshore projects, while Goldwind mostly in intertidal zone projects [3].

Japan is an island country with a very large Exclusive Economic Zone and excellent offshore resources, and as a consequence of the Fukushima disaster in 2011, a strong appetite for renewable energy. The Ministry of Environment (MOE) estimates Japan's theoretical offshore potential at 1573 GW. Currently, Japan has four on-going offshore wind projects at four sites totalling 25.3 MW, including both founded and floating types of turbines. The only new installation in 2012 was the floating turbine demonstration project off Goto Island, near Nagasaki, with a capacity of 100 kW [4].

Smaller projects in South Korea and Taiwan exist. In the USA and South America there are no offshore turbines erected, yet.

3 Ways to Cost Reduction

Offshore turbine manufacturing will enter a phase of intense competition. There is a large number of new market entrants in the last 2 years. Big industrial players such as Alstom, Hyundai, Mitsubishi, and Samsung see offshore wind as attractive. Competition will increase significantly due to the large number of new entrants. Siemens and Vestas are the dominant players, with more than 500 turbines installed. Market entry by large players such as construction companies and utilities leads to growing professionalism in project development.

Offshore projects show significant potential for improvement across the entire project value chain. Offshore project development takes 7–10 years. But there are interface risks for investors, because Engineering-Procurement-Construction-contracts for offshore wind projects are not available and multi-contracting is a high risk. Improved methods exist, like strategic partnerships with a small number of partners, strengthen controls and hiring of experienced personnel from developers. Larger turbines will improve total CAPEX, capacity factors, and O&M costs (see Fig. 1) [5].

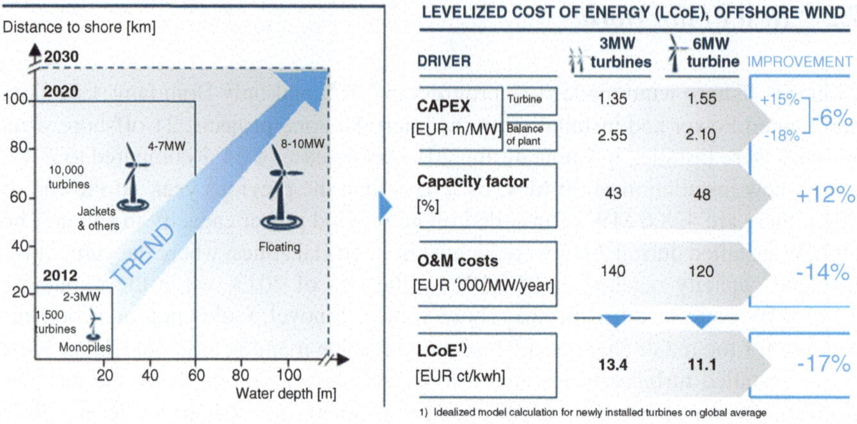

Fig. 1 Wind turbines—size and Levelized Cost of Energy [5]

Monopiles remain the dominant foundation concept, but trend towards deeper water is shifting growth to jackets. Gravity-based foundations (GBF) are currently only used in shallow water; however, new GBF concepts could have potential for renewed future application up to 40 m. Monopiles remain the most widespread foundation type. Limitations in water depth and weight are increasingly being overcome with new concepts. Tripod/-pile high production costs due to complex structure and great weight are likely to limit the use of both concepts. Jackets will increase their share due to their flexibility and low weight (40–50 % less steel than monopiles), commercially worthwhile >35 m. Floating turbines are currently at R&D stage, but could become relevant for countries with steep shores. No commercial use is expected before 2020 [5].

HVDC (High-voltage direct current) connections cause delays and cost overruns in Germany and similar issues may occur in other markets. Bottlenecks are offshore converter stations, since only three suppliers exist (ABB, Siemens, Alstom) and a delivery time up to 50 months is needed. Offshore HVDC cables and cable laying are a problem, as well because only a few suppliers exist and shortages may occur. Furthermore installation vessels for converter stations, only a few vessels can install converter stations >10,000 t. New vessels specifically designed for offshore wind will reduce installation times and costs [5].

Operation & Maintenance (O&M) is a key value driver but concepts for the next generation of wind farms are not yet mature. Efficient, proven O&M concepts are still not available. Excellence in that field is critical to a profitable offshore wind business and O&M is approximately 28 % of lifetime costs. Key O&M variables are the location of the station for service personnel onshore or offshore, service vessel concepts including a potential use of helicopter and an adequate access to vessels for replacing large components. Improvement levers are increased rated power of WTGs reduces O&M costs per kWh, increased reliability of turbines and components to reduce unplanned service activities, geographical clustering of offshore wind farms to create synergies and increased in-house O&M activity by utilities will partly or fully replace by O&M turbine manufacturers [5].

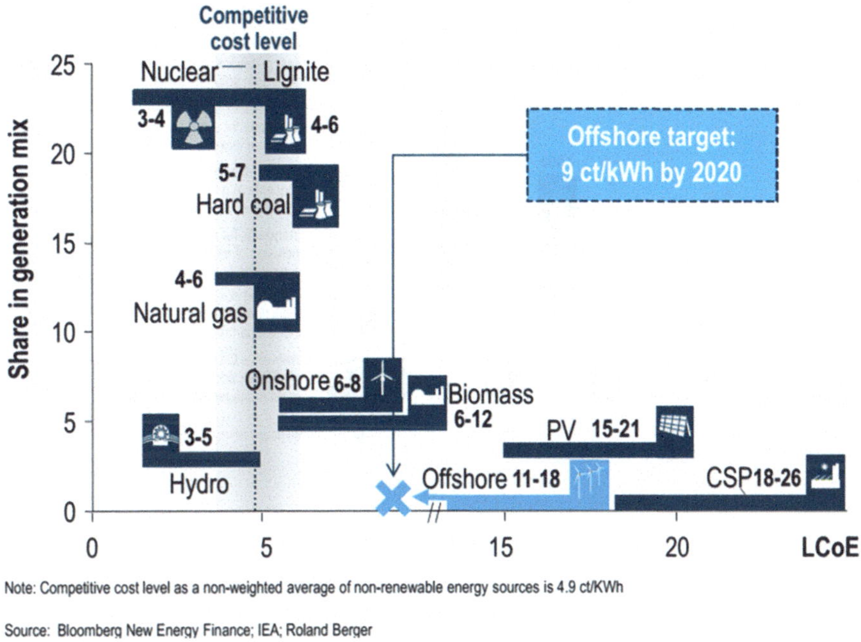

Fig. 2 LCoE 2012 European generation mix in EUR ct/kWh [5]

Utilities are dominant in farm ownership and operation. Financial investors are required to finance the pipelines. Three investment models exist:

Standalone: One utility owns, develops, and operates a project. Utility has full control, but bears all risks. *Lead Investors*: Leading utility with one or more minority investors, utility shares risk and reduces equity requirements. *Joint Venture*: A group of equal players join forces, project development by joint venture company.

The offshore industry needs to raise its cost competitiveness to ensure sustainability (see Fig. 2) [5].

WTG costs are 25 % of lifetime costs—Project elements offer further potential to realize a sustainable cost out (see Fig. 3).

4 Floating Foundations

Monopile substructures remained the most popular EU substructure type in 2013 with 490 installed (79 %). 87 tripod foundations were installed, 14 % of all newly installed substructures, followed by jackets (39, 6 %), tripiles (8, 1 %), and 1 gravity foundation. The weighted average water depth of offshore wind farms where work was carried out in 2013 was 20 m, slightly lower than in 2012 (22 m). The average distance to shore for those same projects was 30 km, almost the same as in 2012 (29 km). Over the years, offshore wind farms have moved further from shore and into deeper waters. At the end of 2013, the average water depth of online wind

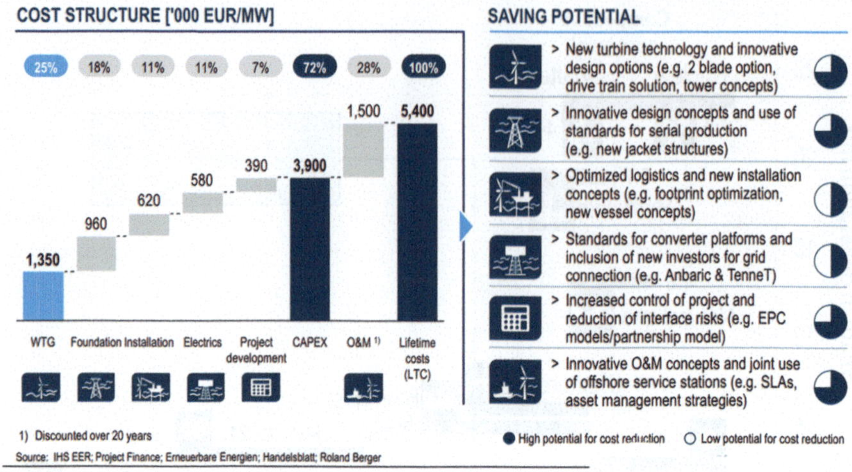

Fig. 3 Cost structure and saving potential [5]

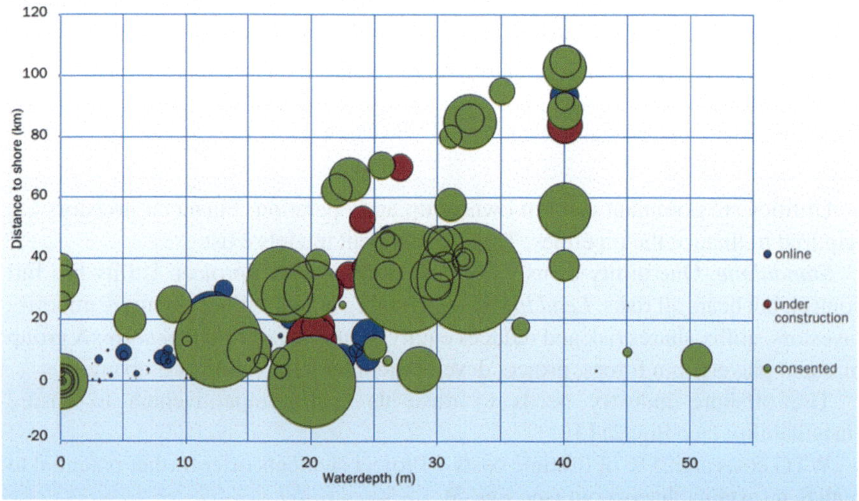

Fig. 4 Average water depth and distance to shore of online, under construction and consented wind farms, in EU end of 2013 [2]

farms was 16 m and the average distance to shore 29 km. Looking at projects under construction, consented or planned, average water depths and distances to shore are likely to increase (see Fig. 4) [2].

Floating wind turbines can be deployed in waters of several hundred meters making the benefits of offshore wind power available to new areas and coastlines. Future developments may open up the possibility of going towards the ultra deep waters, in the 1000 m range and beyond. Consequently, floating offshore wind could be an important contributor to the world's energy mix in the future, with large floating wind farms providing clean, cost-effective electricity for millions of people (see Fig. 5) [6].

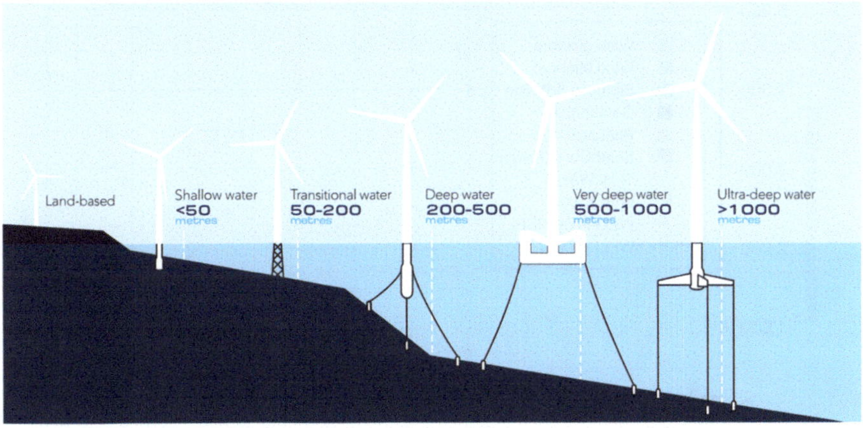

Fig. 5 An overview of applied wind power generation substructure technologies [6]

During this initial stage, development is likely to be concentrated around a small number of countries. The world's first floating wind turbine prototype was installed in 2009 near the Norwegian island of Karmøy, and small-scale demonstrator arrays are now under development around the world. Japan, Portugal, and the USA also have prototypes installed and show very strong interest in this technology. The cost of floating wind turbine technology will, to a large extent, depend on the cost development of fixed offshore wind; hence, a cost analysis will have to include both. Lean manufacturing of substructures, development of turbines, blades, and station-keeping systems are all obvious elements of the future cost reductions for offshore wind technology. A Levelized Cost of Energy model based on learning effects and projections of future market size can then assist in explaining the overall potential for cost compressions for the whole offshore wind industry. However, very few predictions exist for the offshore wind energy development until 2050 and few national targets look this far into the future [6].

5 100 % Renewable Only with Offshore Wind Energy

In a study of GL, Floating Offshore technology is seen as main potential solution for the future:

> Floating wind turbine—the third generation of wind power—is the most recent development in the relatively young offshore wind power business. Currently, only a handful of small demonstration projects are running, accounting for a total generation capacity of just a few megawatts (MW). However, attracted by its potential, governments, large technology developers and other stakeholders are increasingly interested in this technology. As a result, floating wind turbine technology has the potential to mature into a cost-efficient energy source in the coming decade, graduating from today's demonstration projects to large-scale arrays of several thousand MW. [6]

Steve Sawyer, Secretary-General of the Global Wind Energy Council writes in "FLOATING WIND POWER: THE NEXT WAVE?": "These types of platform could justify the build-out of the 10, 12, 15 and even 20MW machines which at this stage remain on the drawing board." [7].

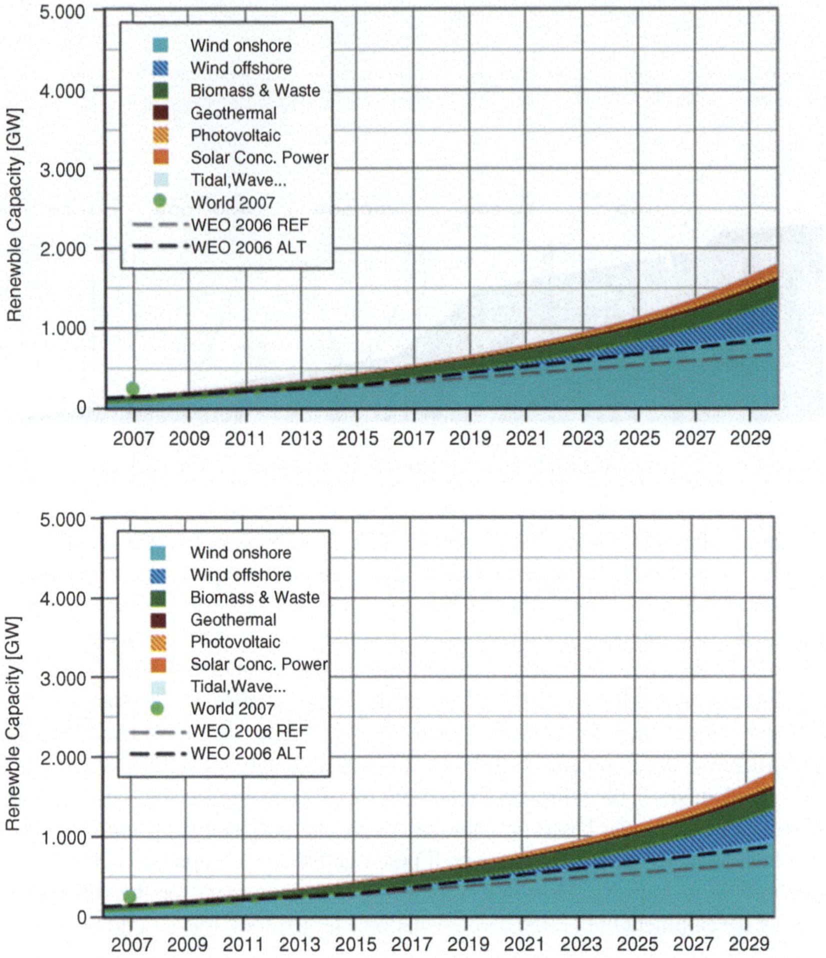

Fig. 6 Development of "new" renewable electricity generating capacities in the world regions in the "High Variant" (*upper figure*) and "Low VariantScenario" (*lower figure*) (EWG; 2008). Data on renewable capacity 2007: (REN 21; 2007) [9]

WWF writes in "The Energy Report—100 % Renewable by 2050": "Wind could meet a quarter of the world's electricity needs by 2050 if current growth rates continue—requiring an additional 1,000,000 on shore and 100,000 offshore turbines. Electricity from offshore is less variable, and turbines can be bigger. Although wind farms have a very visible effect in the landscape, their environmental impact is minimal if they are planned sensitively" [8].

The "Energy Watch Group Global Renewable Energy Scenarios" shows in "Renewable Energy Outlook 2030" two scenarios for the share of the renewable energy in 2030. The "High Variant Scenario" is much more than double the capacity reached in the "Low Variant Scenario". The vast majority of the generating capacity in 2030 in both scenarios is onshore and offshore Wind Energy" (see Fig. 6) [9].

References

1. GWEC (2014) Global wind statistics 2013. (http://www.gwec.net/wp-content/uploads/2014/02/GWEC-PRstats-2013_EN.pdf)
2. EWEA (2014) The European offshore wind industry—key trends and statistics 2013. (http://www.ewea.org/fileadmin/files/library/publications/statistics/European_offshore_statistics_2013.pdf)
3. WWEA (2014) Quarterly Bulletin issue 1. (http://www.wwindea.org/webimages/WWEA_Bulletin-ISSUE_1_2014_reduced.pdf)
4. GWEC (2014) Global wind report 2013. (http://www.gwec.net/publications/global-wind-report-2/global-wind-report-2013/)
5. Roland Berger Strategy Consultants (2013) Offshore wind toward 2020 April 2013 on the pathway to cost competitiveness. (https://www.rolandberger.com/publications/publication_pdf/roland_berger_offshore_wind_study_20130506_1_.pdf)
6. DNV GL (2014) Electrifying the future, Høvik. (http://production.presstogo.com/fileroot6/gallery/DNVGL/files/original/ae11bb7549914e07b60489f7bd8df92f.pdf)
7. Secretary-General of the Global Wind Energy Council (2013) Floating wind power: the next wave?. (http://www.gwec.net/wpcontent/uploads/2013/09/The-Next-Wave-9-2013.pdf)
8. WWF (2011) The energy report: 100% renewable by 2050. (http://assets.panda.org/downloads/101223_energy_report_final_print_2.pdf)
9. Energy Watch Group Global Renewable Energy Scenarios (2008) Renewable energy outlook 2030. (Energy Watch Group Global Renewable Energy Scenarios. Renewable energy outlook 2030)

The Role of Biomass in a 100 % Renewable Energy World

Heinz Kopetz

1 Introduction

1.1 The First Step: More than 50 % Renewables by 2035!

Long run goals like 100 % RES in 2050 make forget that they only can be achieved if strong efforts in this direction start immediately. The intermediate target for the needed change of the global energy system by 2035 in line with the mitigation scenario of IPCC is as follows:

- More than 50 % share of renewables of the global final energy consumption
- a three-fold increase of biomass—biomass starting from a rather high level—as compared to 2010
- a 20-fold increase of wind and solar energy—starting from a rather low level—as compared to 2010
- a halving in the use of fossil fuels as compared to 2010 and
- a reduction in the energy consumption in the developed world!

The technologies and resources are available for this takeoff of renewable energies. What is needed are reliable political framework conditions that attract private capital investment for this energy revolution, and clear and simple rules to drive and guide a sustainable development of biomass. This transformation has to comprise the markets for electricity, heat, and transport. The WBA proposes targets for renewable electricity, heat, and transport until 2035 in line with the mitigation scenario of the IPCC.

H. Kopetz (✉)
WBA—World Bioenergy Association, Vienna, Austria
e-mail: hg.kopetz@netway.at

T.S. Uyar (ed.), *Towards 100% Renewable Energy*, Springer Proceedings in Energy, DOI 10.1007/978-3-319-45659-1_42

Table 1 Proposed electricity generation in TWh in 2035

Energy sources	TWh	
	2010	2035
Fossil sources	14,447	3200
Nuclear	2756	2800
Total RES[a]	4207	32,600
Total generation	21,408	38,600
Hereof renewable (%)	20	84

[a]Electricity from wind, solar geothermal, hydro and biomass

Table 2 Overview of heat energy production 2010, 2035

	2010, IEA	2035, WBA
Energy sources	Mtoe	Mtoe
Fossil sources	3320	1576
Bioenergy	1132	2684
Other RES	38	516
From geothermal	9	86
From solar thermal	29	430
Total renewable heat	1170	3200
Total heat	4490	4776
Hereof renewable (%)	26	67

1.2 Electricity

A rapid deployment of renewable electricity such as electricity from wind, solar, and other technologies with an annual growth rate of 8.5 % is proposed. Such a development in combination with a stable production of nuclear electricity would change completely the structure of the industry. In 2035, electricity demand will be higher than 2010, the share of renewable electricity above 80 % and the fossil fuel-based electricity reduced by more than 75 % (Table 1).

1.3 Heat

Paradoxically the heating sector is often overlooked in the discussion about energy strategies although for the developed world heating makes up about 50 % of the final energy. The main sources of renewable heat are biomass, geothermal energy, and solar energy. WBA proposes a strong push for renewable heat, especially based on biomass and solar thermal technologies (Table 2).

This rapid growth of use of biomass for heat requires specific strategies for different countries. In countries north of 45° latitude biomass in the form of wood chips, other wood residues and agricultural by-products such as straw should be increasingly

Table 3 Structure of energy use in the transport sector (Mtoe)

	2010	2035
Oil/others	2294	2293
Electricity	24	95
Biofuels	59	334
Total RES	83	429
Total	2377	2722
Share of RES (%)	2.7	16

used in district heating plants, plus the use of heat from biomass cogeneration plants. But also the service and manufacturing sector will become a market for this kind of biomass-sourced heat.

In addition a rapidly growing pellets industry will supply pellets for the residential, the manufacturing and service sector and for electricity generation. White pellets and in future also torrefied pellets have a high energy density and can therefore be traded worldwide at low cost as long as the main consumers are along the coast line. In developing countries with a high share of traditional biomass and charcoal the main focus will have to be to improve the efficiency and the technologies used and to drive the move to sustainable production.

1.4 Transport

In the year 2010 the share of renewables in the transport sector, including road, train, ship, and air transport, was low. Out of the total consumption of 2376 Mtoe (99.5 EJ) only 24 Mtoe (1 EJ) was covered by electricity and 60 Mtoe (2.5 EJ) by biofuels, mainly ethanol and biodiesel. As 80 % of electricity was not renewable in 2010 the total amount of renewables in the transport was only 64 Mtoe, corresponding to 2.7 % (Table 3).

But also in the year 2035 WBA does not expect a dominating role of renewables in the transport sector. WBA proposes an increase of 1st generation fuels, of advanced biofuels and of biomethane for transport, to 334 Mtoe in total. Electricity use will grow for trains and subways, and for shorter distance road traffic, but not for long distance road traffic, or for trucks. Biofuels use by ships and air traffic will increase in response to binding targets.

1.5 The Final Goal: 100 % Renewables

The first priority concerns the time period from 2010 to 2035. This is the time span in which the transformation to a new energy system has to be started and partly implemented in order to follow the mitigation pathway. Yet, the described energy concept for 2035 is only an intermediate target on the way to an energy system

Table 4 RA the transformation to 100 % RE

	2010	2035	Vision towards 2050	
	Mtoe	Mtoe	Mtoe	%
Fossil	10,327	5153	238	2.1
Nuclear	719	740	0	0
Hydro	295	593	715	6.3
Bioenergy	1277	3650	4055	35.4
Other RE	112	2504	6441	56.2
Total	12,730	12,640	11,449	100.0
Share of RE in %	13.3	53.4		97.1

based completely on renewables. Such an energy system can be reached in the period between 2035 and 2050. The following Table 4 presents a vision for the sustainable energy system of the future. It should be noted that in such a sustainable system a negligible quantity of fossil fuels might still be needed for specific purposes.

The main characteristics of the road beyond 2035 to a 100 % RE world will be:

- A reduction of the total primary energy demand due to energy savings and a higher efficiency of the total system.
- Phase out of nuclear energy.
- Phase out of fossil fuels except for a small quantity for transport and special uses in industry.
- Strong further growth of solar electricity, solar heat, and wind electricity from 2035 to 2050 supported by additional energy from geothermal resources.
- Small further growth of hydropower and of biomass between 2035 and 2050, biomass mainly for transport fuels
- Decentralized energy supply

2 Enabling Policy Measures

According to this scenario biomass would cover one third of the energy needed in a 100 % renewable world. Following a list of measures needed to push bioenergy accordingly:

- Phasing out all subsidies for Fossil and Nuclear Energy
- Phasing out new investments in Fossil and Nuclear Energy
- Introducing a carbon tax
- Providing financial support from public authorities to construct new district heating systems and install renewable heating systems (biomass, solar thermal) in private houses outside the reach of district heating grids.
- Keep Feed in Tariffs for electricity from biomass/biogas (produced in cogeneration units and CHP plants)

- Keep blending rules for biofuels
- Create incentives to increase the supply of biomass: to use fallow land and unneeded agricultural land to produce biomass, to collect by-products of agriculture, to better use organic waste and bio-wastes, to better manage the forests.

3 The Positive Impacts

The lion share of the energy supply in a 100 % RE world would come from solar and wind. Such a system would have dramatic advantages as compared to the present system.

3.1 Cheap Energy in the Future

One main difference concerns the fuel cost. Fossil energy and biomass have feedstock costs whereas wind, solar, hydro, and geothermal don't have feedstock costs but only operating and capital cost (Table 5).

The proposed transformation of the energy system would mean that the share of energies without costs for the feedstock increases from 3 % now to 25 % in 2035 more than 60 % in the future. This means energy could become really cheap.

3.2 Reduction of the CO_2 Emissions, No Nuclear Risks, No Wars for Energy Resources!

The biggest advantage of the new RE system lies in the avoidance of CO_2 emissions, the various risks associated with nuclear power plants, and of new wars for the dwindling fossil resources. 100 % RE is the way towards an environmental friendly, peaceful, and sustainable energy future.

3.3 New Employment

In addition the transformation to such a system would create millions of jobs as a big part of the total energy system has to be renewed.

Table 5 Changing share of energies without feedstock cost (%)

	2010	2035	2050
Energy carriers without feedstock cost	3	24	63

3.4 *Future Innovations Will Facilitate This Switch to 100 % RE*

In the next 20 years many innovations in the energy field will again take place—just as many innovations were observed to happen over the last 20 years. These innovations will facilitate the transformation to a 100 % RE energy system.

The main challenge until 2035 should be the transformation of the electricity and heating sector to renewables; renewable transport should not be the main concern as long as big quantities of fossil fuels are used for heat or electricity generation. The transformation of the transport sector should follow later, based on new experiences to be gained over the next two decades. It also can be expected that the demand for heating/cooling will go down in the longer run due to the better energy efficiencies of buildings—then a larger share of biomass would be available to be used for the generation of transport fuels after 2035.

4 Conclusion

WBA is convinced: the best answer to the new IPCC report on climate change is an accelerated transition to an energy system without reliance on fossil and nuclear energy resources. Biomass will have to cover around one third of the energy demand. Government policies and investment decisions should be directed towards this goal. A delay in the transformation of the energy system deepens the problem. It leads to additional global warming and would require even bigger adjustment efforts in the future. 100 % renewable energy is possible within less than 40 years and as a first step more than 50 % renewables until 2035 should be reached! The faster we take this road, the better for the future!

References

1. IEA (2012) World energy outlook. IEA, Paris
2. IPCC (2013) First part of 5th assessment report on "Climate Change 2013: The Physical Science Basis—Summary for Policy Makers". IPCC, Stockholm
3. World Bioenergy Association—internal papers (2013) World Bioenergy Association, Stockholm

History of Wind Energy and an Outlook for the Future

Klinger Friedrich and Müller Lukas

1 The Origins of Windmills

There are contradictory speculations about the historical origins of windmills. Some authors maintain that they have discovered the remains of stone windmills in Egypt, near Alexandria, with a supposed age of 3000 years. There is no convincing proof, however, that the Egyptians, Phoenicians, Greeks or Romans really knew windmills. The first reliable information about the existence of windmills from historical sources originates from the year 644 A.D. It tells of windmills from the Persian-Afghan border region of Seistan. A later description, including a sketch, dates back to the year 945 and depicts a windmill with a vertical axis of rotation. It was obviously used for milling grain. Similar, extremely primitive windmills have survived in Afghanistan up to the present time (Fig. 1). Some centuries later, the first news arrived in Europe that the Chinese were also using wind wheels for draining rice fields. Whether the Chinese knew windmills even before the Persians and whether the European mills might have been only an offshoot of the Chinese invention, can no longer be determined with certainty today. It is remarkable, however, that the Chinese wind wheels, too, were simple structures made of bamboo sticks and fabric sails and that they had a vertical axis of rotation (Fig. 2) [1].

Note: The following text and figures are a summary of a chapter of the book: 'Erich Hau: Wind Turbines - Fundamentals, Technologies, Applications, Economics - second edition'. Other sources are appropriately marked as footnotes.

K. Friedrich (✉) • M. Lukas
INNOWIND Forschungsgesellschaft mbH in cooperation with Fraunhofer IWES, Altenkesseler Str. 17/D2, 66115 Saarbrücken, Germany
e-mail: f.klinger@wind-energy-research.de

T.S. Uyar (ed.), *Towards 100% Renewable Energy*, Springer Proceedings in Energy, DOI 10.1007/978-3-319-45659-1_43

Fig. 1 Persian windmill [1]

Fig. 2 Chinese windwheel [1]

2 European Windmills

The windmill with a horizontal axis of rotation, which is the traditional windmill, was probably invented in Europe independently of the vertical-axis windwheels of the Orient. The first verifiable information has its origin in the year 1180 in the Duchy of Normandy. According to this source, the so-called post or trestle mill (Fig. 3) is supposed to have stood there. From this north-western corner of Europe, windmills quickly spread all over North and Eastern Europe as far as Finland and Russia. Numerous post windmills could be found in Germany in the thirteenth century. In addition to the post windmills, which are made entirely of wood, the so-called tower windmills (Fig. 4) make their appearance one or two centuries later. In this type of construction, the windwheel rests on a round tower made of stone. This type of mill mainly spread from the Southwest of France into the Mediterranean

Fig. 3 Post windmill

Fig. 4 Greek tower windmill

Fig. 5 Windmills in Holland, for draining polders [1]

region, which is why it is frequently referred to as the Mediterranean type of windmill. There is no reliable information as to whether the first post and tower windmills could already be yawed into the wind. However, yawing soon became a commonly found property of post windmills. The post windmill in its simple and serviceable form remained in existence right into the twentieth century. In Holland, several decisive improvements were made on windmills in the sixteenth century, leading to a new type of mill, the so-called 'Dutch windmill'. It is not known whether it was the post windmill or the tower windmill, of which some examples were also to be found in the north, which had served as the prototype. The fixed millhouse structure of the Dutch mill, where only the tower cap turned with the windwheel, permitted both the dimensions and the range of applications to be increased. Thus, the historical windmill reached its perfection towards the middle of the nineteenth century [1] (Fig. 5).

3 American Wind Turbines

In the early nineteenth century, when windmill technology was reaching its peak in Europe, numerous windmills were also built in the New World, mainly on the East Coast where the Dutch and British had their settlements. Simultaneously, the great movement to the West started in the USA. The settlers of the great plains of the Mid-West needed water, above all, when they wanted to settle down. In those places which did not have natural surface water, water had to be pumped up from wells. The large windmills were of little help for this purpose. They were too heavy to follow those pioneers rapidly enough. But in the land of unlimited possibilities, solutions were also found for this problem. In about 1850, the mechanic Daniel Halladay from Connecticut found the first solution (Fig. 6). His wind turbine already started

Fig. 6 Hallady wind turbine [1]

turning at low wind speeds, it turned comparatively slowly and developed a high torque at low speeds, exactly the right preconditions for driving a reciprocating water pump. The water pump was driven via a crank mechanism with a long vertical shaft which reached to the foot of the lattice mast. Despite his scepticism, Halladay started manufacturing wind turbines and soon sold large units to the American railroad companies. These had an increasing need for water pumps for re-filling their water tanks en route. With its many joints and bolts, Halladay's wind turbine was a comparatively complex machine. Although it was manufactured until 1929, it remained rather a rarity. The Reverend Leonhard R. Wheeler of Wisconsin found a simpler solution a few years later. Instead of dividing the windwheel into sectors, Wheeler mounted an additional wind vane which was positioned at right angles to the wind direction. This vane was used to turn the entire wind wheel out of the wind. The vane was connected to a weight, so that when the wind speed decreased, the wheel turned back into its original position. Wheeler's concept was manufactured under the name of 'Eclipse' (Fig. 7) and became the standard design of the American wind turbine. In the following years, wind turbines were manufactured in ever greater numbers by an ever increasing number of relatively small firms, especially the model developed by Wheeler, which was built in numerous variants. By 1930, more than six million American wind turbines had been manufactured. For the first

Fig. 7 Eclipse wind turbine [1]

time ever, the utilisation of wind energy was based on an industrially mass-produced article. A remarkable fact and it can hardly be considered a coincidence that it was in the USA where it happened for the first time [1].

4 Electrical Power from the Wind: The First Attempts

The first attempts to generate electric current with the help of wind power were being made in that period of time when the large cities were already being supplied with electricity, but complete coverage of users in rural areas was not yet feasible. The first systematic development aimed at utilising wind power for the generation of electricity took place in Denmark. Like no other, the name Poul La Cour marks the turning-point from historical windmill building to the modern technology of power generating wind turbines. This is the merit of perfecting traditional windmill technology on the basis of the scientific principles worked out by him, and he was a pioneer of electricity generation by means of wind power—all this in the nineteenth century. Professor La Cour built an experimental wind turbine (Fig. 8) driving a 'dynamo' in 1891. The remarkable fact is that he also at once tackled the problem of energy storage. He used the direct current generated by his wind turbine for electrolysis and stored the hydrogen gas thus

Fig. 8 Wind turbine by Poul La Cour [1]

produced. From 1885 to 1902, gas lamps using this method illuminated the school grounds in Askov. As far as the wind wheel was concerned, La Cour's electricity-generating wind turbine strongly followed the model of the traditional windmills. He was possibly the first to carry out tests in a wind tunnel, which he had built himself, and he set up a second, larger test station in 1897. By 1908, it had already built 72 electricity-generating wind turbines, modelled after the test station at Askov, which supplied power to rural settlements. This development was accelerated by the dramatic rise in fuel prices during World War I, so that by 1918, about 120 wind turbines were in operation. The La-Cour-Lykkegard turbines were built in various sizes with power outputs ranging from 10 to 35 kW. The rotor, with a diameter of up to 20 m, had four shutter sails, making it possible to remain below a certain rotational-speed limit. Yawing was carried out by two fantail type side wheels. The electrical generator was installed at the base of the latticed steel tower and was driven by the rotor via a long shaft and intermediary gearbox. Electricity was fed into the small isolated consumer grids via a buffer battery. The overall efficiency of the wind turbines was indicated to be about 22 %. At a good site, the annual energy yield amounted to about 50,000 kWh. The F. L. Smidth company, a manufacturer of machines for the production of cement whose entire export market had collapsed due to the events of the war, turned to building wind turbines. Using the name of 'Aeromotor' (Fig. 9) for their design, Smidth started by developing a wind turbine with a rotor diameter of 17.5 m and a power output of about 50 kW at a wind speed of approx. 11 m/s. Problems with the dynamic characteristics of the two-bladed rotor caused the company to develop a second, larger type with three rotor blades. With a rotor diameter of 24 m, it yielded a power output of about 70 kW at a wind speed of about 10 m/s [1].

Fig. 9 Aeromotor [1]

5 Ambitious Projects in Germany

In Germany, the decisive impulse came from the theoretical camp. Against a background of aircraft aerodynamics, the physicist Albert Betz, director of the Aerodynamische Versuchsanstalt (Aerodynamic Research Institute) in Göttingen, approached the problem of the wind rotor's physics and aerodynamics from a strictly scientific point of view. In an article published in the 'Zeitschrift für das Gesamte Turbinenwesen' (Journal of Turbine Science) in 1920, he proved that the maximum physically possible utilisation of the wind by a disk-shaped, turbine-like wind energy converter is restricted to 59.3 % of the power contained in the air current. In his book Wind Energy and Its Exploitation by Windmills', which was published in 1925, he summarised the results of his research and formulated a theoretical basis for the aerodynamic shaping of wind rotor blades, which has kept its validity to the present day [1].

Fig. 10 Wind power plant by Hermann Honnef [1]

One of the first to work with these new scientific findings was the steel construction engineer Hermann Honnef who developed concepts for absolutely gigantic wind power plants (Fig. 10). Colossal lattice towers were to have carried up to five wind rotors, each with a diameter of 160 m and a power output of 20,000 kW. In retrospect, it must be noted that Honnef's plans were indeed based on mathematical and engineering principles. Their realisation would, however, have most certainly caused considerably more problems than Honnef imagined in 1932. It was no longer the idea of supplying remote farms with electricity which inspired him, he wanted to build large 'wind power plants' which were to generate electricity in combination with conventional power plants at an economical price. In this respect, Honnef was a pioneer of the large wind turbines [1].

In 1937, the engineer Franz Kleinhenz published plans of a large wind turbine project. Unlike Honnef, Kleinhenz knew how to win the co-operation of renowned scientists and industrial firms. His plans took shape in co-operation with the Maschinenfabrik Augsburg-Nürnberg (MAN). Even today, the technical data of the MAN-Kleinhenz project convey an impression of advanced technology: 130 m rotor diameter, three or four rotor blades, 10,000 kW rated power, tip-speed ratio of 5, rotor positioned down-wind, 250 m hub height, directly driven generator with a diameter of 28.5 m or several generators via a mechanical transmission, tower as guyed tubular steel tower, upper part yawing with the rotors. By 1942 the project was ready for implementation, but the war prevented its actual construction [1].

6 The First Large Wind Turbine in the USA

As the electricity supply of the rural areas no longer presented a general problem, plans were made in the USA to deploy large wind turbines within the public utility grid, interconnected to conventional power plants. The American engineer Palmer Cosslett Putnam must be given the credit of being the first to implement these plans. In 1940, he approached the S. Morgan Smith Company, a water turbine manufacturer in York (Pennsylvania), with his concept and some ideas of the technical design of a large electricity-generating wind turbine. Morgan Smith entered into a contract with the Central Vermont Public Service Company concerning the erection of a wind turbine based on Putnam's plans. Palmer C. Putnam won over renowned scientists and technicians of the Massachusetts Institute of Technology (MIT) to co-operate in this project. Among others, Theodore von Kármán was responsible for the aerodynamic design of the rotor. In October 1941, the turbine was installed on Grandpa's Knob, a hill in the state of Vermont. It was the world's first really large wind turbine (Fig. 11) as proven by the technical data:

- 53.3 m rotor diameter
- 1250 kW rated power
- 35.6 m tower height.

The Smith-Putnam wind turbine operated for about 4 years and fed electricity into the utility grid of the Central Vermont Public Service Company for about one thousand operating hours until March 26 in 1945, when a rotor-blade fracture interrupted its operation [1].

7 Wind Turbines in the 1950s

After World War II, the prices of the primary fuels coal and oil dropped again and a period of extremely cheap oil imports began. The availability of fuels for the generation of electricity was no problem at all. The subject of environmental protection had not yet been thought of and if so, not in connection with the production of electricity. Nevertheless, attempts at generating electrical power by means of wind turbines continued in the 1950s in some places, after the shortages of the first post-war years had been mostly overcome [1].

Fig. 11 Smith Putnam wind turbine [1]

Naturally, the Danes were also represented with experimental wind turbines in the 1950s. Basing his concept on the technical model of the Aeromotors, J. Juul built a 200 kW wind turbine with a rotor diameter of 24 m in Gedser in 1957. The Gedser wind turbine (Fig. 12) operated from 1957 to 1966, but then shared the fate of all other wind turbines of this period and was decommissioned. Remarkably, or possibly prudently, it was not disassembled. It was thus the only historical wind turbine which survived to see the renaissance of wind power technology after 1975. Due to an agreement between America's NASA and the Danish authorities, the Gedser wind turbine was recommissioned in 1977 and served as an experimental turbine for several years. The results obtained here, together with the technical documentation from the Hütter W-34 wind turbine (Fig. 13), formed the starting point for NASA's research work in the field of wind power technology from 1975 onward. In the Federal Republic of Germany, the 'Studiengesellschaft Windkraft e.V.' (Society for the Study of Wind Power) was founded in 1949. Ulrich Hütter, who had already distinguished himself with his papers on the theory of wind turbines in 1942, had a leading role in this. On behalf of the Allgaier Werkzeugbau GmbH in Uhingen, Germany, Hütter initially designed a small wind turbine with a 10 m rotor diameter and

Fig. 12 Gedser turbine [1]

8–10 kW rated power. About 90 units of this wind turbine were built and proved to be quite satisfactory. In 1958, Hütter then started to develop a larger wind turbine, the W-34, which was to have a rotor diameter of 34 m and a rated power of 100 kW. In 1958, the wind turbine was erected in Stötten (now Schnittlingen, near Stuttgart) in the Swabian Alb, Germany. The technical concept of Hütter's W-34 has been influencing wind turbine design in numerous features right to the present day. It was particularly the designers of the large experimental wind turbines which marked the first phase of modern-day wind energy technology after 1975, who followed Hütter's ideas, some of them borrowing directly from the technical example of the W-34. Moreover, he gave high priority to a lightweight construction of the wind turbine, ideas which considerably influenced the design of the later German wind turbines after 1980 [1].

Fig. 13 W-34 by Hütter [1]

8 The Large Experimental Turbines of the 1980s

In the 1980s, the state-subsidised and state-initiated programs for developing the wind energy technology were primarily orientated towards the construction of large experimental turbines. Apart from political motives, the opinion prevailing initially that the large utilities should be the potential buyers of these turbines were a decisive argument for concentrating on the development of wind turbines in the Megawatt power range, a development which came much too early from today's point of view. The large experimental turbines were built almost exclusively by large and well-known industrial companies since only these were able to develop and build projects of this magnitude from a standing start, as it were. The names read like a 'Who's Who?' guide through industry: Boeing, General Electric and Westinghouse in the USA, MAN, MBB, Dornier, Voith in Germany or Kvaerner in Sweden. The development began in the USA. From 1975 to 1987, a series of large experimental turbines designated MOD-0 to MOD-5 were erected and tested (and a number of duplicates were built of some of these, e.g., of MOD-0 and MOD-2). The final and largest project, the MOD-5A designed by General Electric, no longer

Fig. 14 MOD-5 (B) [1]

reached completion. The turbine was intended to have a rotor diameter of 122 m and a rated power of 7300 kW. The so-called aileron-controlled two-bladed rotor was provided as a special feature. The project was cancelled in 1993 in favour of the MOD-5 (B) since this design was largely based on the preceding MOD-2 turbines and could be implemented more rapidly and inexpensively with the subsidies by the DOE which were still available. After the MOD-5 (B) (Fig. 14) had been tried out in the Hawaiian Islands, the state-subsidised development for the large experimental turbines came to a halt in the USA. In Denmark, a start was made by a private initiative. In 1975, the 'Tvind Turbine' was erected by a syndicate at an adult education school in Ulfborg. The turbine was built with much enthusiasm and idealism but constructed rather amateurishly in some respects. After that, the Danish utilities built the experimental systems Nibe A and Nibe B. In Germany, the GROWIAN (Fig. 15) project formed the focus of the program.

Looking for an alternate energy source during the oil shortages of the 1970s, German engineers constructed the experimental GROWIAN turbine to generate the maximum amount of electric power with the biggest turbine possible. With blades nearly 50 m long, it was the world's largest wind turbine at the time of its construction. The massive turbine proved difficult to maintain and is no longer in use. Through this experience, engineers learned that medium-sized turbines could harness wind energy at a more cost-effective rate than large turbines.[1]

[1] http://www.getdiyinfo.com/wind-turbine-size-2787981.

Fig. 15 GROWIAN [1]

The GROWIAN Turbine was built by MAN (Nacelle, etc.) and MBB (Blades) North West of Hamburg (Cost: $55 million). The total operation time was 420 h. The Turbine was demolished in 1987.[2]

In addition, however, innovative designs such as the Voith WEC-520 or several systems of the single-bladed design 'Monopteros' were constructed. Some years later—as a second beginning as it were—the Aeolus II turbine followed in co-operation with Sweden and the WKI-60 on the island of Heligoland [1].

9 Nineties till Today

In the 1990s, as aesthetics and durability became more important, turbines were placed atop tubular steel or reinforced concrete towers. Small generators are connected to the tower on the ground and then the tower is raised into position.

[2] http://www.windsofchange.dk/WOC-eurstat.php.

Larger generators are hoisted into position atop the tower and there is a ladder or staircase inside the tower to allow technicians to reach and maintain the generator, while protected from the weather.[3]

European manufacturers like Tacke, Micon, Vestas and Enercon have commercialised turbines with more conventional rotors, but featuring such important innovations as low speed generators and complete variable speed systems incorporating advanced power electronics.[4]

In 2001, with European wind turbine power ratings pushing 2 MW, Denmark's Riso Laboratories touting its new wind turbine airfoil designs (modelled closely after pioneering activities in the USA). The last remaining major area of controversy is the issue of two versus three blades for large wind turbines. Time will tell if one design will win out or if both will be able to exist in specific applications (see footnote 4).

As the twenty-first century began, fossil fuel was still relatively cheap, but rising concerns over energy security, global warming and eventual fossil fuel depletion led to an expansion of interest in all available forms of renewable energy. The fledgling commercial wind power industry began expanding at a robust growth rate of about 30 % per year, driven by the ready availability of large wind resources, and falling costs due to improved technology and wind farm management.

Offshore wind power began to expand beyond fixed-bottom, shallow-water turbines beginning late in the first decade of the 2000s. Europe is the world leader in offshore wind power, with the first offshore wind farm being installed in Denmark in 1991. As of 2010 Siemens and Vestas were turbine suppliers for 90 % of offshore wind power, while Dong Energy, Vattenfall and E.on were the leading offshore operators. As of October 2010, 3.16 GW of offshore wind power capacity was operational, mainly in Northern Europe. According to BTM Consult, more than 16 GW of additional capacity will be installed before the end of 2014 and the United Kingdom and Germany will become the two leading markets. Offshore wind power capacity is expected to reach a total of 75 GW worldwide by 2020, with significant contributions from China and the USA. Offshore turbines require different types of bases for stability, according to the depth of water. To date a number of different solutions exist[5, 6]: (Fig. 16)

10 Outlook

After the slump on the international wind market, further cost reductions and technological optimisation can be expected as wind energy develops. The number of countries in which the technology can already compete with energy generated from

[3] http://www.electricityforum.com/windmills-for-electricity.html.

[4] http://telosnet.com/wind/future.html.

[5] http://www.energyunion.eu/de/blog/offshore_wind.

[6] Source: http://www.offshore-stiftung.com/60005/Uploaded/DGehring|Fundamenteall.jpg.

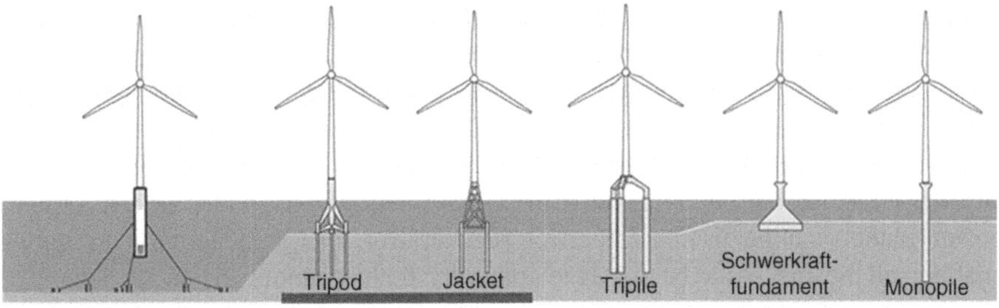

Fig. 16 Comparison of tower concepts for offshore wind turbines

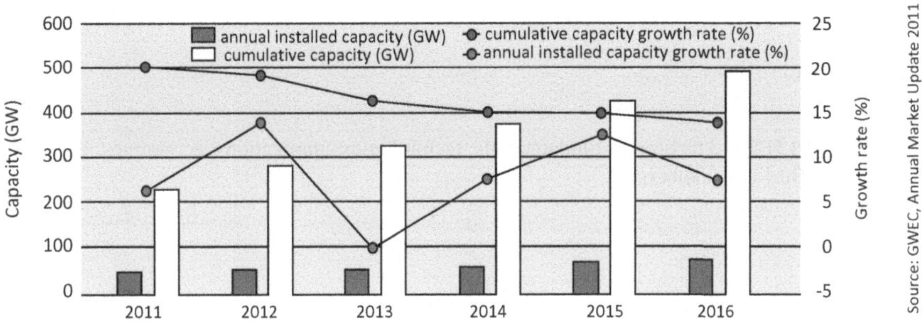

Fig. 17 GWEC, Annual Market Update 2011 worldwide market forecast 2012–2016

fossil fuels is increasing around the world. Many observers expect that half of all the installations worldwide will be installed in emerging markets such as Brazil, China, India, Mexico, Morocco, South Africa and Turkey by 2030. It is also notable that national and multilateral development banks are increasingly diverting investments in wind energy to these emerging countries.

Further international expansion of wind energy over the years to come will depend on the regulatory framework in energy policy and urban planning, to name but two areas. Essential prerequisites include designation of suitable areas for onshore and offshore wind farms, the abolition of restrictive height limits, expansion of the grid infrastructure, funding for storage technologies and the creation of incentives for repowering in order to utilise high-yield sites even more efficiently. If wind energy is to live up to its full potential in decades to come, new sources of finance must also be developed which can be used, to some extent, to offset the ongoing reluctance to lend in the OECD. Offshore wind farms will become increasingly important over the next few years. The German government is expecting installed offshore output in Germany to reach up to 25 GW by 2030, which would mean that marine wind farms in Germany would cover up to 15 % of electricity demand in the longer term (Fig. 17).

The increasing transnational trade in electricity across Europe, the shift of the energy generation focus away from conventional power plants and the expansion of renewable energies, in particular of wind energy, have made it necessary to modify the power grid infrastructure, with a focus on optimising the existing network and making it more flexible. Measures to expand the power grid and improve its utilisation, through temperature monitoring, for example, are currently in preparation in Germany. The use of new storage technologies including compressed air storage, better load management in the domestic and industrial sectors and the networking of decentralised power generation into what are referred to as virtual power plants all offer considerable potential for the optimal integration of wind energy. Virtual power plants can be used to connect regenerative energy generation systems, enabling all the turbines to be managed optimally, both economically and technically. Moreover, further research and development in wind energy will concentrate on reducing the negative environmental impact (noise and light emissions).[7]

Reference

1. Hau E (2013) Wind turbines—fundamentals, technologies, applications, economics, 2nd edn. Springer, Berlin Heidelberg

[7] Federal Ministry of Economics and Technology in Germany: http://www.renewables-made-in-germany.com/en/renewables-made-in-germany-start/wind-energy/wind-energy/outlook.html.

Electric Power System Transition and the "Polluter Pays Principle"

Aviel Verbruggen

Abbreviations

DNO	Distribution Network Operator
IGOP	Independent Generator of Own Power
RE	Renewable Electricity
TSO	Transmission System Operator

1 Introduction

The inherited fossil fuel, nuclear and large hydro based power systems are more and more evaluated as non-sustainable.[1] By 2050 their transition to 100 % renewable electricity (RE) supplies is considered possible, desirable, and necessary for preserving climate stability [3]. Power systems are multi-leveled and their bottom-up development in the industrialized world to continentally integrated systems, with very large generation stations as central nodes, took about a century. Except large hydro dam plants, RE generation units are mostly medium to small scale and distributed at locations where natural resources are available [4]. The transition to 100 % RE is not free from tensions between more centralized RE deployments (e.g., off-shore wind parks) and priority to numerous distributed independent generators of own power

[1]This contribution is based on a lecture presented at IRENEC-2012 in Maltepe-Istanbul [1], reworked and updated in 2013 (this text), and further complemented in cooperation with 11 colleagues in 2014–2015 [2]. The reader may observe that the study and discussion on the topics covered are far from finished.

A. Verbruggen (✉)
University of Antwerp, Prinsstraat 13, 2000 Antwerp, Belgium
e-mail: aviel.verbruggen@uantwerpen.be

T.S. Uyar (ed.), *Towards 100% Renewable Energy*, Springer Proceedings in Energy, DOI 10.1007/978-3-319-45659-1_44

(IGOP).[2] The former start from top-down centralized operation (successful in the established systems); the latter want to anchor and grow the future 100 % RE systems around distributed generation, with a central role for household PV and micro-generation [6], nested in smart grids [8]. Both approaches must obey physical, economic, and social laws to supply power in an effective, efficient and fair way, within continentally wide networks. Shifting back the gravity point of power systems from their tops to the floors where end-users prevail is unavoidable when sustainable power systems are intended. But in an interconnected system the frequency is regulated at a single value (50 Hz in Europe) for the entire area, and every participant has to synchronize, that is a modern power system is always multi-leveled.

The future growth of renewable electricity (RE) supplies will ever more disrupt inherited and incumbent power generation, transmission, and distribution systems. Disruption increases with RE variability, randomness, distance from load centers, and constraints on flexible dispatching by system operators. Dispatching constraints are related to physical conditions, technical factors, and to economic-institutional attributes like ownership, elasticity of demand, and regulatory rules or conventions. Because of their independent autonomy and by swapping bi-directional power exchanges, IGOP are most disrupting incumbent systems. Scholars seem to assign merit to active participation by residential end-users in addressing intricate electricity system balancing issues [7, 9]. The societal benefit/cost ratio of such active participation is low, and should be taken into account when mechanisms for supporting IGOP are proposed.

The technical aspects of system stability (frequency), balancing demand and supply (load management and load following), and adequacy (sufficient capacity to reliably meet (peak) loads in the future) received most attention (e.g., [10–12]). Added is work on the financial trade-offs behind the investments and operations by the German federal ministry BMU as architect of the German RE support system, and by academics, e.g., Schaber et al. [13] and Gawel and Purkus [14]. Costs can be shifted around [11] what makes accurate assessments precarious. Most literature on the integration of renewable electricity in power systems focuses on generation and transmission (e.g., [11, 13]). The role of distribution networks and of future smart grids changes with the growth of distributed generation [5, 7, 9, 15, 16].

This paper adds to the burgeoning literature in two ways. First, it provides more clarity on some important gaps between economic value and price. Second, it develops an unfamiliar, yet comprehensive perspective on footing the bill of the transition to 100 % renewable electricity.

The flow of arguments is as follows. In Sect. 2, important gaps between economic value and price in the energy systems transition debate are addressed for clarifying the boundaries on public energy strategies and therefore also on practical regulation. This comprehensive helicopter perspective frames the substitution of RE, in particular renewable IGOP, for existing power systems. The proper reference point for fixing the analysis and policy is no longer the prevailing power systems and practices, but the future, fully RE based, electricity generation and supply systems.

[2] IGOP as general and neutral term [5] is preferred above e.g. "prosumers" [6] or "co-providers" [7]. The adjective independent is added to distinguish from joint ventures between incumbent power companies and industries that house on site the shared (often cogeneration) power plant.

The transition from present systems to the due future electricity systems is a long-term and expensive undertaking. With the end-state as the valid reference point, and applying the widely adopted "polluter pays principle," the agents accountable for the inherited and present non-sustainable power systems are also accountable for the transition expenditures. The main components and relationships of today's liberalized electric power systems in Europe are described in Sect. 3. Next to an overview of the bulk and retail activities, the section situates the Independent Generators of Own Power at a different place from the other generators. Other than available power generation dichotomies, Sect. 4 identifies two classes of power suppliers: "Commanded Generation Plants," built to serve power loads of customers, versus plants operated by "Independent Generators of Own Power." Both classes include renewable electricity providers. For being successful the 100 % transition to RE will rest on the unfettered growth of IGOP. What makes IGOP special is their autonomy and the bi-directional power exchanges with the central power systems. In Sect. 5 is shown that IGOP will not or poorly develop when a short-run market pricing scope is imposed, what would reflect the observed value-price gaps of Sect. 2. On major aspects making electricity an attractive energy for end-users, IGOP do not deliver simply. Policy implications of the proposed spread of liabilities over actors during the power sector transition are overviewed in Sect. 6. The concluding Sect. 7 discusses whether politics should give priority to present market conditions and rules, or should opt 100 % for the transition to a 100 % RE sector.

2 Gaps Between Economic[3] Value and Price

At the 2009 Copenhagen conference, political leaders of the major world economies joined the EU in targeting +2 °C as maximum global ambient temperature increase. Since then, global politics has added one more schism to the sustainable development one, agreed upon at the 1992 Rio conference. The intentions and declarations are fine, but their conversion in reality requires unseen changes [18], while deep change actions and practices are not evidenced. Several disciplines study the rifts, and provide elements of explanation (e.g., [19]).

In energy policy terms, the +2 °C limit means the full transition of fossil fuel based economies to economies thriving on renewable energy sources [3]. The economics of such transition are challenging, facing two major economic value-price gaps (Fig. 1). First, the direct use value of fossil fuels is huge because they are versatile, dense in energy content, easily stored and transported, etc. [20]. Their negative impacts on human health, environment, and in particular greenhouse gas concentration in the atmosphere, are unpaid externalities. Fossil fuels are sold at low prices because only winning and processing expenses are accounted for.

[3]The focus is on economic value, i.e. on direct use value and on option value for preserving future direct use values. Not considered are other components of "total economic value" [17], for example serendipity, existence, and bequest values.

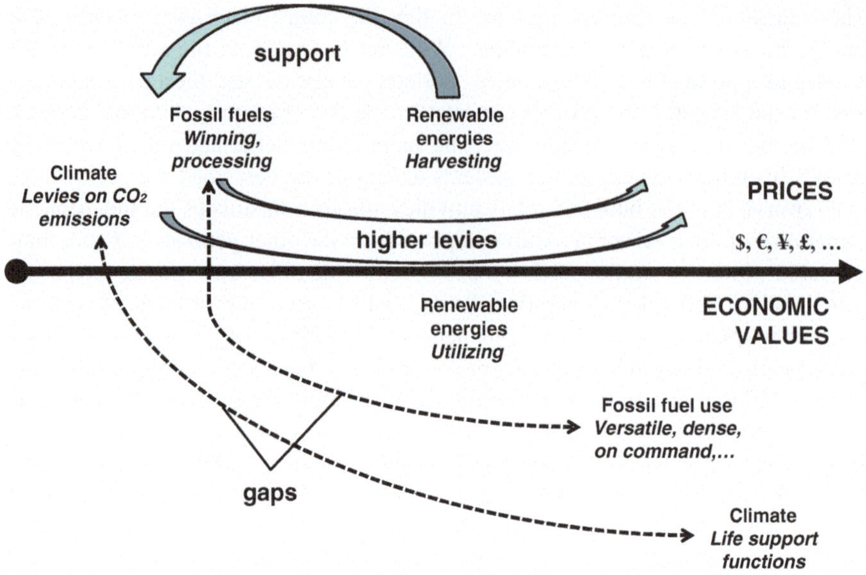

Fig. 1 Economic value-price gaps for climate, fossil fuels, and renewable energies

The wide gap between direct use value and low prices[4] explains their irresistible use, fuelling the impressive economic growth in some parts of the world since World War II. Second, respecting the +2 °C limit implies most of the fossil fuels resources will be left in their natural deposits. Foregoing the huge fossil fuel bequest for preserving sufficient climate stability means the risks of wrecking the climate are evaluated higher by global politics than utilizing the fossil fuel bequest. The gap between the economic value of the natural climate as a major life support system and the (almost zero) price of global greenhouse gas emissions is even larger than in the fossil fuel case. As a corollary, the global CO_2 emissions continue to grow, increasing the risks of irreversible disruption of the climate. Figure 1 indicates that the economic value-price gap is small for renewable energies: the distance between harvesting and utilizing renewable energies is mostly short, and even reversals in the gap (i.e., harvesting price higher than utilizing value) may occur under local and short-term conditions. For meeting modern energy services of exigent consumers, temporal and spatial spot dependent renewable energies are of lower value than fossil fuels available on command.

Bridging the value-price gaps is feasible by increasing the prices of fossil fuel use and by setting increasingly higher prices on emitting greenhouse gases, with both pricing policies concurring. Most fossil fuel bequests are national states' property or privatized. Handled as private goods, huge rents and profits for the owners are generated, especially when prices exceed several times the expenses of supplies. Continuous high (and mounting) fossil fuel prices trigger alternative energies substitutions. The substitutions would multiply when the bulk of fossil fuel rents would be appropriated by

[4]Fossil fuels moreover are subsidized. The International Energy Agency (World Energy Outlook) estimates US$406 bn over 2010, and US$523 bn over 2011.

end-user countries through levies with spending revenues on R&D and subsidies for renewable energies (Fig. 1). The global climate is a pure public good. The revenues from pricing emissions belong to the treasuries of public authorities, and are welcomed for covering mitigation and adaptation expenses. Presumably the most effective mitigation strategy is the transition of today mainly fossil fuel based electric power systems to very low-carbon ones, practically spoken: 100 % renewable resources.

The climate and fossil fuel value-price gaps overarch and direct many costs-expenses and benefits-revenues gaps, thoroughly affecting the transition processes. Economic theory defines costs as forgone opportunities of all—private and public—factor use. Economic agents' decisions are based on expenses as the product of prices applied or accounted for factor use. The transformers from full costs to actual expenses are opaque, incomplete, and biased by the addition of rents [21]. In a similar way revenues received by economic agents may be very different from obtained—private and public—benefits. Schaber et al. [13] "quantify changes in power producer revenue due to variable renewable energy generation" as performance indicator, not clearly comparable with the more standard public economics indicator "overall welfare gain." Mingling concepts lead to confusing vocabulary and statements in the transition endeavor. Confusion mostly strengthens the position of the established or incumbent state of affairs. Especially economists consider "what is" as the relevant reference point for assessing costs and benefits of propositions or measures deviating from that point. A reference position is implicitly assumed as the best. In case of climate change and non-sustainable energy supply systems, the "status quo" reference is evidently a very inferior situation. A society that fully thrives on renewable energy is a better reference than the "status quo", although its development is work in progress and its final shape is ex ante more difficult to describe.

Identification of the proper reference point on the transition track is highly important for the assignment of tasks and allocation of the incurred expenses. First, the superior endpoint of the transition track rightly claims to be vested as reference. Explicitly describing the future state a society should reach also is a major component of "back-casting" analysis [22]. Second, reversals in thinking, analysis, vocabulary, evaluations, etc. are necessary when shifting the reference position from the present to the future. Renewable energies cannot be labeled disturbing or disruptive any longer, but every progress in their development and deployment is a step towards the goal to reach. One also will thoughtfully and differently "integrate" RE in current power systems, when keeping clearly in mind the task of turning current systems in 100 % RE based power systems in a proximate future. Third, but not least, loading the expenses of the transition on the solutions and change agents of the superior future is not really helpful to advance that future; remains the bill is better footed to the lagging incumbents. This implies, for instance, that the costs of integrating RE supplies in existing central power systems and the expenditures for adapting the systems fall largely or entirely on incumbent power sector interests. This approach opposes claims for charging costs of disturbing incumbent production and transmission systems on RE supplies when the latter make inroads on established power systems.

The "polluter pays" principle (PPP) legitimates the imposition of obligations on incumbent power companies to pay for the costs of transitioning from existing high-carbon and high-risks systems, inherited from the fossil and nuclear era, to future RE systems. In 1972 the OECD agreed that polluters should pay the costs of abating the own environmental pollution, for example by installation of filters, sanitation plants and other add-on techniques. This narrow interpretation of polluter pays intended to avoid that governments would (continue to) subsidize polluting industries for building treatment plants, scrubbers, waste incinerators, etc. Rather "the polluter should bear the expenses of carrying out the above-mentioned measures decided by public authorities to ensure that the environment is in an acceptable state" [23]. The PPP extends the responsibility of polluters when, additionally to abatement expenses, they pay for the damages their residual pollution is causing or may be causing. Eventually the PPP may also scope the impacts of historical pollution, although "allocating responsibility raises a series of practical and ethical questions. The attribution of "blame" should arguably depend on some knowledge that harm is being caused" [24].

Another extension is the "precautionary polluter pays principle" where potential polluters are mandated to take insurance or preventive measures for pollution that may occur in the future. For example, requiring full-indemnity insurance for the harm and costs that any specific power plant may occasion would increase the price of polluting fossil fuel plants. Arguably, the requirement will preclude the construction and operation of nuclear power plants, because the global re-insurance sector rejects underwriting nuclear accident risks [20].

Polluter pays is also known as "extended producer responsibility." This makes actors responsible for the effects and impacts of their actions, not necessarily limited to directly objected and observed harm to living people whom property rights exclusivity is trespassed [25]. Applying extended responsibility to incumbent electric power interests equals charging them with objective liability, making one accountable for causing societal harm without establishing guilt. It signifies the application of payment for actions in the past where the power sector planners and investors being held accountable did not or could not have foreseen the consequences. The practical application of objective liability is utterly difficult and contentious, and every measure "needs to somehow reflect degrees of responsibility for the causes of the problem" [24].

In focusing on the main principles, this article is not attempting specific solutions for implementing the extended producer responsibility. When discussing policy implications in Sect. 6, some specific issues of footing the transition bill in Flanders are examined.

3 Extended Participants in EU's Liberalized Electric Power Supply Systems

Liberalization of electric power systems started during the 1980s [26] and ever since affected national power systems on a global scale [27]. In February 1997 the EU published a directive on the internal electricity market, but its design and implementation delivered a variety of mixed market structures all over Europe [28]. Figure 2

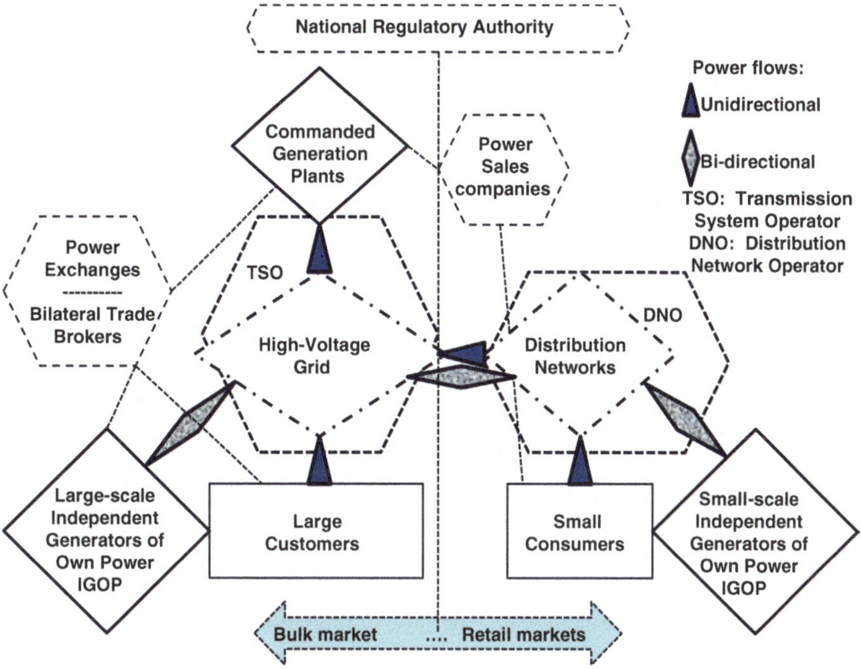

Fig. 2 Components and relationships in liberalized electric power supply systems in Europe

shows the main components of, and participants in, present-day electric power supply systems in Europe, and some of their relationships. The left side of Fig. 2 represents the bulk electricity market; the right side the retail markets within a given geographical area.

Before liberalization, electricity supply was a fully vertical integrated industrial activity covering generation, transmission, and distribution. The latter main economic activities were organized as separate entities within the vertical column, and guided by the central intelligence of a multi-leveled system operation unit. Investment and operational decisions were optimized with the help of scientifically based algorithms. Central supervision provided stability and neat balancing of supply of and demand for the non-storable power flows. Internal system operators continuously maintained system equilibrium. They were entitled with full command (unit commitment, dispatching) over the generation capacities, and covered a national or subnational territory. Exchanges with adjacent control areas were limited, with some arbitraging of peak loads and last resort back-up power supplies by colleagues [3].

Electricity sector liberalization intended to substitute free market principles for vertically integrated supply structures. However, realizing workable competition in such tightly managed systems was contingent on a logical sequence of prerequisites, viz. proper harmonization of rules and conditions for participants in the to become "competitive" markets, transparency of the institutions and activities, unbundling of the main functions (generation, transmission, distribution), and firm guidance and supervision by excelling independent regulators [5]. The three EU regulatory packages (1997, 2003, 2009) could not fully impose the prerequisites on the member states; one had to continue with very different institutional and political contexts. The three packages could not yet iron the inherited uneven systems to a single internal electricity market, and competition remains partial and incomplete with influential roles by remaining state companies (e.g., EDF, Vattenfall AB).

Figure 2 shows an unbundled structure of power generation activities, the high-voltage grid transmitting power to bulk demand nodes, and distribution companies operating low-voltage networks to serve the retail demands. Liberalization forced unbundling of the organizational entities that are processing physical power flows, and added several new entities, such as power exchanges, bilateral trade brokers, power sales companies (also called: suppliers), embroiled as intermediaries in contracting power transactions. The new institutions function on legal and financial terms, not intensely interfering in physical electricity flows; they are shown as hexagons with dashed borders in Fig. 2. At the top of Fig. 2 is shown the national regulatory authority, supervising the electricity sector. System operators can function independently of any physical power supply activity. In Europe they mostly are merged with grid owners and operators, and named transmission system operators (TSO) that also assume responsibility for overall balancing power generation with loads. In large areas, TSO decentralize to subdivisions and to distribution network operators (DNO). System operation is growing more challenging, because of limitations on the authority and on the flexibility in unit commitment and dispatching of many low-carbon power plants (nuclear stations, flow renewable energy sources), exponential growth in number of new power producers [6], and more technical or institutional constraints on operating individual plants. Technical constraints are, for example, ramping rates in loading and de-loading generation units. Institutional constraints are related to ownership, legal or contractual privileges (e.g., "must run" or interruptible), reliability priority for supplying particular end-users (e.g., hospitals), and similar factors.

4 Two Main Classes of Power Generators

Power generators are classified according to specific purposes, with terms and definitions often unsettled, for example: central versus distributed; independent versus incumbent; (variable) renewable energies versus (on command) fueled plants; small-scale versus large-scale. Figure 2 identifies two main classes. At the top of the figure are mentioned "Commanded Generation Plants," permitting full institutional dispatching of their capacity on contract with the TSO, i.e. delivering power when requested or withholding generation when demand for power is low. It encompasses the production facilities of previously vertically integrated incumbent power companies, mostly consisting of several stations and units of a wide variety and range of capacities, including (very) large-scale plants. This class also includes independent power producers that exclusively generate power for selling to customers through the integrated power system. Liberalization and unbundling are anyhow blurring the differences between incumbent and independent ordered generation. Both may deploy conventional nuclear or fossil fuel technologies, combined heat and power,

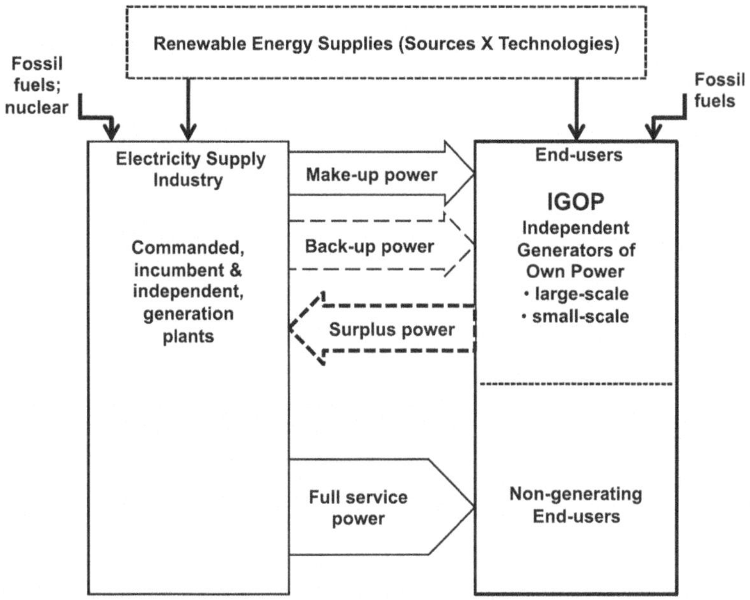

Fig. 3 Two main classes of power generators: standard generation plants versus independent generators of own power (IGOP)

or renewable supplies (now mostly biomass, hydro, wind; in the future, presumably more concentrated solar power). Generally, the independent plants are more small-scale and distributed than the plants of incumbent power companies.

The other class of power generators (bottom of Fig. 2) consists of—large and small—"Independent Generators of Own Power (IGOP)." They are often named "on site" generation because they are placed at the premises of large customers (industrial plants, commercial sites) or of households and small businesses (PV at building roofs, small-scale cogeneration). IGOP use fossil fuels (often cogeneration or combined heat and power units) or renewable sources and technologies. They build and run power plants to serve primarily the own loads but in interaction with the—high-voltage or low-voltage—power grids. Grid connection is preferable for attaining the best reliability/cost ratio, due to the non-storable character of electric power.

Commanded (incumbent or independent; central or distributed) generation plants are single-directionally linked to the power system: they only deliver power. IGOP (large-scale and small-scale) are bi-directionally linked (Figs. 2 and 3). IGOP mostly switch roles from (net) supplier to (net) consumer of electricity, forth and back. This aspect created the name "prosumer." When technically feasible and financially opportune IGOP first serve the own loads and eventually send surplus power to the grid. When the own loads exceed the power output of the IGOP plants, electricity is imported from the grid as "make-up" or as "back-up." The distinction between the latter flows is important when electricity tariffs include a high fixed term (price per monthly kW-peak) argued as coverage of high investment outlays in base-load plants. Power use by energy intensive industrial sites is generally labeled as base-load power, with electricity tariffs including high payment for the monthly requested (quarter-hour peak) capacity. It is not grounded to apply this tariff on a demand spike for back-up purposes of short duration [29].

Surplus power from IGOP delivered to the grid is disturbing power system balancing when overall low load is already challenging TSO in keeping commanded, inflexible (often large-scale), generation capacities on line. The short-run price of the kWh may then fall to zero, or become negative when payment is needed to purge superfluous power. Structurally, the system is too heavily locked in large-scale fossil and nuclear production plants resisting to be reduced in output. Also the system is short in buffering facilities where power can be converted in storable energy, which can be reconverted back to electricity. Technically, spiky fluctuations are wearing and tearing electricity supply equipment. Financially, longer periods of running below full-load capacity erode the bottom line accounts of generation plants.

Assigning a central role to IGOP will increase the size of surplus power deliveries to the grid. When, for example, cogeneration IGOP is dimensioned on heat demand of some industries, significant electric power capacity may be installed when the terms for delivery to the grid of surplus power are financially guaranteed. Also private persons and small companies with large, well-oriented roof surface may yearly generate double or more PV electricity than the own activities absorb.

A completed transition to 100 % renewable based power systems may find IGOP as the most common and predominant type of power supplies. Growing importance of IGOP is inevitable, and regulation can play a stimulating role not a choking one. Germany is applying premium schemes to safeguard positive stimulation. Gawel and Purkus [14] provide an evaluation of the schemes, and confirm that "limited possibilities of wind and PV installations to react to short-term price signals impose fundamental constraints on the (premium) instrument's ability to improve their system integration," with as logical next step that "systemic concepts are required, which draw on all components of the energy system (…)." They express "doubts whether the current electricity market design is suitable at all for integrating large scales of RES." This quote and several other positions in their analysis reveal that they look at the transition challenges from the present, incumbent perspective, while concluding another view is necessary. The latter finding is argued in this article.

Geelen et al. [7] sympathize an active role as co-providers for end-users connected to smart grids, but observe "little is known yet on how to shape active participation of residential end-users in smart grids and thus how to support them in achieving the role of co-provider." They join Verbong et al. [9] in observing that technology and financial incentives dominate the discussions and the development. They consider "the focus on technology and the protection of vested interests" as main threat to smart grids. Nykamp et al. [16] investigate various regulation designs on effectively stimulating DNO investment in innovative smart grid solutions (local storage, voltage regulation).

If one adopts the future 100 % RE system as reference position (as argued in Sect. 2), we propose to deploy a different approach by regulators, as follows: DNO are hold liable for all cases where small scale IGOP investments or activity are curtailed; TSO assume similar liability for large scale IGOP. Penalties paid by DNO and TSO (and charged on electricity consumers) for such shortcomings are redirected to fund innovative smart grid investments, storage facilities, dedicated fast ramping, decentralized generation units, optimization of IGOP activities and their power grid interactions, etc. Remedying the problems is directly linked to penalizing their symptoms.

5 Integrating IGOP in Power Systems

Table 1 provides an overview of five variables (column 1) affecting the transient quality and therefore the spot market value of the supplied kWh. In columns 2 and 3 is assessed how commanded power generators and how IGOP perform on the five variables. On time and speed, commanded generation by far outperforms IGOP; this is due to their respective roles in dispatching and optimizing integrated power supplies. The differences sublimate in the attribute "liability to serve." Commanded generation plants adopt full liability if ready to supply when the system requires their contribution, and if abstaining from delivery when there is no demand for their power. "Full" liability is attenuated by specific terms in the relationship between commanded plants and the TSO. In principle, IGOP assume no liability to serve: they are not engaged to supply when the TSO would want it, and they deliver power to the grid when technical feasible and matching their financial self-interest. This reduces the market value of IGOP power compared to the value of power from

Table 1 Market value of a kWh supplied depends on five variables, implemented differently by IGOP and by commanded generation plants

Market value of a kWh supplied depends on	Commanded generation plant	Independent generator of own power (IGOP)
Time of delivery (synchronous with system base to peak load fluctuations)	Delivery at command if unit was committed; variable RE contribute when sources deliver on time of request	Delivery not at command; net power offered according to source supplies (renewable) and own demand for power, and for heat (cogeneration)
Speed of delivery (immediate, within seconds, minutes, hours)	Plants ready for dispatching but limited by ramping rates and flexibility; some plants specialized in high flexibility	Most IGOP capacity not available for dispatching
Liability to serve	Produce power on demand—shunt power production if not demanded	Deliver power in surplus of own needs when profitable; IGOP switch roles producer–consumer
Place of delivery	Central large-scale stations supply bulk power; renewable sources often distant from the grid (e.g., off-shore wind; hydro dams)	Distributed locations near load centers, creating meshed deliveries; participants in smart grids
Reliability	Source, technology, project, environment, ... specific	Source, technology, project, environment, ... specific

ordered generation plants. By accepting some liabilities to manage their plant availability, IGOP enhance the market value of their kWh supplies [14].

Mostly IGOP is well located, avoiding transmission activities and corresponding losses. The advantages of being located near the power load centers are difficult to measure and to quantify, because they depend heavily on the momentary simultaneity between IGOP surplus delivery to the grid and demand for power by consumers on the local grid. Because IGOP may request full back-up power when the plant is down, the compensation advantages may have only minor impact on grid capacity and thus on a significant share of the investments. The capacity effect is also dependent on the type of IGOP, considering, for example, the difference between intermittent PV and industrial combined heat and power units.

There are no arguments why reliability of particular plants should change because of other ownership, e.g. PV wherever installed, is technically very reliable, presumably the most reliable power generation technology forever.

From Table 1 follows that electricity forthcoming from IGOP scores a lower market value than electricity from commanded generation plants, the latter being exclusively dedicated to serve the market. Leftover to the established systems and institutions, it is unlikely that IGOP may win the uphill market battle against incumbent power generators that run ordered plants. Nykamp et al. [16] and Gawel and Purkus [14] reveal the difficulties in developing proper financial incentive mechanisms to overcome the prevailing market structures and rules. Moreover, the mechanisms must stay transparent and provide certainty for implying many millions more of small RE generators. There is no future in expecting that more than a small percentage of end-users ever can be engaged in the intricacies of electric power systems, a neither economic nor social beneficial time passing for the vast majority of people. The corollary is high exigencies to regulations for bringing IGOP from its present non-competitive position to the default electric power generation option. Only effective, efficient, and fair, but also simple and transparent regulations may engage millions of building owners and small companies.

6 Policy Implications of the Proposed Spread of Liabilities

Before developing the detailed, and often tricky, regulations of electric power sector transition to 100 % renewable energy supplies, an encompassing helicopter vision is recommended. Every modern power system is widely branched and multi-leveled, however continuously integrated by a single, common frequency control. Within the technical constraints there is ample choice on the degree of centralization of the power supply industry.

A first policy decision is to state explicitly the priority assigned either to centralized power supply plants, or to distributed IGOP (Independent Generators of Own Power). Priority to IGOP is the more sustainable option, but also the more distant one from inherited power sector systems and practices. Acting along the priority for

IGOP implies innovative technical, financial, and regulatory approaches and mechanisms, which development and deployment necessitates further policy decisions.

The second policy decision is shifting the targeted benchmark or reference point, now anchored at past or present electric power sector structures and practices, to a reference reflecting the 100 % renewable electric system as envisioned by the first policy decision. A clear view on the future vantage point is most helpful in backcasting the suitable steps for an effective, efficient, and fair transition.

The third policy decision is the application of the polluter pays principle in its advanced version of "extended polluter responsibility," for allocating the expenses of the conversion of the power systems from fossil fueled and nuclear, to very low-carbon and low-risk power supplies. Once societies and their political representatives have recognized the significant economic value-price gaps in climate, fossil fuels, and renewable energy supplies, they opt for the full (and urgent) transition to 100 % renewable electricity systems. Realizing this principled choice in practice means significant reversals in ongoing activities and practices. In comparison with ordered power supplies that are dispatched on command by a system operator, IGOP power has a lower value due to weaker commitment to supply liability in meeting the intense and exigent demands by affluent electricity consumers in wealthy countries. Left over to thriving electricity trade practice, IGOP has little chance to fulfill its historical role in the 100 % renewable energy transition. Public intervention for changing the trade rules is urgently on the agenda.

Germany is the industrialized country with the strongest commitment to turn over its electricity supply sector from a mainly fossil fuel and nuclear fission based one, to a 100 % renewable electricity version. Integrating RE is mandated for grid operators in Germany, except when the connection exceeds "reasonably economical" costs. Minimizing expenses is a justified efficiency goal, when all costs are fully identified, and measured from the proper future reference point.

The danger of volatile policies is real when the encompassing meta-vision on the transition is lacking. For example, the Flemish government and regulator are now taking back money from PV investors, after attributing too high subsidies as guaranteed certificate payments (in fact: guaranteed premiums on PV output) in the period 2008–2012. Since January 2013 onwards, household PV plants in Flanders are practically no longer subsidized. On the contrary, all PV owners pay a yearly fixed fee (per kWp installed inverter capacity: €56–83 in 2013, depending on the distribution utility), or they can install a smart meter to measure and bill their power exchanges with the grid [30, 31]. This policy U-turn has significantly retarded the transition to a low-carbon and low-risk electricity supply sector.

7 Conclusion: Opposite Perspectives on Integrating Renewable Supplies

Respecting the +2° limit of the Copenhagen Accord (in December 2015, reaffirmed in the Paris Agreement), implies that the world's electricity generation systems turn to zero or very low carbon sources [3]. When accepting that RE, in particular RE built and operated by IGOP, have to become the default generation option in a 100 % RE system, a comprehensive helicopter view on the transition and integration issues is requested. In case of direct short-term competition between established power systems and IGOP challengers, the latter will fail to develop (Sect. 5). Two arguments for overcoming the fallacy of direct spot competition merit consideration. The minor argument is that today's power supply systems are distant from market competitive optima as hailed in the economics literature. Most economic analysis of power systems starts from the hypothesis that electricity companies obey competitive market rules. For example, it is assumed that most electricity is traded at power exchange prices, and that the spot prices at the exchanges reflect the true short-run marginal supply costs of power. This way of pricing would approach the theoretical bliss under two conditions: all power is traded at the transient short-run marginal costs that include all the public and private costs of used economic resources, and the generation system is optimally composed (so that the equality of short-run and long-run marginal costs prevails, guaranteeing that short-run prices also cover the fixed expenses on the plants). In reality, both conditions are not fulfilled. Large quantities of electricity are traded under long-term contracts, mostly including substantial capacity payments. Consecutively, the national power supply systems in Europe (that together make up the European system) are far from textbook optimal composition, formerly targeted by vertically integrated monopolies. The long-living fixed assets systems have been perturbed, e.g., by the EU liberalization packages, and by unexpected growth of RE capacities. Public authorities and regulators deliberately acting to promote and support RE (like in Germany is the case) are molding the 100 % RE systems of the future as the reference to construct. They base the policy on the other, major argument: today's power supply systems are completely grown adverse because of unpaid externalities and unpaid risks of its large-scale fossil fueled and nuclear plants. The energy systems of the industrialized world were driven in a non-sustainable direction by two major economic value-price gaps (Sect. 2). Fossil fuels and the global climate have been and are still highly over-used (or abused) because their prices were and still are far below their value. The political target of a maximum global temperature rise of +2 °C is set, implying the task of transiting to low carbon energy economies. The electricity sector is seen as the first major energy subsystem that can and should realize the transition. Understanding well the structure, composition, working, and participants in this sector is a prerequisite for finding the most effective and efficient solutions (Sect. 3). The position and characteristics of IGOP are highlighted, without full dissection of the various cases and situations that one meets in practice. New mechanisms to support IGOP must stay transparent and provide certainty for implying many millions more of small RE generators. There is no future in expecting that more than a small percentage of end-users ever can be engaged in the intricacies of electric power systems, moreover neither economic nor social beneficial. This condition challenges governments, and their regulatory authorities, to stay on top of the power sectors (Fig. 2) and of their evolution. For supporting IGOP, objective information is needed about asymmetries in supply liability, remuneration of IGOP surplus power supplies to the grid, pricing of back-up power from the grid to independent generators. The issues are often not well understood and generally contentious, but of high importance for the effective, efficient, and fair transition from non-sustainable to 100 % renewable power systems.

References

1. Verbruggen A (2012) Balancing incumbent and opposite perspectives on key issues in the 100% renewable electricity transition. Second international 100% renewable energy conference and exhibition (IRENEC 2012), Maltepe-Istanbul
2. Verbruggen A, Di Nucci MR, Fischedick M, Haas R, Hvelplund F, Lauber V, Lorenzoni A, Mez L, Nilsson LJ, del Rio Gonzalez P, Schleich J, Toke D (2015) Europe's electricity regime: restoration or thorough transition. Int J Sustain Energy Plann Manag 5:57–68
3. IPCC (2012) Renewable energy sources and climate change mitigation. Special Report of the Intergovernmental Panel on Climate Change. Cambridge University Press. Chapter 8 deals with "Integration of Renewable Energy into Present and Future Energy Systems". Available at www.ipcc.ch
4. Twidell J, Weir T (1997) Renewable energy resources. E&FN Spon, London
5. Verbruggen A (1997) A normative structure for the European electricity market. Energy Policy 25(3):281–292
6. Schleicher-Tappeser R (2012) How renewables will change electricity markets in the next five years. Energy Policy 48:64–75
7. Geelen D, Reinders A, Keyson D (2013) Empowering the end-user in smart grids: recommendations for the design of products and services. Energy Policy 61:151–161. doi:10.1016/j.enpol.2013.05.107
8. Eurelectric (2013) Active distribution system management: a key tool for the smooth integration of distributed generation. Eurelectric, Brussels
9. Verbong GPJ, Beemsterboer S, Sengers F (2013) Smart grids or smart users? Involving users in developing a low carbon electricity economy. Energy Policy 52:117–125
10. George M, Banerjee R (2011) A methodology for analysis of impacts of grid integration of renewable energy. Energy Policy 39:1265–1276
11. IEA (2011) Harnessing variable renewables: a guide to the balancing challenge. Organization for Economic Cooperation and Development/International Energy Agency, Paris
12. Mason IG, Page SC, Williamson AG (2013) Security of supply, energy spillage control and peaking options within a 100% renewable electricity system for New Zealand. Energy Policy 60:324–333
13. Schaber K, Steinke F, Hamacher T (2012) Transmission grid extension for the integration of variable renewable energies in Europe: who benefits where? Energy Policy 43:123–135
14. Gawel E, Purkus A (2013) Promoting the market and system integration of renewable energies through premium schemes—a case study of the German market premium. Energy Policy 61:599–609. doi:10.1016/j.enpol.2013.06.117
15. De Joode J, Jansen JC, van der Welle AJ, Scheepers MJJ (2009) Increasing penetration of renewable and distributed electricity generation and the need for different network regulation. Energy Policy 37:2907–2915

16. Nykamp S, Andor M, Hurink JL (2012) 'Standard' incentive regulation hinders the integration of renewable energy generation. Energy Policy 47:222–237
17. Pearce DW, Turner RK (1990) Economics of natural resources and the environment. Harvester Wheatsheaf, New York
18. WCED (1987) Our common future. The World Commission on Environment and Development. Oxford University Press, Oxford
19. Meadowcroft J, Langhelle O, Ruud A (eds) (2012) Governance, democracy, and sustainable development: moving beyond the impasse. Edward Elgar, Cheltenham
20. Verbruggen A (2008) Renewable and nuclear power: a common future? Energy Policy 36:4036–4047
21. Verbruggen A, Fischedick M, Moomaw W, Weir T, Nadai A, Nilsson LJ, Nyboer J, Sathaye J (2010) Renewable energy costs, potentials, barriers: conceptual issues. Energy Policy 38: 850–861
22. Robinson J (1982) Energy backcasting: a proposed method of policy analysis. Energy Policy 10(4):337–344
23. OECD (1972) Guiding principles concerning international economic aspects of environmental policies. Organisation for Economic Cooperation and Development, Paris
24. Heyward M (2007) Equity and international climate change negotiations: a matter of perspective. Clim Pol 7:518–534
25. Cordato RE (2001) The polluter pays principle: a proper guide for environmental policy. Institute for Research on the Economics of Taxation, Washington, DC
26. Joskow PL, Schmalensee R (1983) Markets for power. MIT Press, Cambridge
27. Besant-Jones JE (2006) Reforming power markets in developing countries: what have we learned? The World Bank, Washington, DC, Energy and Mining Sector Board Discussion Paper 19
28. Glachant J-M, Finon D (2003) Competition in European electricity markets. A cross-country comparison. Edward Elgar, Cheltenham
29. Verbruggen A (1990) Pricing independent power production. Int J Global Energy Issues 2(1):41–49
30. Eandis (2013) Veel gestelde vragen. www.eandis.be/eandis/faq.htm?datald=9123708
31. Infrax (2013) Netbijdrage voor lokale elektriciteitsproducenten. www.infrax.be/nl/over-infrax/nieuws/2013/20121212-netbijdrage-producenten

The Importance of Diversity for Renewables and Their Control in Future Electrical Infrastructure

Hasan Basri Çetinkaya, Serhat Uzun, Hasan Göksun Virlan, and Halit Can Altay

1 Introduction

The grid integration of renewables, which have variable energy sources, will have a significant impact on the future network. Nowadays, the transmission system integration has gained quite a pace. The integration of generation from distribution system will accelerate in the coming years. Today's distribution grids were not designed according to integration of generation. Therefore, the integration of this generation can be said to create a quiet revolution in the energy structure. The future reliability of supply will depend on how much of the existing structure become smart which will be consistent with this supply integration.

Today the electricity enters in our lives in every field and no one will accept the increasing frequency of energy cuts as well as the decreasing quality of the supply. For the integration of generation, it is very important to design the system regarding the type of generation and the location in the grid with taking necessary control and protection measures. In such kind of system the optimization and efficiency of energy production will also gain importance.

The main disadvantage for renewables is the fluctuating source. Because the source is uncertain and uncontrollable, it is hard to get a stabilized power curve. Renewable energy sources must increasingly provide balancing power [1]. This balancing power can be improved mainly by three issues. The first one is using maximum resource diversity in order not to depend on a single source. The second one is to control the power of renewables as the source is fluctuating and the third one is to have energy storage to maximize the efficiency and stability.

H.B. Çetinkaya (✉) • S. Uzun • H.G. Virlan • H.C. Altay
SIEMENS A.Ş., Energy Management/Digital Grid,
Yakacik Cd. No:111 34870, Kartal, Istanbul, Turkey
e-mail: hasan.cetinkaya@siemens.com; serhat.uzun@siemens.com; goksun.virlan@siemens.com; can.altay@siemens.com

T.S. Uyar (ed.), *Towards 100% Renewable Energy*, Springer Proceedings in Energy, DOI 10.1007/978-3-319-45659-1_45

2 Challenges of Infrastructure for Renewables

Transmission system is flexible for generation due to its structure, but for distribution system the situation differs. Distribution system is not designed for generation integration. When generations are integrated to distribution system, load flows, short circuit levels and protection philosophy changes. The change amount is related with the amount of integration and may create huge voltage/frequency oscillations that may shut down the system [2]. The first expected problem is the voltage rise that may occur close the generation units. The cause of this voltage rise is related with the voltage control at TEIAS (Transmission System Operator in Turkey) transformer station. The voltage of medium voltage side is normally controlled via tap changers at power transformer. In most cases these taps are arranged according to the lowest voltage that may occur at distribution busses that are fed from transformer station. This may create a voltage that is a little bit higher than the nominal at the main station. A large generation that may integrate from a far location has led to increased voltage at its terminal due to controlled voltage at the main transformer station. This voltage rise may also limit the reactive power that is supplied from the generation unit. Measures should be taken and system equipment should be protected against abnormal voltage levels (Fig. 1).

The second issue that should be correctly addressed is the short circuit contributions from dispersed generations. This contribution amount depends on the type, technology, and the nominal power of the generation unit. Adverse events can be avoided in the design stage. Another problem that may arise after integration of generation is islanding. Islanding may occur if generation and loads are islanded together after a distribution system fault. At that time, power quality only depends on the generation units that are running in island. Today's code are not expecting a control of frequency and/or voltage for small scale renewable integration which means it is possible to have dangerous voltage and/or frequency oscillation inside

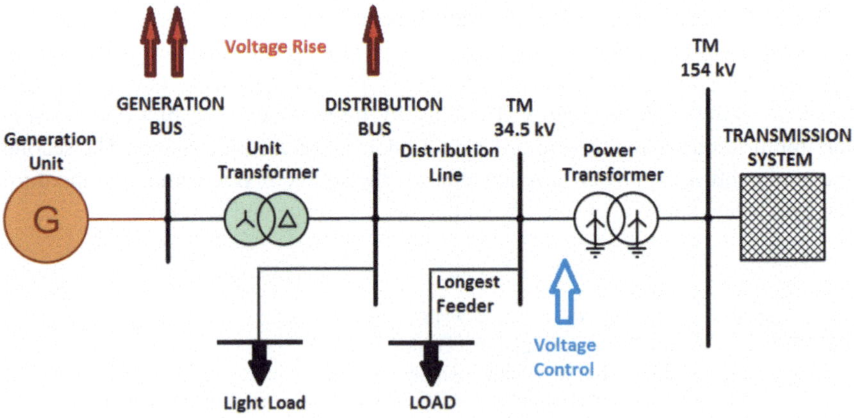

Fig. 1 Voltage rise that may happen due to generation integrations at distribution system

the island. The code is expecting disconnection of generation units at the time of islanding. However as generation integration amount rises, the control for these generations will be required. When control is satisfied, the "Microgrids" will begin to occur. At this stage, the island structure should be able to read very well and necessary control measures should be taken. Storage technology improvements will also be really important for "Microgrid" structure.

3 Diversity for Renewables

The resource diversity and the high integration ratio of renewables are really important to have a more smooth power curve. Although wind, solar, tidal, and wave resources will always be intermittent when they are considered in isolation and at one location, several methods exist to reduce intermittency of delivered power. These include combining geographically disperse intermittent resources of the same type, using storage, and combining different renewables with complementary intermittencies [3] (Fig. 2).

There exist studies which show 100 % renewable supply is possible when sources are combined in a true mix according to load curves of cities, countries. It is known that multiple sites, combining different resources will reduce variability and an optimal renewables portfolio can be identified according to load demand [4, 5]. In most cases, a true mix of renewables is not sufficient to get a smooth power curve. Storage and the control of power outputs should also be required.

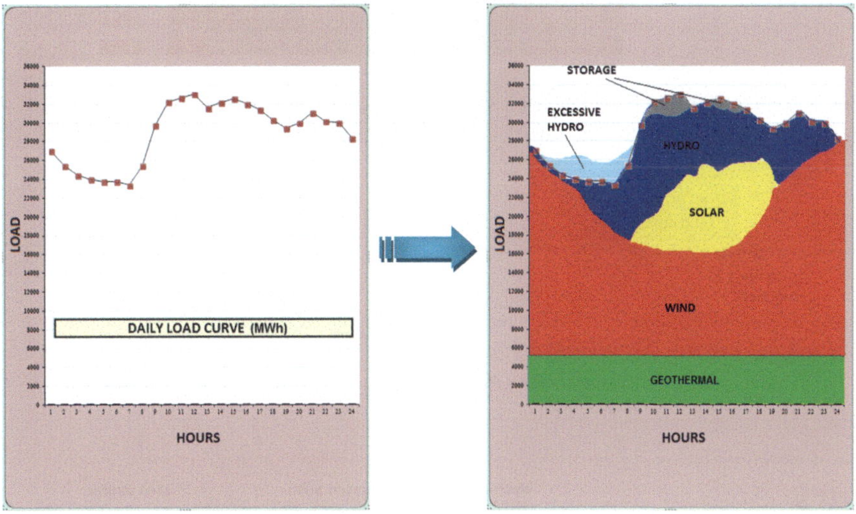

Fig. 2 Daily power curve of Turkey in May 2014 (www.teias.gov.tr) and representative illustration of 100 % renewable supply

4 Controllable "Smart" Infrastructure

The new era of electricity requires a smart structure for transmission and distribution system. Control is a key enabling technology for the deployment of renewable energy systems. Solar and wind power require advanced control techniques for high-performance and reliable operation [6]. The protection system and required parameters should always be tracked in such a dynamic system. There exist applications like SIGUARD® PSA [7] for protection security assessments in order to improve grid reliability (Fig. 3).

In future power systems, there should also be a management system which controls and optimizes dispersed generations regarding the technical and economic aspects. One of the example of this system is Decentralized Energy Management System (DEMS®) [8] (Fig. 4).

Fig. 3 SIGUARD® PSA™ output that is showing protection situation online

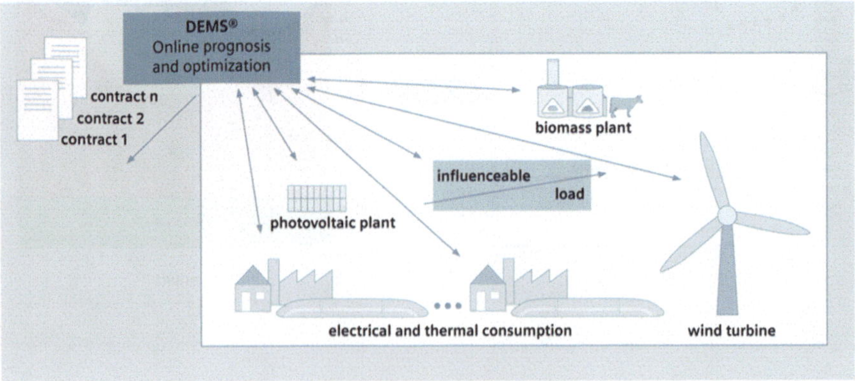

Fig. 4 Decentralized Energy Management System (DEMS®)

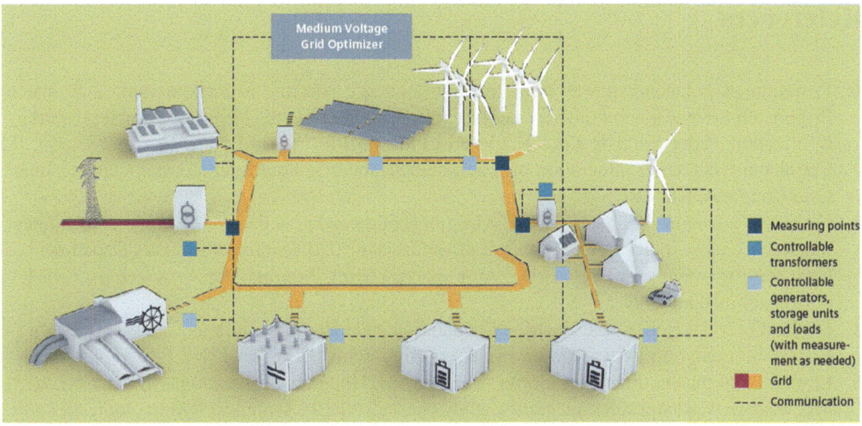

Fig. 5 Medium voltage grid optimizer

Grid management is facing challenges like an unclear, fluctuating direction of load flow and, more and more often, critical voltage violations. There is a growing risk of voltage range infringement and thus malfunctions or even damaged equipment on the consumer side. At the same time, the danger of overloads on lines, transformers, and other equipment are growing, which can even result in grid failure. Processes in the distribution grids must be made visible at all times in order to reliably assess the status and take efficient countermeasures before critical situations arise. For this reason smart grid optimization tools should be used (Fig. 5).

Using these tools, load flow values and load flow directions are reliably monitored. Voltage violations and equipment overloads are detected quickly and accurately. Balancing measures primarily for maintaining grid stability and for protecting equipment can be initiated at an early point. Distribution losses can be effectively reduced. An optional automatic mode allows transformer tap changers, capacitor banks, loads and generators, including battery storage, to be controlled without operator intervention.

5 Conclusion

The strategy vision of Turkey for 2023 and also the energy vision of the world clearly show that future energy infrastructure will be different from today. Huge amount of renewable will be integrated from transmission and distribution systems. If the right mix can be made, renewable and intermittent technologies can be close to electricity demand patterns. This will reduce the need for backup, and makes renewables a serious alternative to conventional power sources. It is also essential to disperse the generators as widely as possible. By increasing the separation between sites, it can be ensured that power is always being generated somewhere and so smooth out the supply curve. Control is the complementary issue for 100 % renewable generation.

References

1. Provision of balancing power from renewable energy sources. Kombikraftwerk 2 Research Project. http://www.kombikraftwerk.de/fileadmin/Kombikraftwerk_2/English/Kombikraftwerk2_Information_Paper_Field-Test.pdf. Retrieved 13 May 2014
2. Çetinkaya HB (2014) Integration of renewables to the grid. 2nd International Istanbul Smart Grid Congress, Istanbul
3. Hoste GRG, Dvorak MJ, Jacobson MZ. Matching hourly and peak demand by combining different renewable energy sources. A case study for California in 2020. Stanford University Department of Civil and Environmental Engineering, Atmosphere/Energy Program. http://www.etopia.be/IMG/pdf/HosteFinalDraft.pdf
4. Sinden G (2005) Renewable electricity generation and electricity storage. Cambridge Energy–Energy Storage. http://www.cambridgeenergy.com/archive/2005-06-22/CEF-Sinden.pdf
5. Lew D, Brinkman G, Ibanez E, Florita A, Heaney M, Hodge B-M, Hummon M, Stark G (2013) The western wind and solar integration study phase 2. NREL, Technical Report. www.nrel.gov/docs/fy13osti/55588.pdf
6. Samad T, Annaswamy AM (2011) The impact of control technology. IEEE Control Systems Society, New York
7. SIGUARD® PSA. Protection security assessment. www.siemens.com/siguard
8. Virtual Power Plants by Siemens. DEMS®-decentralized energy management system. http://w3.usa.siemens.com/smartgrid/us/en/distributech/Documents/DEMS_VPPs.pdf

Cellular Power Grids for a 100 % Renewable Energy Supply

Eberhard Waffenschmidt

1 Introduction

It is proven that an energy supply with 100 % renewable energies is possible in Germany [1], Europe and the whole world. However, this requires a decentralized, distributed energy generation. It will probably not be sufficient to generate and store the energy only in those regions, where the cheapest generation is possible. Instead, each region needs to contribute on its own to the energy supply by using its potential. Every roof needs a photovoltaic (PV) system and every village its own wind park. Also storage needs to be decentralized, located at the generation sites to reduce the grid load.

Such decentralized generation and storage offers new opportunities for a reliable structure of the electrical power grid. It is proposed to subdivide the power grid into much smaller cells than in a traditional power grid (Fig. 1, left). Since all of these cells contain distributed generators, they could be able to operate on their own, if necessary. In case of a global black-out, the individual cells could be able to survive.

There are already some proposals for similar structures. The report of a VDE working group 'Active Grids' [2] raises the idea of such a cellular grid. The initiative 'C/sells' [3], part of smartgrids-BW also aims for such a cellular grid structure.

This publication aims to derive the issues related to such an approach and identify further topics of research. Some examples for parts of related aspects are presented. The paper focuses on Germany, but the general conclusions can be taken for any region in the world. Figure 1 (right) illustrates the tasks of such a cell of a cellular grid. The issues can be grouped into four groups: generation (with renewable energies), control, fault management and organization. Of these four groups, this paper mainly considers generation and control. Before, general issues are discussed.

E. Waffenschmidt (✉)
CIRE—Cologne Institute for Renewable Energies, Cologne University of Applied Sciences, Betzdorfer Str. 2, 50679 Cologne, Germany
e-mail: eberhard.waffenschmidt@fh-koeln.de

T.S. Uyar (ed.), *Towards 100% Renewable Energy*, Springer Proceedings in Energy, DOI 10.1007/978-3-319-45659-1_46

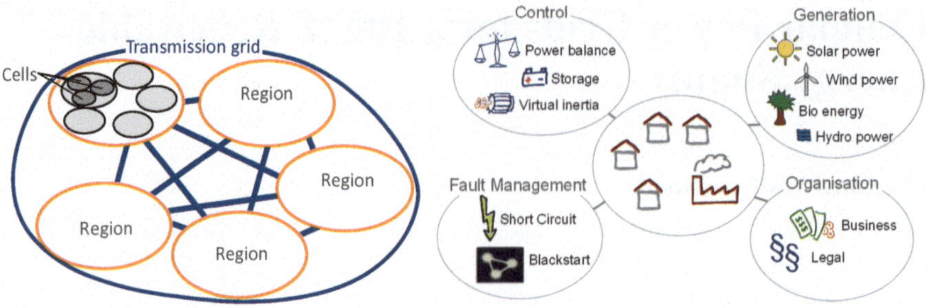

Fig. 1 Structure of a cellular grid (*left*) and tasks for a cell of a cellular grid (*right*)

2 Autarky Versus Autonomy

If the regions are interconnected to each other, a complete *autarky* for each cell is not useful. Autarky means that a cell is able to operate completely independent of external connections. However, it is proposed to select the cell size such that *autonomy* can be achieved. Autonomy is understood as being able to generate the required energy demand within a period of time (e.g. a year). The cells should be able to interchange power, because some regions might have an excess of power, while other may have a power demand at the same time.

A figure of merit to what extend the individual cells must become autarkic is the grade of autarky. It is defined as the ratio of the amount of self-generated power used in a cell to the power demand in this cell, averaged over a period of time (e.g. a year). A grade of autarky of 100 % means full autarky.

3 Cell Size

A higher grade of autarky obviously reduces the need for infrastructure for power transmission between the cells. On the other hand, it increases the need for storage in the cells. A way to reduce the need for storage is to compensate the fluctuating generation by different modalities like wind and solar. Figure 2 shows the grade of autarky which can be achieved with a pure solar or wind supply and with a combination of both [4].

The figure shows it for an investigated community in northern Germany as a function of the ratio between wind and PV generation. The blue curve relates to a grade of autonomy of 100 %, while the orange curve relates to an excess generation of 260 % (which is the case for the investigated community, see red dot). Both curves don't include storage. Further details are explained in [4]. It is obvious that a supply with only PV results in a much lower grade of autarky as a supply with wind. A combination increases it by about 8–10 %. In [4] it is shown that the storage demand reduces nearly by one order of magnitude in this exemplary community with a combined generation with wind and PV compared to a generation with PV only. Concluding, a mix of modalities improves the grade of autarky in a cell. An additional possibility is to use demand side management (DSM). Both requirements lead to conclusions about a reasonable cell size:

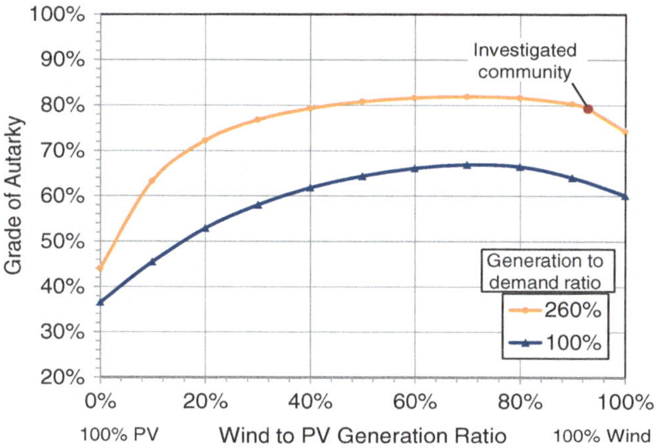

Fig. 2 Grade of autarky as a function of the ratio between the annually generated wind energy and PV energy for two different excess ratios [4]

A cell size relating to an individual home or street is certainly too small, because then usually only one modality, typically PV, is available. Furthermore, the possibilities for DMS are limited. This would require a much larger storage size. Thus a size which allows a good combination of different power generation modalities and the use of DSM would be reasonable for the cellular grid concept. This would relate to the area of about a community or a similar size. It is a matter of ongoing research to investigate the suitable size of a cell. The aim is to minimize the necessary power exchange between the cells to avoid extensions of the transmission grid.

4 Autarkic Emergency Operation

In cases of severe grid states up to black-out, the individual cells should be able to operate autarkic. The full functionality may not always be guaranteed in such a case, but it would be the goal to provide at least a minimal supply for the most important loads. However, such an operation has several implications.

4.1 Conflict Between Continuous Operation and Safety Shut Down

The actual grid is not designed for such a case. Instead, all distributed generators must shut down in case of a black-out [5] because of safety reasons in case of maintenance. As a further problem, connecting the islanded grid after continuous operation to the main grid may be done only with proper synchronization to avoid damage [6]. As a solution, the generators could be switched off remotely. The control, however, needs to be safe. In Germany, newer PV systems are obliged to have a remote control connected to the grid operator. However, only a fraction of existing PV systems are reliably remote controlled [7]. Another option would be have at least one 'master' generator (e.g. a biogas generator or a combined heat and power (CHP) system), which is able to 'build' the grid. Then, existing infrastructure can easily be integrated. It remains an open question, which are the required minimal properties for such a master generator.

4.2 *Balance of Power*

As a further challenge the balance of power must be able to be maintained in each individual cell. This requires an intelligent control not only of the generators but also of the selected loads.

4.2.1 Existing Situation

Unfortunately, most distributed generators are not supposed to modulate their feed-in power, but to feed-in always the maximum available power. Larger generators should be remotely controlled by the grid operator, but as mentioned already above, only a fraction can actually be controlled [7]. In addition, there are no loads, which modulate their power demand based on the available power generation. In case of lack of energy the grid operator can only switch on or off parts of the grid. In Germany, legal regulations demand switching off grid areas in a non-discriminating way. It is not allowed to prefer critical infrastructure like hospitals or fire brigades. In Germany this principle is under discussion and related associations demand a change in legislation.

4.2.2 Short Term Solutions

As short term solution, at least one 'master' generator can be established. It must be able to modulate its generation in a wide range and to control the grid frequency. Preferably, it is realized using a synchronous generator or by an inverter, capable of providing similar properties. In case of an excess generation, it can force 'slave' generators to switch off or reduce their power generation by emulating an emergency situation by increasing the grid frequency beyond the normal range. In Germany, if the grid frequency exceeds 50.2 Hz, the generator must reduce its feed-in power linearly with increasing grid frequency. Alternatively, it shuts down at an arbitrary frequency between 50.2 and 51.5 Hz, such that the sum of the feed-in power of all generators follows the required curve [5]. The master generator can then set the frequency to the required value in order to balance the power in the grid.

In case of lack of energy, the only available option today is to shut off areas of the grid in the cell. If legally possible, a priority plan should be made to be able to supply most important areas containing critical infrastructure.

4.2.3 Medium and Long Term Solutions

In a long term perspective each of the participating generators should be able to modulate its feed-in power according to the grid state. In order to avoid additional communication, a control based on physical grid variables is proposed. Preferably, feed-in power should be controlled by a frequency droop control, similar as today for large generators in power plants.

To achieve a seamless transfer between the 'micro-grid' mode and 'normal' mode it is preferred to use the same control mode in both cases. During 'normal' mode the grid frequency range will probably be much smaller, requiring less power control of each individual generator, while in 'micro-grid' mode the frequency range will be broader resulting in a much stronger power modulation.

On the load side, an automatically switching or modulation of loads should be implemented. Here, DSM methods could be applied. The details are still an open question and require further research.

Finally, storages will definitely help maintaining the power balance. Distributed storages, especially those connected to PV systems, are emerging. Attention should be paid to their inverters. They should as soon as possible be able to build and control the grid as distributed 'masters'. Since they are still in an early phase of introduction, related standards can still be set.

4.3 Grid Control

Most of the renewable energy is fed in by electronic inverters, which do not contribute to grid control (like frequency and voltage control). Especially rotating masses (inertia) of conventional generators are missing. Therefore, a suitable grid control like virtual inertia must be applied to all related inverters. Several research projects have shown such a solution for PV inverters [8, 9]. Enercon offers wind turbines with virtual inertia control, if demanded by the grid operator (private communication). The research project Kombikraftwerk 2 [10] is going to demonstrate a stable grid operation only with renewable generation in Germany, and Younicos AG [11] is going to show a stable grid operation for the island Graciosa only with wind, PV and storage. Own research showed that it should be possible to use the elcap storage, which is available in feed-in inverters to compensate the pulsating ac power, additionally as storage for inertia control [12]. Then, the existing hardware can be used and only the power control needs to be modified.

These examples show that grid control with feed-in only by electronic inverters is feasible. However, an operation in a cellular grid cell has not been demonstrated up to now.

5 Examples

There are more and more communities and regions in Germany, which aim for a 100 % renewable energy supply of their community or region or have it realized already. The project '100ee region' aims to support those regions by giving consultancy and to interconnect them. The related website gives a good overview of 100 % renewable regions in Germany [13].

There are several examples of regions in Germany where at least some of the aspects of a cellular grid are investigated and demonstrated. Some exemplary regions: On the grid connected North-Sea island Pellworm the project 'SmartRegion Pellworm' [14, 15] aims to demonstrate a high grade of autarky with a combination of wind, PV and biogas power and with using electrical storage. The project Smart Country [16, 17], led by power provider RWE, relates to a region in Western Germany, Rheinland-Pfalz, in the Eifelkreis Bitburg-Prüm. The region wants to be supplied 100 % by renewable energies by 2030. The project aims to demonstrate the power balance by using controllable biogas power. Even further the project IREN2 in Wildpoldsried, Bavaria [18], is demonstrating a real autarkic operation of a micro grid, very similar as proposed in the cellular grid approach. In this project an island grid operable microgrid is in development since July 2014. All necessary components are considered: Controllable low voltage transformers, controllable generators for electrical power and heat (PV systems, biomass and biodiesel generators), energy storages (battery, thermal storage), measurement and communication systems, controllable power consumers and in addition components for forcing and testing load steps.

6 Towards 100 % Renewable Energy

Solving these issues for an individual cell will rise questions and give answers, which will appear sooner or later not only in individual cells, but will also be relevant for the total power grid. Therefore, with this bottom-up approach we will learn step by step how to operate a complete global power grid with fluctuating, decentralized renewable energies. Then, despite the fluctuating nature of renewable energies, our power grid will be more reliable and resilient than today.

References

1. Waffenschmidt E, von Fabeck W (2007) Aggregate scenario for an energy supply exclusively from domestic Renewable Energy resources. Second international renewable energy storage conference (IRES II), Bonn, 19–21 Nov 2007
2. Adam K, Benedix C, Berg V, Blug C, Brochtrop M, Waffenschmidt E, Buchholz B, et al (2013) Aktive Energienetze im Kontext der Energiewende. Studie der Energietechnischen Gesellschaft im VDE (ETG), Frankfurt
3. C/sells – das Energiesystem der Zukunft. http://www.smartgrids-bw.net/csells/. Accessed 1 May 2015
4. Waffenschmidt E, Reinartz D (2015) The optimal combination of energy storages, photovoltaic systems and wind turbines at local level. 9th international renewable energy storage conference (IRES 2015), Düsseldorf, 9–11 Mar 2015
5. Erzeugungsanlagen am Niederspannungsnetz – Technische Mindestanforderungen für Anschluss und Parallelbetrieb von Erzeugungsanlagen am Niederspannungsnetz. VDE-Anwendungsregel VDE-AR-N 4105, Aug 2011
6. Laudahn S (2015) Aktive Inselnetzerkennung und Netzstützungsfunktionen: Widerspruch oder ideale Ergänzung? Otti-Konferenz "Zukünftige Stromnetze für Erneuerbare Energien", Berlin, 27–28 Jan 2015
7. Döring M (2015) Herausforderungen bei der Weiterentwicklung von Netzanschlussbedingungen im Rahmen der Energiewende. Otti-Konferenz "Zukünftige Stromnetze für Erneuerbare Energien", Berlin, 27–28 Jan 2015
8. VSYNC Project. www.vsync.eu. Accessed 1 May 2015
9. Rodriguez P (2013) Control of grid-interactive power converters: the synchronous power controller. ECPE workshop "Power Electronics in the Electrical Network (Renewables, Energy Storage, Grid Stability)", Kassel, 12–13 Mar 2013
10. Kombikraftwerk 2: Das regenerative Kombikraftwerk. http://www.kombikraftwerk.de/. Accessed 3 Nov 2013
11. Graciosa—the road so far. http://www.younicos.com/de/projekte/graciosa/video_graciosa.html. Accessed 30 Nov 2013
12. Waffenschmidt E (2015) Instantaneous grid power control with PV inverters using DC-link capacitors. Submitted as extended abstract to the 6th international symposium on power electronics for distributed generation systems (PEDG 2015), Aachen, 22–25 June 2015
13. 100ee Erneuerbare Energie Regionen. IdE—Institut dezentrale Energietechnologien. http://www.100-ee.de/. Accessed 1 May 2015
14. Nicolai S (2015) SmartRegion Pellworm—results from the field test of an hybrid storage system. 9th international renewable energy storage conference (IRES 2015), Düsseldorf, 9–11 Mar 2015
15. SmartRegion Pellworm – Energie-Region mit Zukunftsanschluss. http://www.smartregion-pellworm.de/home.html. Accessed 30 Nov 2013
16. RWE: Vorweg gehen mit Smart Country. http://www.rwe.com/web/cms/de/2034950/smart-country/modellprojekt/. Accessed 30 Apr 2015
17. BINE Informationsdienst: Projektinfo: Ländliches Stromnetz smart geregelt. http://www.bine.info/publikationen/projektinfos/publikation/laendliches-stromnetz-smart-geregelt/. Accessed 30 Apr 2015
18. Zukunftsfähige Netze für die Integration Regenerativer Energiesysteme. IREN2 Project. http://www.egrid.de/projekte/iren2/. Accessed 1 May 2015

Solar Atlas: One Quarter of 1 % of Turkey's Surface Area Is Sufficient to Meet All the Power Demand with PV by 2050!

Jean-Philippe Denruyter and Mustafa Özgür Berke

1 Introduction

According to 2011 figures, only 6.5 % of the primary energy demand in Turkey was met by renewable energy resources such as hydropower, wind, geothermal and solar [1]. In the power sector, renewables account for 25 % of total generation [2]. The energy vision of Turkey for 2023 includes renewable electricity targets making use of all the hydropower potential, installing 20,000 MW wind, 3000 MW solar and 600 MW geothermal capacity. The targets on the fossil fuel side are much more ambitious, such as utilizing all domestic lignite resources.

The current share of renewable energy resources in the energy mix of Turkey and the 2023 renewable energy targets are below the levels that are required for Turkey to attain its target of decreasing dependency on external energy resources and have a meaningful contribution to global climate change mitigation efforts. Despite Turkey's significant solar energy potential, the share of solar energy in electricity generation is essentially non-existent, and the share it is projected to have in 2030 is just 5 % [3].

The way we produce and consume energy is not sustainable. The road to sustainable energy has a number of components. Aggressive energy conservation coupled with achieving the 100 % renewable energy target is one of these. Ensuring the sustainability of renewable energy resources is an equally imperative factor.

One of the questions concerning the sustainability of solar power focuses on the obvious question of "how much land is required for solar power generation?" "Solar

J.-P. Denruyter (✉)
Av. du Mont-Blanc 1196 Gland Switzerland, Büyük Postane Cad. No.19 Kat. 5, Bahçekapı, Istanbul, Turkey
e-mail: jpdenruyter@wwf.panda.org

M.Ö. Berke
WWF-Turkey, Büyük Postane Cad. No.19 Kat. 5, Bahçekapı, Istanbul, Turkey
e-mail: mberke@wwf.org.tr

T.S. Uyar (ed.), *Towards 100% Renewable Energy*, Springer Proceedings in Energy, DOI 10.1007/978-3-319-45659-1_47

energy requires too much land" is also the first line of arguments presented by proponents of conventional energy. This paper is an attempt to provide a contribution to this debate. There are two main questions that we sought answers for:

- How much land would be required if Turkey was providing all its electricity demand from solar PV for the year 2011?
- How much land will be required if Turkey decides to meet 100 % of its electricity demand from solar PV in 2050?

This paper is not suggesting that 100 % of future electricity consumption should be met with solar PV solely. However, this extreme case provides a very clear understanding of solar PV land use.

2 Methodology

"The Energy Report" by WWF [3] shows that meeting the global energy demand by 2050 with renewables is possible, using current technologies. Attaining this target on global, national and regional scales would require a combination of energy conservation and widespread deployment of renewable energy generation. In reality, the shares of various renewable energy resources in the 100 % renewable energy mix would differ according to numerous parameters. This paper focuses on one of these resources and depicts a world in which 100 % of the electricity generation is being met by solar PV systems in 2011 and 2050.

The calculations in the paper assume PV efficiency of 15 %, which is conservative concerning the current state of commercial PV technology. It is this same rate, which is also used for calculations for year 2050.

It is assumed that not every bit of the land dedicated to PV electricity production will actually host modules: about 20 % of land will be necessary for the plants' roads, shadow reduction, service stations, etc. In order to reflect this in the calculations, we "de-rate" the land required so that our land estimates reflect all land dedicated broadly to PV generation, including these roads and shading area. Therefore, the de-rating factor is 80 %.

The global annual irradiation value for Turkey is taken as 4.91 kWh/m^2/day. The resulting reasonable annual PV generation per m^2 is calculated as 0.215 MWh/m^2.

The boldest assumptions of the paper are related energy demand in the year 2050. Though PV efficiency is still assumed to be 15 % for 2050, the fundamental assumption for 2050 is that there will be energy equality across borders (i.e., per capita electricity consumption will be the same all over the world), and the 100 % Renewable Electricity Target will have been attained, in line with the findings of the Energy Report of WWF [8]. The Energy Report of WWF has calculated total global electricity supply for 2050 as 127.4 ej/year or 3.5389e + 10 MWh/year. Per capita electricity consumption in 2050 is calculated by dividing this total by UN Population Department's projection for 2050 global population (9,191,287,000).[1] The ensuing

[1] These figures were pulled from Population Division of the Department of Economic and Social Affairs of the United Nations Secretariat, World Population Prospects: The 2006 Revision and

figure is 3.85 MWh/year/person. Turkey's projected energy consumption in 2050 is the projected population in 2050 (90,551,000 according to UN projections) [4] multiplied by 3.85 MWh/year/person.

3 Solar Power in Turkey

According to the Ministry of Energy and Natural Resources, Turkey's technical solar energy potential is 35 mtoe/year or 380 billion kWh/year [5]. Turkey is ranked 2nd in solar water heating capacity following China. However, despite this potential and proven record of success in solar water heating, the solar photovoltaic market in Turkey is quite small. As of 2011, official figures rate solar PV at 1 MW of installed capacity. The first tender for solar field investments is expected to be finalized in mid-2013, though limited with a cap of 600 MW for licenses.

Potential investors in Turkey's solar energy sector are attracted by Turkey's possible participation in the European Electricity Market, following the 2009 synchronization between the Turkish power system and the interconnected power systems of continental Europe. Turkey is also expected to play a key role in cross-border renewable projects "Desertec" and "Mediterranean Solar Plan,"[2] both as a solar energy supplier and a bridge between the Middle East and Europe.

4 Findings

Most of Turkey's southern half has annual global irradiation values above 5.0 kWh/m^2/day. Solar development along the Aegean and Mediterranean coasts appears to present the best opportunity for aligning population centers and large PV plants.

This paper suggests that a total of 790 km^2 of solar panels (20 km^2 short of the total surface area of Atatürk Dam) would generate enough electricity to meet Turkey's total current electricity generation. As population and per capita electricity consumption are expected to rise in the coming decades even under the 100 % renewable energy scenario, 1600 km^2 of land dedicated to solar production would be required to meet 100 % of Turkey's total electricity needs in 2050.

The solar map (Fig. 1) displays the area that would have to be covered by solar panels to meet all the power demand for 2010 (red square) and 2050 (blue square).

World Urbanization Prospects: The 2007 Revision, http://esa.un.org/unup, Tuesday, June 5, 2012; 4:43:45 a.m.

[2] The Mediterranean Solar Plan targets the construction by 2020 of 20 GW of renewable energy installed capacity. For detailed information: http://www.medea.be/en/themes/euro-mediterranean-cooperation/mediterranean-solar-plan-msp/ DESERTEC concept is based on the idea for the large-scale utilization of renewable energy in deserts and arid regions. The objective of DESERTEC is to provide 15 % of power demand in Europe. For detailed information: www.desertec.org.

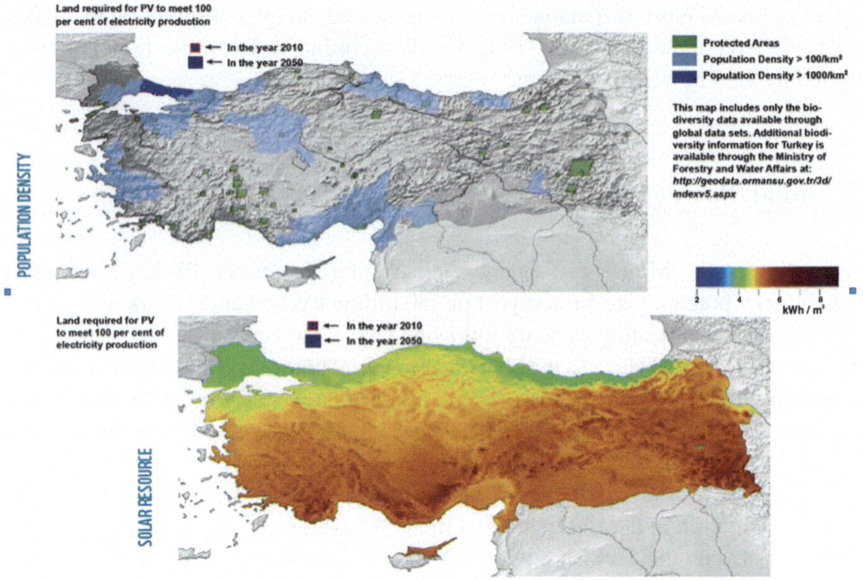

Fig. 1 Land required to meet 100 % of electricity production

It suggests that, even based on very conservative estimates where we disregard any development in solar PV efficiency, less than one quarter of 1 % (0.21 %) of Turkey's land would need to host solar PV generation in order to meet 100 % of Turkey's projected electricity needs in 2050. A large portion of these land requirements would actually be located on existing infrastructure like roofs.

Even if Turkey does not meet ambitious energy efficiency objectives and exceeds electricity consumption that is forecasted in this paper, total amount of land use would still be limited. Total land use by PV would still be below 0.5 %, in case electricity demand compared to the forecasts doubled.

5 Conclusion

Today solar PV electricity provides 0.1 % of total global electricity generation. However, PV has seen an average annual growth rate of more than 40 % since 2000. According to Bloomberg New Energy Finance, PV has already reached grid parity in a number of countries and given enough scale, it is expected to reach grid parity in Turkey by 2014 [6]. Now a well-established, commercially available and reliable technology, it has significant potential for long-term growth in nearly all world regions now and in the coming decades.

Solar power generation requires land. But just how much land would be required for Turkey to meet all of its electricity demand from solar PV? If Turkey chooses to meet 100 % of its electricity with solar PV by 2050, this would only require 0.21 % of its total surface area.

In a realistic portfolio of renewable energy generation technologies, PV will very likely require far less land than displayed in this paper. Grid integration, storage and balancing are significant issues to address for renewable energy to be successful. These topics are beyond the scope of this paper but are addressed more in depth in WWF's Energy Report [3].

Solar PV is taking its first steps in Turkey. Experience elsewhere indicates that at the local level, there has been some concern that PV development could conflict with livelihood and conservation goals. While these concerns are important to consider, research has consistently found that, when developed responsibly, ground-mounted and roof-mounted solar PV power plants provide considerable economic and environmental benefits. In order to ensure the sustainability of solar energy, PV solar manufacturers, project developers, policy makers and other stakeholders need to convene a multi-party initiative to establish sector-wide guidelines for responsible community engagement and land management.

This paper demonstrates that PV technology, when well-planned, does not conflict with conservation goals. On a macro level, no country or region must choose between solar PV and space for humans and nature. On the contrary, as climate change threatens humans and the environment, it is more important than ever to work for the efficient and wide-scale adoption of well sited, responsibly and effectively operated renewable energy generation facilities. Environmental protection and renewable energy can and must develop in parallel.

References

1. Yıldız T (2012) Budget presentation for year 2013. Power Point Presentation. Ministry of Energy and Natural Resources, Ankara, 15 Dec 2012. http://www.enerji.gov.tr/yayinlar_raporlar/2013_Genel_Kurul_Konusmasi.pdf. Retrieved 3 Jan 2013
2. Ministry of Energy and Natural Resources (2012) Energy outlook in Turkey and the world. Power Point Presentation, Ankara. http://www.enerji.gov.tr/yayinlar_raporlar/Dunyada_ve_Turkiyede_Enerji_Gorunumu.pdf. Retrieved 11 Nov 2012
3. The energy report: 100% renewable energy by 2050 (2011) WWF. http://wwf.panda.org/what_we_do/footprint/climate_carbon_energy/energy_solutions/renewable_energy/sustainable_energy_report/
4. World urbanization prospects: the 2007 revision. http://esa.un.org/unup/p2k0data.asp
5. Yazar Y (2013) Promotion of renewable energy in a liberalizing Turkish energy sector (2002-2012). Power Point Presentation. General Directorate of Renewable Energy, Ministry of Energy and Natural Resources. Wind Workshop, Ankara, 26 Mar 2013. http://events.ewea.org/events/workshops/wp-content/uploads/2013/03/EWEA-TUREB-Workshop-27-3-2013-Yusuf-Yazar-YEGM.pdf
6. Bazilian M et al (2012) Re-considering the economics of photovoltaic power. Bloomberg New Energy Finance. http://www.ourenergypolicy.org/wp-content/uploads/2012/05/BNEF_re_considering_the_economics_of_photovoltaic_power.pdf. Retrieved 23 Dec 2012
7. Demircan Z (2013) What energy mix for Turkey in 2030. Power Point Presentation. General Directorate of Energy Affairs, Ministry of Energy and Natural Resources. Wind Workshop, Ankara, 26 Mar 2013. http://events.ewea.org/events/workshops/wp-content/uploads/2013/03/EWEA-TUREB-Workshop-27-3-2013-Zafer-Demircan.pdf. Retrieved 28 Mar 2013
8. WWF (2012) Solar PV Atlas: Solar Power in Harmony with Nature - Towards 100 per cent renewable energy. http://awsassets.panda.org/downloads/solar_atlas__low_res__final_8_jan_2013__1_.pdf

The manufacturer's authorised representative in the EU is Springer Nature Customer Service Centre GmbH, Europaplatz 3, 69115 Heidelberg, Germany. If you have any concerns regarding our products, please contact ProductSafety@springernature.com

Printed and bound by CPI Group (UK) Ltd, Croydon, CR0 4YY
30/09/2026
02225657-0014